S0-ACY-059

IF FOUND, please notify and arrange return to owner.

Pilot's Name _____

Address _____

City _____ State ____ Zip Code ____

Telephone (_____) _____

Additional copies of Gleim's *Aviation Weather and Weather Services* are available from

Gleim Publications, Inc.
P.O. Box 12848
University Station
Gainesville, Florida 32604
(904) 375-0772
(800) 87-GLEIM

The price is $18.95 (subject to change without notice). Orders must be prepaid. Use the order form on page 433. Add applicable sales tax to shipments within Florida.

Gleim Publications, Inc. guarantees the immediate refund of all resalable materials returned within 30 days. Shipping and handling charges are nonrefundable.

REVIEWERS AND CONTRIBUTORS

Maria M. Bolanos, B.A., University of Florida, has 10 years of production experience in scientific and technical publications. Ms. Bolanos coordinated the production of the text and reviewed the final manuscript.

Gillian Hillis, B.A., University of Florida, is our editor. Ms. Hillis drafted portions of the text, reviewed the entire manuscript, and revised it for readability.

Barry A. Jones, CFII, B.S. in Air Commerce/Flight Technology, Florida Institute of Technology, is a charter pilot and flight instructor with Gulf Atlantic Airways in Gainesville, FL. Mr. Jones updated the text, incorporated numerous revisions, assisted in assembling the manuscript, and provided technical assistance throughout the project.

Heiko E. Kallenbach, ATP, CFII, MEI, MFA, Carnegie Mellon University, is a flight instructor in single and multiengine airplanes in Gainesville, FL. Mr. Kallenbach assisted in editing portions of the text.

John F. Rebstock, B.S., School of Accounting, University of Florida, assisted in editing the text, reviewed the entire edition, and composed the page layout.

The CFIs who have worked with me throughout the years to develop and improve my pilot training materials.

A PERSONAL THANKS

This manual would not have been possible without the extraordinary efforts and dedication of Jim Collis, Chris Eichelberger, and Connie Steen, who typed the entire manuscript and all revisions, as well as prepared the camera-ready pages.

The author also appreciates the proofreading and production assistance of Michael Amideo, Alison Barrett, Laura David, Kim Houellemont, Gregory Mullins, Steve Palasay, and Charmaine Smith.

Finally, I appreciate the encouragement, support, and tolerance of my family throughout this project.

FIRST EDITION

AVIATION WEATHER AND WEATHER SERVICES

by Irvin N. Gleim, Ph.D., CFII

ABOUT THE AUTHOR

Irvin N. Gleim earned his private pilot certificate in 1965 from the Institute of Aviation at the University of Illinois, where he subsequently received his Ph.D. He is a commercial pilot and flight instructor (instrument) with multiengine and seaplane ratings, and is a member of the Aircraft Owners and Pilots Association, American Bonanza Society, Civil Air Patrol, Experimental Aircraft Association, and Seaplane Pilots Association. He is also author of Practical Test Prep books for the private, instrument, commercial, flight instructor, multiengine, and seaplane certificates/ratings, and study guides for the private/recreational, instrument, commercial, flight/ground instructor, fundamentals of instructing, and airline transport pilot FAA written tests.

Dr. Gleim has also written articles for professional accounting and business law journals, and is the author of the most widely used review manuals for the CIA exam (Certified Internal Auditor), the CMA exam (Certified Management Accountant), and the CPA exam (Certified Public Accountant). He is Professor Emeritus, Fisher School of Accounting, University of Florida, and is a CIA, CMA, and CPA.

iv

Gleim Publications, Inc.
P.O. Box 12848
University Station
Gainesville, Florida 32604
(904) 375-0772
(800) 87-GLEIM

Library of Congress Catalog Card No. 93-91537
ISBN 0-917539-39-7
First Printing: August 1993

HELP !!

Please send any corrections and suggestions for subsequent editions to me, Irvin N. Gleim, c/o Gleim Publications, Inc. • P.O. Box 12848 • University Station • Gainesville, Florida • 32604. The last page in this book has been reserved for you to make your comments and suggestions. It should be torn out and mailed to me.

Also, please bring this book to the attention of flight instructors, fixed base operators, and others interested in flying. Wide distribution of our books and increased interest in flying depend on your assistance and good word. Thank you.

TABLE OF CONTENTS

AUTHOR'S MESSAGE

The purpose of this book is to provide an up-to-date compilation of all the FAA's weather publications in one easy-to-understand and easy-to-use book. This book is largely a reformation of *Aviation Weather* (AC 00-6A-1975) and *Aviation Weather Services* (AC 00-45D-1993)*, but other ACs are covered as well.

Parts I and II contain 16 chapters which are a restatement and amplification of each of the 16 chapters in AC 00-6A: *Aviation Weather*. We have broken *Aviation Weather* into Parts I and II to differentiate the advanced topics in Part II from the more basic topics in Part I.

The FAA's AC 00-6A: *Aviation Weather* is 192 pages in length. After editing the material in *Aviation Weather* and presenting it in larger type and in outline format, our outline is 210 pages in length. Our outline is much easier to read and study. We explain how to study each chapter, including which material to scan vs. the important material.

More importantly, you learn more from our outline. Your mind is not a word processor. It does not deal in terms of sentences and paragraphs. Your mind works with concepts, and our outline presents concepts and their relationship to other concepts.

Similarly, the FAA's AC 00-45D: *Aviation Weather Services* (AWS) is 136 pages in length. Our Part III of this book, which outlines AWS, is 171 pages in length. Our outline includes all tables, diagrams, weather maps, charts, etc., which appear in AWS (AC 00-45D).

NOTE: We provide guidance on which topics are important and need to be studied vs. topics which are not necessary for most pilots to study. We also have 100% coverage of all material in the FAA's AC 00-6A and AC 00-45D. Our index is very thorough (beginning on page 435) so you can research any topic, concept, etc., you desire.

---

* AC 00-45D was not yet published when the first printing of this book went to press. We have researched and updated the topics in AC 00-45C with the expectation that this book will summarize AC 00-45D.

PART I
AVIATION WEATHER

Aviation Weather was published jointly by the FAA and National Weather Service (NWS) in 1975. The book had previously included weather services (e.g., then available reports and forecasts); however, these prior editions quickly had gone out of date. In 1975, *Aviation Weather* was accompanied by *Aviation Weather Services* (AC 00-45A). Subsequent editions of *Aviation Weather Services* were published in 1979 and 1985, and AC 00-45D was issued in 1993*, as was the first edition of this book: *Aviation Weather and Weather Services* by Gleim.

Note: Gleim Publications, Inc. does not have the FAA's problem in staying up-to-date because we revise and reprint our books frequently. Thus, we can and will provide pilots with all current relevant weather information in one book.

Aviation Weather (AC 00-6A) contains 16 chapters which are outlined on pages 3 through 210. Most figures in *Aviation Weather* are reproduced. Only those that do not add to or facilitate under-standing of the topics explained in the outline are omitted. Conversely, additional information and explanations have been added where they facilitate learning and understanding of weather by pilots. You, as a pilot, should learn and understand aviation weather as presented in Chapters 1 through 12 in Part I of this book.

Part I: Aviation Weather
1. The Earth's Atmosphere
2. Temperature
3. Atmospheric Pressure and Altimetry
4. Wind
5. Moisture, Cloud Formation, and Precipitation
6. Stable and Unstable Air
7. Clouds
8. Air Masses and Fronts
9. Turbulence
10. Icing
11. Thunderstorms
12. Common IFR Producers

> * AC 00-45D was not yet published when the first printing of this book went to press. We have researched and updated the topics in AC 00-45C with the expectation that this book will summarize AC 00-45D.

Chapters 13 through 16, which are presented as Part II of this book, cover specialized topics which may be beyond your present interest.

Part II: Aviation Weather -- Over and Beyond
13. High Altitude Weather
14. Arctic Weather
15. Tropical Weather
16. Soaring Weather

You do **not** have to read the FAA's *Aviation Weather* (AC 00-6A) in conjunction with this outline. We have presented all of the information that is in *Aviation Weather*. Our contribution is to present the information in a format that will facilitate **knowledge transfer** to you. Our outline is easy to study because you seek a working knowledge of *weather* as it relates to flying.

Our outline is effective. Remember that human minds are not word processing computers that deal with sentences and paragraphs. Human minds deal with concepts and how they relate to other concepts -- which is exactly what our outline presentation facilitates for you.

Of the 16 chapters in *Aviation Weather*, the FAA has "In Closing" sections at the back of eight chapters. In addition to a summary, the FAA's "In Closing" section provides additional comments on and insight into chapter discussion. We have provided an "In Closing" section at the end of each chapter which is a summary of the FAA's "In Closing" section if one exists and our summary of the chapter.

HOW TO STUDY EACH CHAPTER

1. Each chapter begins with a listing of the chapter's topics.

2. Before beginning to study (i.e., learn from) the outline, think about the subject of the chapter and how the chapter's topics relate to the overall subject of the outline. Anticipate what will be said in the outline about these topics (even if you have never studied the topics). This step should require only 3 to 5 minutes.

3. Study the outline. Studying differs from reading in that studying, by definition, results in understanding. In other words, by studying you are internalizing weather information so you can use it as a pilot in the future.

4. In each chapter, ask yourself

    a. What each topic is about and how it affects your present and future flying.

    b. What was covered on each page as you complete each page.

    c. How well the "In Closing" section of the outline summarized and gave you further insight into the chapter.

    d. How you would have improved the chapter if you had to explain the topics in the chapter to your flight instructor or another pilot.

5. Relate each chapter's coverage to the expected coverage in other chapters by looking back to the Table of Contents on page v.

6. Make notes as appropriate on pages 441 and 442 to be sent to your author upon completion of your study of *Aviation Weather and Weather Services*.

CHAPTER ONE
THE EARTH'S ATMOSPHERE

> Please take a few minutes to study each of the concepts listed above and anticipate/imagine what they are and how they relate to the other listed concepts.

A. **Introduction** -- Our restless **atmosphere** is almost constantly in motion as it strives to reach equilibrium.

　　1. Because the sun heats the atmosphere unequally, differences in pressure result, which cause a series of never-ending air movements.

　　2. These air movements set up chain reactions which culminate in a continuing variety of weather.

　　3. Virtually all of our activities are affected by weather, but aviation is affected most of all.

B. **Composition.** Air is a mixture of several gases.

　　1. When completely dry, it is about 78% nitrogen and 21% oxygen.

　　2. The remaining 1% is other gases such as argon, carbon dioxide, neon, helium, and others.

　　3. However, air is never completely dry. It always contains some water vapor, in amounts varying from *almost* zero to about 5% by volume.

　　　　a. Regardless of the amount of water vapor in the atmosphere, the percentages of the dry gases remain the same.

C. **Vertical Structure.** The atmosphere is classified into layers, or spheres, by the characteristics exhibited in these layers. See Figure 2 below.

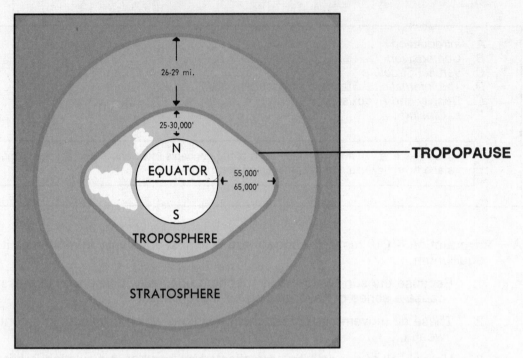

Figure 2. Earth's Atmosphere to 29 Miles High

1.   The **troposphere** is the layer from the surface to an average altitude of about 7 mi.

   a.   It is characterized by an overall decrease of temperature with increasing altitude.

   b.   The height of the troposphere varies with latitude and seasons.

   c.   It slopes from about 20,000 ft. over the poles to about 65,000 ft. over the Equator (4-12 mi.);

   d.   It is higher in summer than in winter.

2.   At the top of the troposphere is the **tropopause**, a transition zone marking the boundary between the troposphere and the layer above, the stratosphere.

   a.   The thickness of the tropopause varies from several discrete layers in some regions to entirely broken up in others.

   b.   The tropopause acts to trap most of the water vapor in the troposphere.

   c.   It is characterized by decreasing winds and constant temperatures with increases in altitude.

3.   Above the tropopause is the **stratosphere**, i.e., beginning at 4-12 mi.

   a.   This layer is characterized by relatively small changes in temperature with height except for a warming trend near the top (26-29 mi.) and a near absence of water vapor.

D. **The International Standard Atmosphere (ISA)** is a fixed standard of reference for engineers and meteorologists. To arrive at a standard, the average conditions throughout the atmosphere for all latitudes, seasons, and altitudes have been used.

    1. ISA consists of a specified sea-level temperature and pressure: 15°C (59°F) and 29.92 in. Hg (1013.2 millibars)

        a. And specific rates of decrease of temperature and pressure with height: 2°C per 1,000 ft. and 1 in. Hg per 1,000 ft.

    2. These standards are used for calibrating pressure altimeters and developing aircraft performance data.

E. **Density and Hypoxia**

    1. Air is matter and has weight.

        a. Since it is a gas, it is compressible.
        b. Thus, as its pressure increases, its **density** also increases.

    2. The pressure that the atmosphere exerts on a given surface is the result of the weight of the air above that surface.

        a. Thus, air near the surface of the Earth exerts more pressure, and is more dense than air at high altitudes.

        b. There is a decrease of air density and pressure with height.

    3. The rate at which the lungs absorb oxygen depends on the pressure exerted by oxygen in the air. Oxygen makes up about 20% of the atmosphere.

        a. Since air pressure decreases as altitude increases, the oxygen pressure also decreases.

        b. A pilot continuously gaining altitude or making a prolonged flight at high altitude without supplemental oxygen will likely suffer from a deficiency of oxygen.

        c. This deficiency of oxygen results in **hypoxia**, which may cause

            1) A feeling of exhaustion,
            2) Impairment of vision and judgment, and
            3) Unconsciousness.

    4. FAR 91.211 requires you to use aviation breathing oxygen for any flight in excess of 30 min. at cabin pressure altitudes above 12,500 ft. MSL up to 14,000 ft. MSL.

        a. You are required to use oxygen continuously when piloting an unpressurized airplane 14,000 ft. MSL or above.

        b. Your passengers are required to have continuous oxygen provided at altitudes of 15,000 ft. MSL and above.

        c. FARs concerning commercial flight require that supplemental oxygen be provided to pilots and used at even lower cabin pressure altitudes.

        d. Hypoxia may affect night vision adversely at altitudes as low as 5,000 ft. MSL.

        e. If you feel fatigued, drowsy, etc., descend and see if you begin to feel better.

F. **In Closing**

1. Troposphere extends upward to an average altitude of about 7 mi.

    a. Tropopause is a transition zone between the troposphere and stratosphere.
    b. Stratosphere is next with small changes in temperature.

2. International standard atmosphere (ISA)

    a. Sea level 59°F (15°C) and 29.92 in. Hg (1013.2 mb)

    b. Average lapse rate (i.e., decrease of temperature and pressure with altitude) is 2°C per 1,000 ft. and 1 in. Hg per 1,000 ft.

3. Air pressure decreases with altitude and insufficient oxygen is available to pilots beginning about 10,000 to 12,000 ft. MSL.

    a. Oxygen deficiency is called hypoxia.

END OF CHAPTER

CHAPTER TWO
TEMPERATURE

> Please take a few minutes to study each of the concepts listed above and anticipate/imagine what they are and how they relate to the other listed concepts.

A. **Introduction** -- Temperature affects every aspect of weather and aviation.

    1.   Temperature variation is the cause of most weather phenomena.

    2.   It is a major factor in the determination of aircraft performance.

B. **Temperature and Heat**

    1.   The temperature of a particular substance, such as air or water, is a measurement of the average kinetic energy of the molecules in that substance, relative to some set standard.

        a.   Heat is energy which is being transferred from one object to another because of the temperature difference between them.

        b.   After being transferred, this heat is stored as internal energy.

    2.   A specific amount of heat absorbed by or removed from a substance raises or lowers its temperature a definite amount.

        a.   Each substance has its unique temperature change for the specific change in heat.

        b.   EXAMPLE: If a land surface and a water surface have the same temperature and an equal amount of heat is added, the land surface becomes hotter than the water surface. Conversely, with equal heat loss, the land becomes colder than the water.

    3.   The Earth receives energy from the sun in the form of solar radiation.

        a.   The Earth, in turn, radiates energy, and this outgoing radiation is "terrestrial radiation."

        b.   The average heat gained from incoming solar radiation must equal heat lost through terrestrial radiation in order to keep the Earth from getting progressively hotter or colder.

            1)   This balance is world-wide; we must consider regional and local imbalances which create temperature variations.

C. **Temperature Scales.** Two commonly used temperature scales are Celsius (Centigrade) and Fahrenheit.

    1.   The Celsius scale is used exclusively for upper air temperatures and is rapidly becoming the world standard for surface temperatures also.

        a.   Fahrenheit is used to report surface temperatures in the U.S.

2.   Two common temperature references are the melting point of pure ice and the boiling point of pure water at sea level.

    a.   The boiling point of water is 100°C or 212°F.

    b.   The melting point of ice is 0°C or 32°F.

        1)   NOTE: Although 0°C (32°F) is generally the freezing point of water as well as the melting point of ice, this is not always so.

           a)   Under certain conditions, supercooled (not yet frozen) droplets of water can often be found in clouds at temperatures down to −15°C (5°F). This will be further explained in Part I, Chapter 5, Moisture, Cloud Formation, and Precipitation, on page 45.

           b)   Supercooled water freezes on impact with an airplane.

    c.   Figure 3 below compares the two scales.

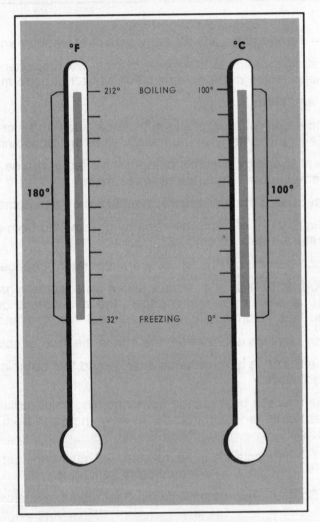

Figure 3. Fahrenheit vs. Celsius

3.   The difference between the boiling point of water and the melting point of ice (the temperature range between which water is a liquid) is 100°C or 180°F.

    a.   The ratio between degrees Celsius and Fahrenheit is 100/180 or 5/9.

    b.   Since 0°F is 32 Fahrenheit degrees colder than 0°C, you must apply this difference when comparing temperatures on the two scales.

c.   Conversion formulae:

$$°C = \frac{5}{9}(F - 32) \quad or \quad °F = \frac{9}{5}C + 32$$

d.   EXAMPLES:

°Celsius	°Fahrenheit	°Fahrenheit	°Celsius
0	+ 32	0	−18
10	50	10	−12
20	68	20	− 7
30	86	30	− 1
40	104	40	+ 4
50	122	50	10
60	140	60	16
70	158	70	21
80	176	80	27
90	194	90	32
100	212	100	38

e.   Most flight computers provide for direct conversion of temperature from one scale to the other.

D. **Temperature Variations.** The amount of solar radiation received by any region varies with time of day, with seasons, with latitude, with differences in topographical surface, and with altitude. These differences in solar radiation create temperature variations.

1.   **Diurnal variation** is the change in temperature from day to night brought about by the daily rotation of the Earth.

   a.   The Earth receives heat during the day by solar radiation and continuously dissipates heat by terrestrial radiation.

   b.   During the day, solar radiation exceeds terrestrial radiation and the surface becomes warmer.

   c.   At night, solar radiation ceases, but terrestrial radiation continues and cools the surface.

      1)   Cooling continues after sunrise until solar radiation again exceeds terrestrial radiation.

         a)   Minimum temperature usually occurs after sunrise, sometimes by as much as 1 hr.

      2)   The continued cooling after sunrise is one reason that fog sometimes forms shortly after the sun is above the horizon.

2. **Seasonal Variation.** Since the axis of the Earth tilts to the plane of orbit, the sun is more nearly overhead in one hemisphere than in the other, depending upon the season.

    a. The Northern Hemisphere is warmer in June, July, and August because it receives more solar energy than does the Southern Hemisphere.

    b. The Southern Hemisphere receives more solar radiation and is warmer during December, January, and February.

    c. Figures 4 and 5 on page 11 show these seasonal surface temperature variations.

3. **Variation with Latitude.** Since the Earth is essentially spherical, the sun is more nearly overhead in equatorial regions than at higher latitudes.

    a. Equatorial regions receive the most solar radiation and are warmest.

    b. Slanting rays of the sun at higher latitudes deliver less energy over a given area with the least being received at the poles.

    c. Thus, temperature varies with latitude from the warm Equator to the cold poles.

        1) You can see this average temperature variation in Figures 4 and 5 on page 11.

4. **Variations with Topography.** Not related to the movement or shape of the Earth are temperature variations induced by water and terrain.

    a. Large, deep water bodies tend to minimize temperature changes, whereas continents favor large changes.

        1) Wet soil such as in swamps and marshes is almost as effective as water in suppressing temperature changes.

        2) Thick vegetation tends to control temperature changes since it contains some water and also insulates against heat transfer between the ground and the atmosphere.

        3) Arid, barren surfaces permit the greatest temperature changes.

    b. These topographical influences are both diurnal and seasonal.

        1) EXAMPLE: The difference between a daily maximum and minimum may be 10° or less over water, near a shore line, or over a swamp or marsh, while a difference of 50° or more is common over rocky or sandy deserts.

        2) Figures 4 and 5 on page 11 also show the seasonal topographical variation.

            a) Note that in the Northern Hemisphere in July, temperatures are warmer over continents than over oceans; in January they are colder over continents than over oceans.

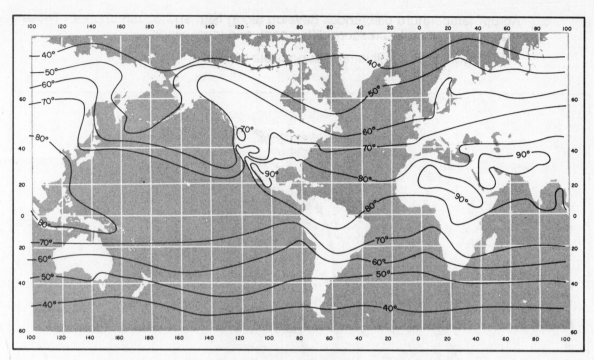

Figure 4. World-Wide Average Surface Temperatures (°F) in July

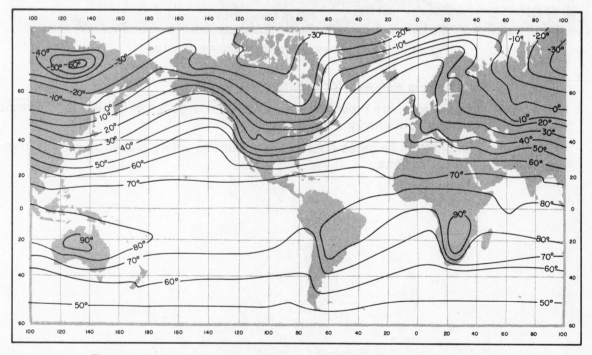

Figure 5. World-Wide Average Surface Temperatures (°F) in January

c.   Abrupt temperature differences develop along lake and ocean shores. These variations generate pressure differences and local winds.

    1)   Figure 6 below illustrates a possible effect.

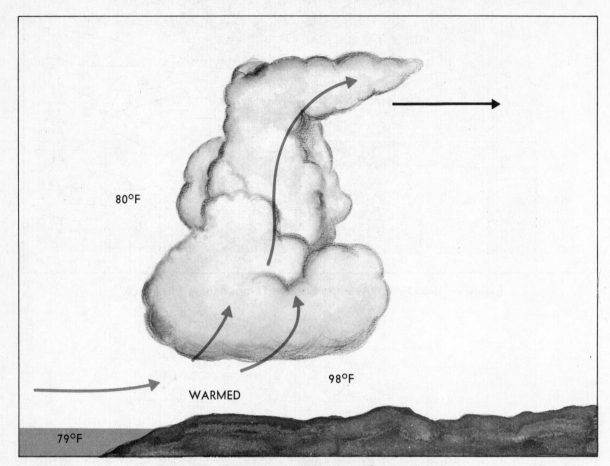

Figure 6. Temperature Differences Create Weather

d.   Prevailing wind is also a factor in temperature controls.

    1)   In an area where prevailing winds are from large water bodies, e.g., islands, temperature changes are rather small.

    2)   Temperature changes are more pronounced where prevailing wind is from dry, barren regions.

5. **Variation with Altitude.** Temperature normally decreases with increasing altitude throughout the troposphere. This is due primarily to solar radiation heating the surface, and the surface, in turn, warming the air above it by terrestrial radiation.

    a. This decrease of temperature with altitude is defined as lapse rate.

        1) The average decrease of temperature -- average lapse rate -- in the troposphere is 2°C per 1,000 ft.

    b. An increase in temperature with altitude is defined as an inversion, i.e., the lapse rate is inverted.

        1) An inversion often develops near the ground on clear, cool nights when wind is light.

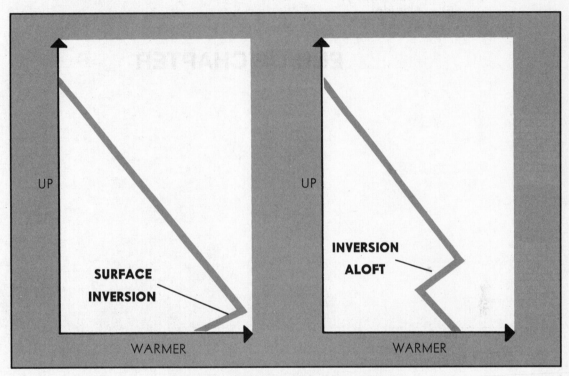

Figure 7. Temperature Inversion Graphs

            a) The ground radiates and cools much faster than the overlying air.

            b) Air in contact with the ground becomes cold while the temperature a few hundred feet above changes very little.

            c) Thus, temperature increases with height.

        2) Inversions may also occur at any altitude when conditions are favorable.

            a) EXAMPLE: A current of warm air aloft overrunning cold air near the surface produces an inversion aloft.

        3) Inversions are common in the stratosphere.

E. **In Closing**

   1.   The two commonly used scales for measuring temperature are Celsius and Fahrenheit.

   2.   Land absorbs and radiates heat much faster than water.

   3.   Variations in solar radiation create temperature variations which are the primary cause of weather.

      a.   The amount of solar radiation received varies with

         1)   Day-night.
         2)   Season.
         3)   Latitude.
         4)   Topography.
         5)   Altitude.

END OF CHAPTER

CHAPTER THREE
ATMOSPHERIC PRESSURE AND ALTIMETRY

> Please take a few minutes to study each of the concepts listed above and anticipate/imagine what they are and how they relate to the other listed concepts.

A. **Atmospheric Pressure** -- the force per unit area exerted by the weight of the atmosphere.

   1. The instrument designed for measuring pressure is the **barometer**.

      a. Weather services and the aviation community use two types of barometers in measuring pressure -- mercurial and aneroid.

   2. The **mercurial barometer** diagrammed in Figure 8 on page 16 consists of an open dish of mercury into which we place the open end of an evacuated glass tube.

      a. Atmospheric pressure forces mercury to rise in the tube.
      b. Near sea level, the column of mercury rises on the average to a height of 29.92 in.
      c. The height of the mercury column is a measure of atmospheric pressure.

   3. An **aneroid barometer** (illustrated in Figure 9 on page 16) uses a flexible metal cell and a registering mechanism.

      a. The cell is partially evacuated of air and contracts or expands as the pressure outside it changes.

         1) One end of the cell is fixed, while the other end moves the registering mechanism.

         2) This mechanism drives an indicator hand along a scale graduated in pressure units.

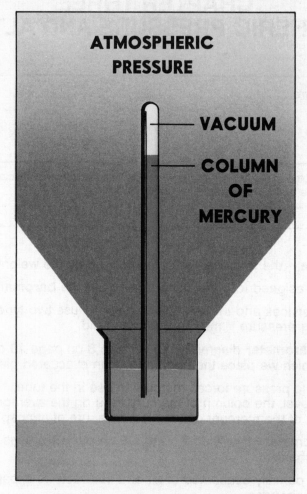

Figure 8. The Mercurial Barometer

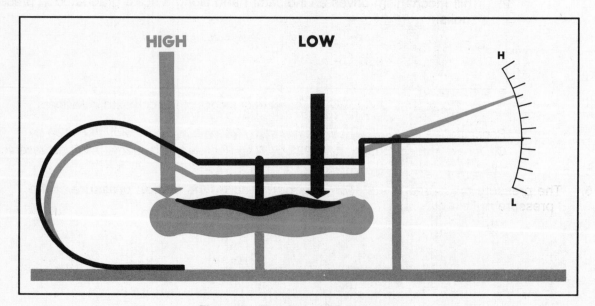

Figure 9. The Aneroid Barometer

4. Two commonly used pressure units are inches of mercury (in. Hg) or millibars (mb).

   a. Inches of mercury refers to the height to which a column of mercury would be raised in the basic mercurial barometer.

   b. The term millibar precisely expresses pressure as a force per unit area.

      1) The millibar is rapidly becoming the universal pressure unit.

   c. Note the relationship between inches and millibars in Figure 9(a) below.

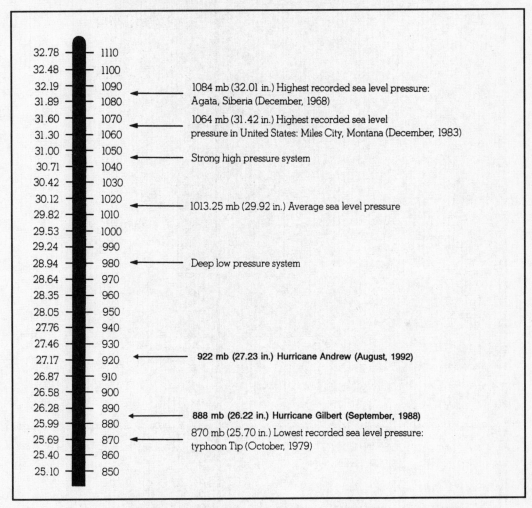

Figure 9(a). Atmospheric Pressure in Inches of Mercury and in Millibars

5. The pressure measured at a station or airport is called the **station pressure** or the pressure at field elevation.

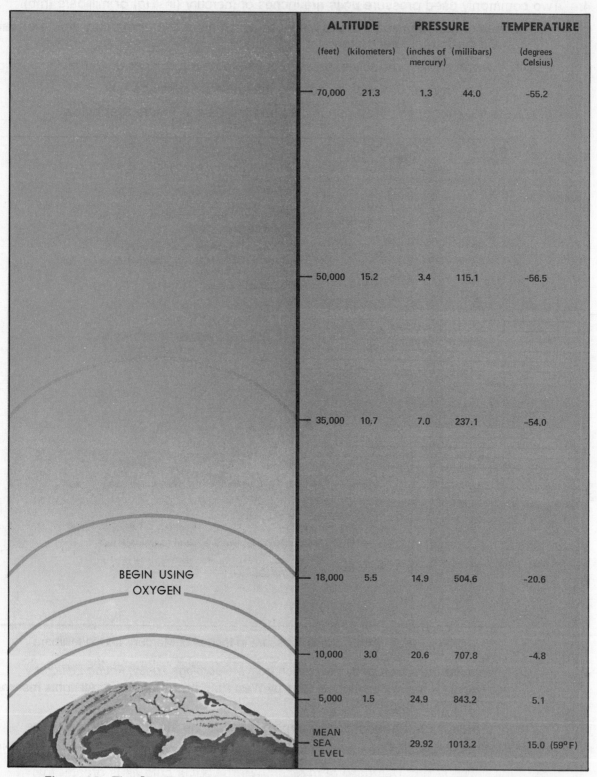

ALTITUDE		PRESSURE		TEMPERATURE
(feet)	(kilometers)	(inches of mercury)	(millibars)	(degrees Celsius)
70,000	21.3	1.3	44.0	–55.2
50,000	15.2	3.4	115.1	–56.5
35,000	10.7	7.0	237.1	–54.0
18,000	5.5	14.9	504.6	–20.6
10,000	3.0	20.6	707.8	–4.8
5,000	1.5	24.9	843.2	5.1
MEAN SEA LEVEL		29.92	1013.2	15.0 (59°F)

BEGIN USING OXYGEN

Figure 10. The Standard Atmosphere -- Decrease in Pressure with Increasing Heights

B. **Pressure Variation.** Pressure varies primarily with altitude and temperature of the air.

   1. **Altitude.** As we move upward through the atmosphere, the amount, and thus the weight, of the air above us becomes less and less.

      a. If we carry a barometer with us, we can measure a decrease in pressure as weight of the air above decreases.

         1) Within the lower few thousand feet of the troposphere, pressure decreases roughly 1 in. for each 1,000-ft. increase in altitude.

         2) The higher we go, the slower is the rate of decrease with height.

      b. Figure 10 on page 18 shows the pressure decrease with height in the standard atmosphere.

   2. **Temperature.** Like most substances, air expands as it becomes warmer and contracts as it cools.

      a. Figure 11 below shows three equal columns of air -- one colder than standard, one at standard temperature, and one warmer than standard.

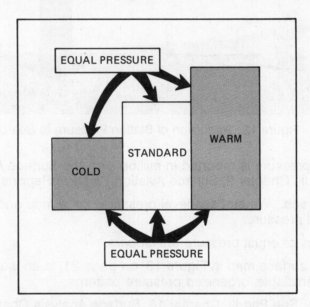

Figure 11. Effect of Temperature on Pressure

         1) Because each column contains the same amount of air, pressure is equal at the bottom of each column and equal at the top of each column.

         2) Thus, pressure decrease upward through each column is the same.

      b. However, vertical expansion of the warm column has made it higher than the column at standard temperature.

         1) Shrinkage of the cold column has made it shorter.

      c. Thus, the rate of decrease of pressure with height in warm air is less than standard.

         1) The rate of decrease of pressure with height in cold air is greater than standard.

3.   **Sea Level Pressure.** To readily compare pressure between stations at different altitudes, we must adjust them to some common level, i.e., mean sea level (MSL).

   a.   In Figure 12 below, pressure measured at a 5,000-ft. station is 25 in. Hg; pressure changes about 1 in. Hg for each 1,000 ft. or a total of 5 in. Hg. Sea level pressure at this station is thus approximately 25 + 5, or 30 in. Hg.

   1)   The weather observer takes temperature and other effects into account, but this simplified example explains the basic principle of sea level pressure.

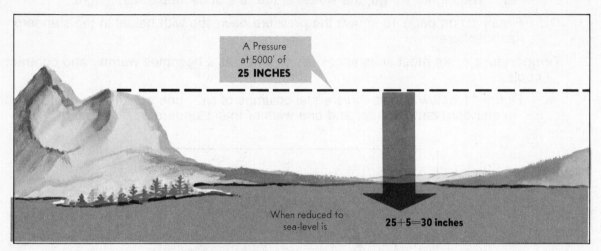

A Pressure at 5000' of **25 INCHES**

When reduced to sea-level is

25+5=30 inches

Figure 12. Reduction of Station Pressure to Sea Level

   b.   Sea level pressure is reported in millibars on the Surface Aviation Observation report, see Part III, Chapter 2, Surface Aviation Weather Reports, beginning on page 225.

4.   **Pressure Analyses.** We plot sea level pressures on a map and draw lines connecting points of equal pressure.

   a.   These lines of equal pressure are *isobars*.

   1)   The surface map in Figure 13, on page 21, is an isobaric analysis showing identifiable, organized pressure patterns.

      a)   See Part III, Chapter 16, Surface Analysis Chart, beginning on page 297.

   2)   The five pressure systems shown are defined as follows:

      a)   **Low** -- an area of pressure surrounded on all sides by higher pressure, also called a cyclone. In the Northern Hemisphere, a cyclone is a mass of air which rotates counter-clockwise (i.e., a low-pressure system), viewed from above.

      b)   **High** -- a center of pressure surrounded on all sides by lower pressure, also called an anticyclone. In the Northern Hemisphere, an anticyclone is a mass of air which rotates clockwise (i.e., a high-pressure system), viewed from above.

      c)   **Trough** -- an elongated area of low pressure with the lowest pressure along a line marking maximum cyclonic curvature.

      d)   **Ridge** -- an elongated area of high pressure with the highest pressure along a line marking maximum anticyclonic curvature.

e) **Col** -- the neutral area between two highs or two lows. It also is the intersection of a trough and a ridge. The col on a pressure surface is analogous to a mountain pass on a topographic surface.

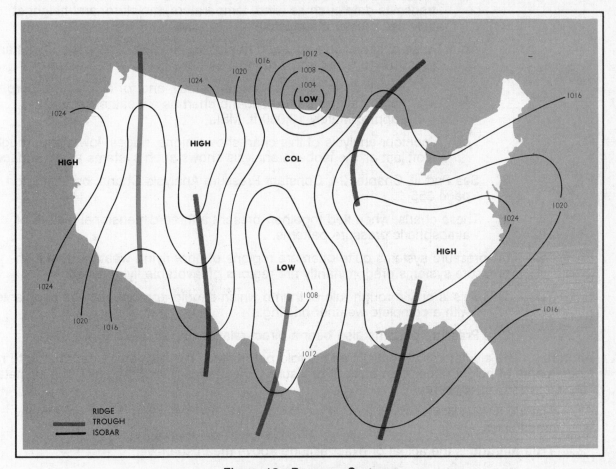

Figure 13. Pressure Systems

b. Upper air weather maps reveal these same types of pressure patterns aloft for several levels.

1) They also show temperature, moisture, and wind at each level.

c. An upper air map is a *constant pressure analysis*. Constant pressure simply refers to a specific pressure (e.g., 700 mb).

1) Everywhere above the Earth's surface, pressure decreases with height. At some height, it decreases to this constant pressure of 700 mb (about 10,000 ft. MSL, depending upon temperature).

a) Thus, there is a "surface" throughout the atmosphere at which pressure is 700 mb.

b) This is called the 700-mb constant pressure surface.

c) However, the *height* of this surface is *not* constant.

i) Rising pressure pushes the surface upward into highs and ridges.

ii) Falling pressure lowers the height of the surface into lows and troughs.

iii) These systems migrate continuously as "waves" on the pressure surface.

        2)   The National Weather Service (NWS) and military weather services take routinely scheduled upper air observations, sometimes called soundings.

            a)   A balloon carries aloft an instrument (called a radiosonde) which transmits data such as wind, temperature, moisture, and height at selected pressure surfaces.

            b)   These observations are used to plot the heights of a desired constant pressure surface (e.g., 700 mb).

                i)   Variations in these heights are small, and for all practical purposes, you may regard the 700-mb chart as a weather map at approximately 10,000 ft. MSL.

            c)   A contour analysis of this chart shows highs, ridges, lows, and troughs aloft just as the isobaric analysis shows such systems at the surface.

        3)   See Part III, Chapter 24, Constant Pressure Analysis Charts, beginning on page 355.

        4)   These charts, when tied together, present a three-dimensional picture of atmospheric pressure patterns.

    d.   Low-pressure systems quite often are regions of poor flying weather, and high-pressure systems predominantly are regions of favorable flying weather.

        1)   This is only a rough rule of thumb which should appropriately be supplanted with a complete weather briefing.

        2)   Pressure patterns also bear a direct relationship to wind.

C.   **Altimetry.** The altimeter is essentially an aneroid barometer. The altimeter is graduated to read increments of height rather than units of pressure. The standard for graduating the altimeter is the standard atmosphere.

    1.   Altitude seems like a simple term; it means height. But in aviation, it can have many meanings.

    2.   **True altitude** is the actual or exact altitude above mean sea level.

        a.   Since existing conditions in a real atmosphere are seldom standard, altitude indications on the altimeter are seldom actual or true altitudes.

    3.   **Indicated altitude** is the altitude above mean sea level indicated on the altimeter when set at the local altimeter setting.

        a.   The height indicated on the altimeter changes with changes in surface pressure.

            1)   Figure 15 on page 23 shows this effect.

            2)   A movable scale on the altimeter permits you to adjust for variations in surface pressure.

            3)   **Altimeter setting** is the value to which the scale of the pressure altimeter is set so as to indicate true altitude at field elevation.

                a)   This value is the atmospheric pressure adjusted to sea level in the region in which you are flying.

            4)   You must keep your altimeter setting current by adjusting it frequently in flight to the setting reported by the nearest tower or weather reporting station.

            5)   If an altimeter setting is not available before takeoff, you can set the altimeter to read field elevation, i.e., true altitude.

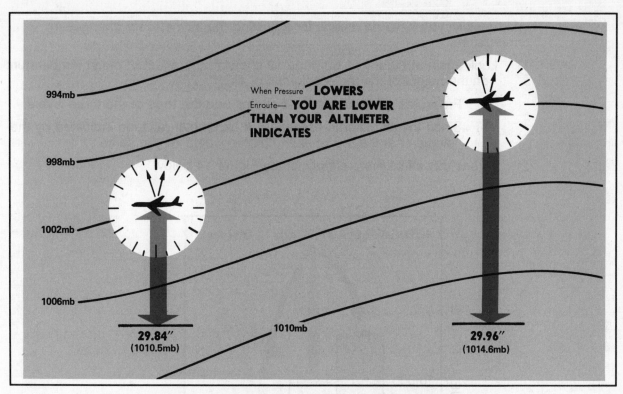

Figure 15. Effect of Pressure on Indicated Altitude

b.   Unfortunately, you have no means for adjusting the altimeter for changes in temperature.

   1)   Look again at Figure 11 on page 19 showing the effect of mean temperature on the height of the three columns of air.

      a)   Pressures are equal at the bottoms and the tops of the three layers.

      b)   Since the altimeter is essentially a barometer, altitude indicated by the altimeter at the top of each column would be the same.

   2)   To see this effect more clearly, study Figure 14 below.

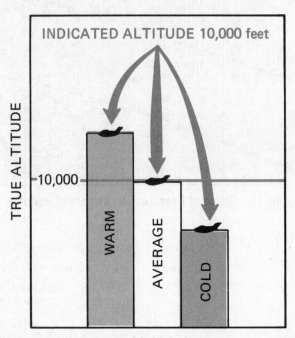

Figure 14. Effect of Temperature on Indicated Altitude

      a)   Note that in the warm air, true altitude is higher than indicated.
      b)   In the cold air, true altitude is lower than indicated.

4. **Corrected (approximately true) altitude** is indicated altitude corrected for the temperature of the air column below the aircraft.

    a. If it were possible for a pilot to determine mean temperature of the column of air between the aircraft and the surface, flight computers would be designed to use this mean temperature in computing true altitude.

        1) However, the only guide a pilot has to temperature below him/her is the outside air temperature at his/her altitude.

        2) Thus, the flight computer uses outside air temperature to correct indicated altitude to approximate true altitude.

            a) It is close enough to true altitude to be used for terrain clearance provided you have your altimeter set to the value reported from a nearby reporting station.

            b) Pilots have met with disaster because they failed to allow for the difference between indicated and true altitude. In cold weather when you must clear high terrain, take time to compute true altitude.

    b. EXAMPLE: If you are flying at an indicated altitude of 5,000 ft. MSL with an altimeter setting of 29.92 (standard pressure), your pressure altitude is also 5,000 ft. Once you determine pressure altitude and know the outside air temperature (OAT), use your flight computer to determine the corrected altitude.

        1) If your OAT is +5°C (standard temperature at 5,000 ft. MSL), your corrected altitude is 5,000 ft. MSL.

        2) However, if your OAT is −15°C (20° below standard), your corrected altitude is only 4,640 ft. MSL.

5. **Pressure altitude** is the altitude in the standard atmosphere where pressure is the same as where you are currently flying.

    a. In the standard atmosphere, pressure at sea level is 29.92 in. Hg and decreases approximately 1 in. per 1,000 ft. of altitude.

    b. Thus, for example, if the pressure at your altitude of flight is 27.92 in. Hg, you are flying at a pressure altitude of approximately 2,000 ft. higher than your indicated altitude.

    c. You can determine pressure altitude by setting your altimeter to the standard altimeter setting of 29.92 in.

    d. All flights at and above 18,000 ft. MSL, i.e., in Class A airspace (formerly known as the Positive Control Area), are flown at pressure altitudes.

6.   **Density altitude** is the altitude in the standard atmosphere where air density is the same as where you are.

    a.   Pressure, temperature, and to a lesser extent, humidity determine air density.

        1)   On a hot day, the air becomes "thinner" or lighter, and its density where you are is equivalent to a higher altitude in the standard atmosphere -- thus the term "high density altitude."

        2)   On a cold day, the air becomes heavy; its density is the same as that at an altitude in the standard atmosphere lower than your altitude -- "low density altitude."

    b.   Density altitude is not a height reference; rather, it is an index to aircraft performance.

        1)   Low density altitude increases performance.

        2)   High density altitude is a real hazard since it reduces aircraft performance. It affects performance in three ways.

            a)   It reduces power because the engine takes in less air to support combustion.

            b)   It reduces thrust because the propeller gets less grip on the light air or a jet has less mass of gases to spit out the exhaust.

            c)   It reduces lift because the light air exerts less force on the airfoils.

    c.   You cannot detect the effect of high density altitude on your airspeed indicator. Your aircraft lifts off, climbs, cruises, glides, and lands at the prescribed indicated airspeeds.

        1)   But at a specified indicated airspeed, your true airspeed and your groundspeed increase proportionally as density altitude becomes higher.

    d.   The net results are that high density altitude lengthens your takeoff and landing rolls and reduces your rate of climb.

        1)   Before liftoff, you must attain a faster groundspeed, and therefore, you need more runway; your reduced power and thrust add a need for still more runway.

        2)   You land at a faster groundspeed and, therefore, need more room to stop.

        3)   At a prescribed indicated airspeed, you are flying at a faster true airspeed, and therefore, you cover more distance in a given time, which means climbing at a more shallow angle.

            a)   Add to this the problems of reduced power and rate of climb, and you are in double jeopardy in your climb.

            b)   Figure 17 on page 27 shows the effect of density altitude on takeoff distance and rate of climb.

                i)   It shows the takeoff roll increasing from 1,300 ft. to 1,800 ft. and the rate of climb decreasing from 1,300 fpm to 1,000 fpm.

e. High density altitude also can be a problem at cruising altitudes. When air is abnormally warm, the high density altitude lowers your service ceiling.

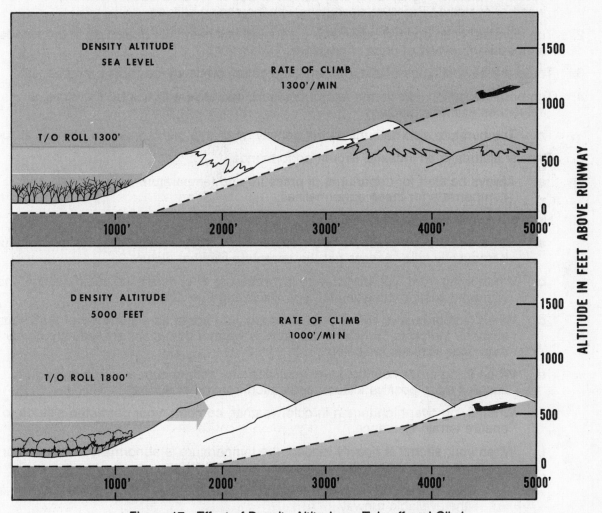

Figure 17. Effect of Density Altitude on Takeoff and Climb

f. To compute density altitude:

1) Set your altimeter at 29.92 in. Hg or 1013.2 mb and read pressure altitude from your altimeter.

a) Read outside air temperature, and then use your flight computer or a graph to get density altitude.

b) EXAMPLE: If temperature at 10,000 ft. pressure altitude is 20°C, density altitude is 12,700 ft. (Check this on your flight computer.) Your aircraft will perform as though it were at 12,700 ft. MSL indicated with a standard temperature of −8°C.

2) At an airport served by a weather observing station, you usually can get density altitude for the airport from the weather observer.

3) A graph for computing density altitude is in Part III, Chapter 26, on page 377.

D. **In Closing**

1. Pressure patterns (highs, lows, etc.) can be a clue to weather causes and movement of weather systems. Pressure decreases with increasing altitude.

2. The altimeter is an aneroid barometer graduated in increments of altitude in the standard atmosphere instead of units of pressure.

3. There are several types of altitude: true, indicated, pressure, corrected, and density.

4. Temperature greatly affects the rate of pressure decrease with height; therefore, it influences altimeter readings.

   a. Temperature also determines the density of air at a given pressure (density altitude).

5. Density altitude is an index to aircraft performance.

   a. Always be alert for departures of pressure and temperature from normals and compensate for these abnormalities.

6. Here are a few operational reminders:

   a. Beware of the low pressure-bad weather, high pressure-good weather rule of thumb. It frequently fails. Always get the **complete** weather picture.

   b. When flying from high pressure to low pressure at constant indicated altitude and without adjusting the altimeter, you are losing true altitude.

   c. When temperature is colder than standard, you are at an altitude *lower* than your altimeter indicates. When temperature is warmer than standard, you are *higher* than your altimeter indicates.

   d. When flying cross country, keep your altimeter setting current. This procedure assures more positive altitude separation from other aircraft.

   e. When flying over high terrain in cold weather, compute your corrected altitude to ensure terrain clearance.

   f. When your aircraft is heavily loaded, the temperature is abnormally warm, and/or the pressure is abnormally low, compute density altitude.

      1) Then check your aircraft manual to ensure that you can become airborne from the available runway.

      2) Check further to determine that your rate of climb permits clearance of obstacles beyond the end of the runway.

      3) This procedure is advisable for any airport regardless of altitude.

   g. When planning takeoff or landing at a high altitude airport regardless of load, determine density altitude.

      1) The procedure is especially critical when temperature is abnormally warm or pressure abnormally low.

      2) Make certain you have sufficient runway for takeoff or landing roll.

      3) Make sure you can clear obstacles beyond the end of the runway after takeoff or in event of a go-around.

END OF CHAPTER

CHAPTER FOUR
WIND

> Please take a few minutes to study each of the concepts listed above and anticipate/imagine what they are and how they relate to the other listed concepts.

A. **Introduction** -- Differences in temperature create differences in pressure. These pressure differences drive a complex system of winds in a never-ending attempt to reach equilibrium.

B. **Convection**

1. When two surfaces are heated unequally, they heat the overlying air unevenly, resulting in a circulatory motion called convection. Figure 18 on page 30 shows the convective process.

   a. The warmer air expands and becomes lighter or less dense than the cooler air.

   b. The dense, cooler air sinks to the ground by its greater mass, forcing the warmer air upward.

   c. The rising air spreads and cools, eventually descending to complete the convective process.

   d. As long as the uneven heating persists, a continuous convective current is maintained.

2. The horizontal air flow in a convective current is wind. Convection of both large and small scales accounts for systems ranging from hemispheric circulations down to local eddies.

   a. This horizontal flow, wind, is sometimes called **advection**.

   b. However, the term advection more commonly applies to the transport of atmospheric properties by the wind, i.e., warm advection, cold advection, advection of water vapor, etc.

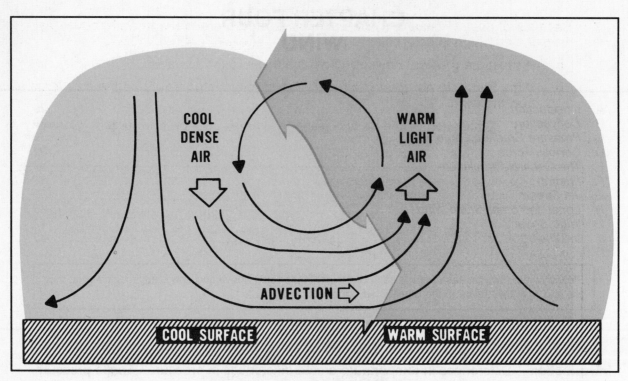

Figure 18. The Convective Process

C. **Pressure Gradient Force**

1. Whenever a pressure difference (or gradient) develops over an area, the air begins to move directly from the higher pressure to the lower pressure.

a. The pressure gradient is the rate of decrease of pressure per unit of distance at a fixed time.

b. The force that causes the air to move is called the pressure gradient force.

2. Recall that areas of high and low pressure are encircled by isobars, or lines of equal pressure. (See Figure 13 on page 21.)

a. The pressure gradient force moves the air directly across the isobars.

b. Closely spaced isobars indicate a strong pressure gradient, or a sharp change in pressure over a short distance.

1) Thus, the pressure gradient force is also strong.

c. The stronger the pressure gradient force, the stronger the wind

d. From a pressure analysis, you can get a general idea of wind speed from the isobar spacing.

1) Closely spaced isobars mean strong winds.
2) Widely spaced isobars mean lighter winds.

3.   Because of uneven heating of the Earth, surface pressure is low in warm equatorial regions and high in cold polar regions.

   a.   A pressure gradient develops from the poles to the Equator.

   b.   If the Earth did not rotate, this pressure gradient force would be the only force acting on the wind.

      1)   Circulation would be two giant hemispheric convective currents as shown in Figure 19 below.

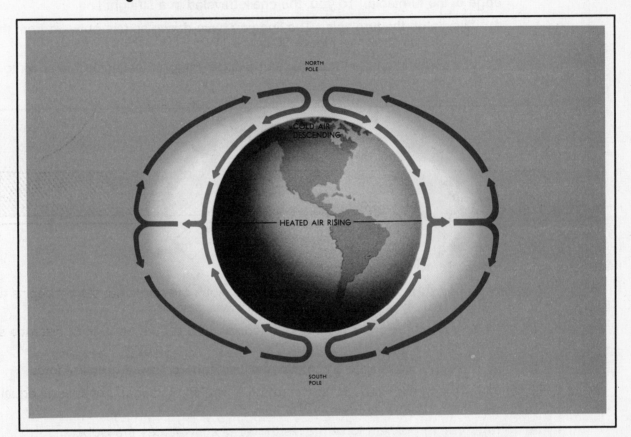

Figure 19. Circulation Pattern of a Nonrotating Earth

      2)   Cold air would sink at the poles, wind would blow straight from the poles to the Equator, warm air at the Equator would be forced upward, and high level winds would blow directly toward the poles.

   c.   However, the Earth does rotate, and because of its rotation, this simple circulation is greatly distorted.

D. Coriolis Force

1.   A moving mass (e.g., air) travels in a straight line until acted upon by some outside force (Newton's First Law of Motion).

    a.   However, if you viewed the moving mass from a rotating platform, the path of the moving mass relative to your platform would appear to be deflected or curved.

    b.   EXAMPLE: Imagine the turntable of a record player rotating counterclockwise. Use a piece of chalk and a ruler to draw a "straight" line from the center to the outer edge of the turntable. To you, the chalk traveled in a straight line.

        1)   Now stop the turntable. The line you have drawn spirals outward from the center as shown in Figure 20 below.

        2)   To a viewer on the turntable, some apparent force deflected the chalk to the right.

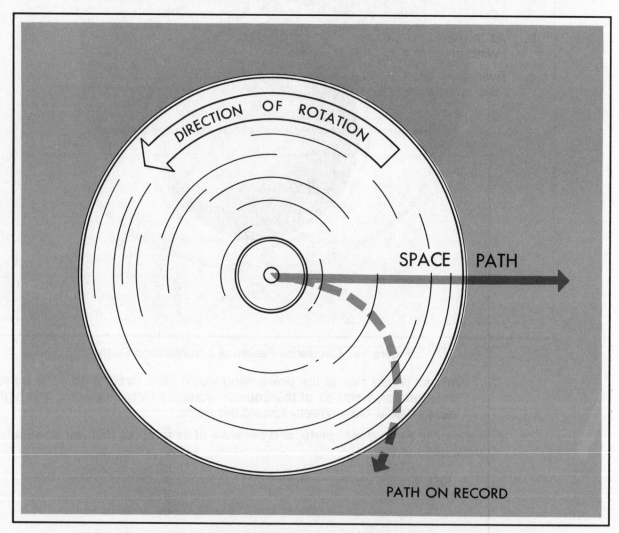

Figure 20. Apparent Deflective Force Due to Rotation of a Horizontal Platform

2.   Similarly, air moving across the surface of the Earth seems to us to be deflected by some force.

    a.   This principle was first explained by a Frenchman, Gaspard Coriolis, and carries his name -- the Coriolis force.

    b.   Because the Earth is spherical, the Coriolis force is much more complex than the simple turntable example.

3.   The Coriolis force affects the paths of aircraft, missiles, flying birds, ocean currents, and most important to the study of weather, air currents.

    a.   It deflects air to the right in the Northern Hemisphere and to the left in the Southern Hemisphere.

    b.   This book concentrates on deflection to the right in the Northern Hemisphere.

4.   The Coriolis force varies directly with wind speed and latitude.

    a.   The stronger the wind speed, the greater the deflection.
    b.   Coriolis force varies with latitude from zero at the Equator to maximum at the poles.

5.   When a pressure gradient force is first established, wind begins to blow from higher to lower pressure directly across the isobars. In Figure 21 below it would be from bottom to top.

    a.   However, the instant the air begins moving from high pressure to low pressure, the Coriolis force deflects it to the right, curving its path.

    b.   As the speed of this moving air increases, the Coriolis force increases, deflecting the wind more and more to the right.

    c.   Eventually, the wind speed increases to a point where the Coriolis force balances the pressure gradient force, as shown in Figure 21 below.

       1)   At this point, the wind has been deflected 90° and is now parallel to the isobars (depicted as "Resultant Wind" in Figure 21).

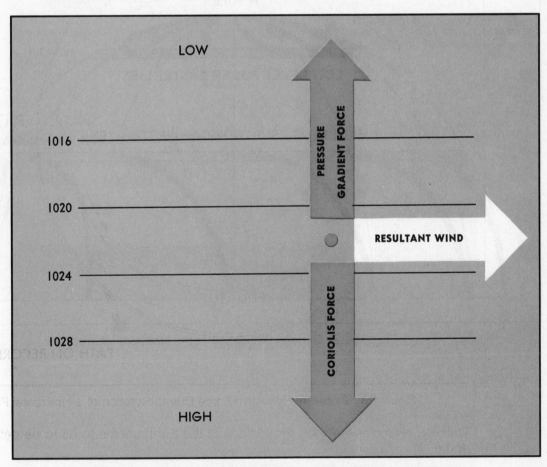

Figure 21. Effect of Coriolis Force on Wind Relative to Isobars

E. The General Circulation

1. As air is forced aloft at the Equator and begins its high-level trek northward, the Coriolis force turns it to the right or to the east as shown in Figure 22 below.

    a. Wind becomes westerly (i.e., moves from west to east) at about 30° latitude temporarily blocking further northward movement, as shown in Figure 22 below.

    b. Similarly, as air over the poles begins its low-level journey southward toward the Equator, it likewise is deflected to the right and becomes an east wind, halting for a while its southerly progress at about 60° latitude, also shown in Figure 22.

2. As a result, air piles up between 30° and 60° latitude in both hemispheres.

    a. The added weight of the air increases the pressure into semipermanent high pressure belts.

        1) Figures 23 and 24 on page 35 are maps of mean surface pressure for the months of July and January.

            a) The maps show clearly the subtropical high pressure belts near 30° latitude in both the Northern and Southern Hemispheres.

    b. These high pressure belts create a temporary impasse disrupting the simple convective transfer between the Equator and the poles.

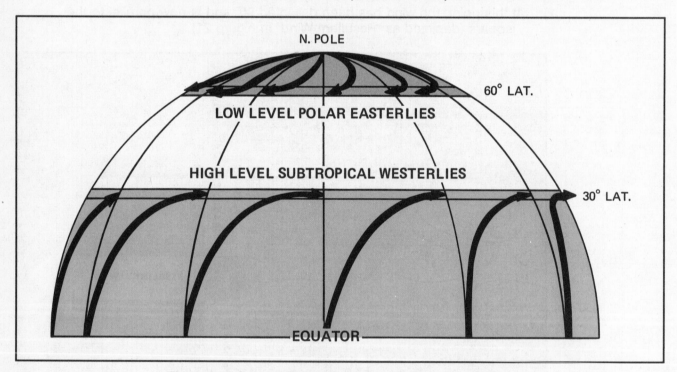

Figure 22. General Circulation in the Northern Hemisphere

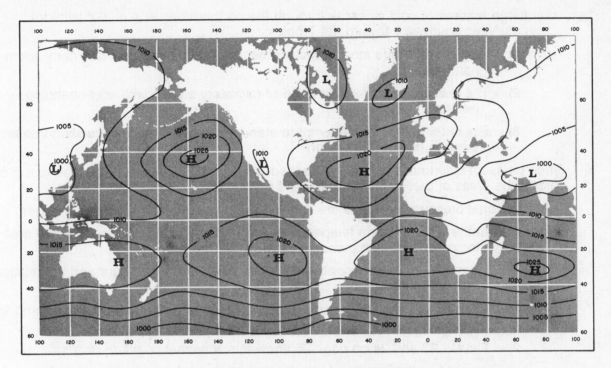

Figure 23. Mean World-Wide Surface Pressure Distribution in July

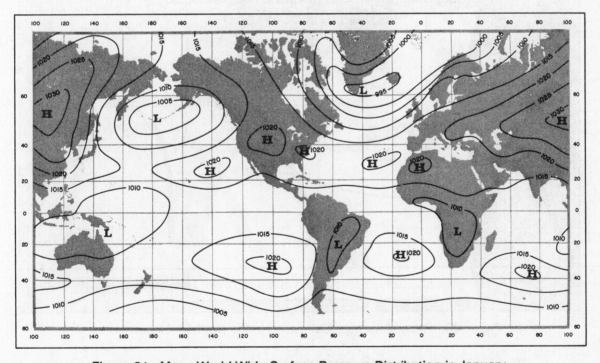

Figure 24. Mean World-Wide Surface Pressure Distribution in January

   c.   Large masses of cold air break through the northern barrier (i.e., 60° latitude) plunging southward toward the Tropics.

       1)   Large mid-latitude storms develop between cold outbreaks and carry warm air northward.

       2)   The result is a mid-latitude band of migratory storms with ever-changing weather.

       3)   Figure 25 below is an attempt to standardize this chaotic circulation into an average general circulation.

3.   Since pressure differences cause wind, seasonal pressure variations determine to a great extent the areas of these cold air outbreaks and mid-latitude storms.

   a.   Seasonal pressure variations are largely due to seasonal temperature changes.

       1)   At the surface, warm temperatures largely determine low pressure and cold temperatures, high pressure.

   b.   During summer, warm continents tend to be areas of low pressure and the relatively cool oceans, high pressure.

       1)   In winter, the reverse is true -- high pressure over the cold continents and low pressure over the relatively warm oceans.

       2)   Figures 23 and 24 on page 35 show this seasonal pressure reversal.

Figure 25. General Average Circulation in the Northern Hemisphere

   c.   Cold outbreaks are strongest in the winter and are predominantly from the colder continental areas.

       1)   Outbreaks are weaker in the summer, and more likely to originate from the cooler water surfaces.

       2)   Since these outbreaks are masses of cool, dense air, they characteristically are high-pressure areas.

4.   As the air moves outward from high pressure, it is deflected to the right by the Coriolis force.

   a.   Thus, the wind around a high moves clockwise.

      1)   The high pressure with its associated wind system is called an *anticyclone*.

   b.   The storms that develop between high pressure systems are characterized by low pressure.

      1)   As wind moves inward toward low pressure, it is also deflected to the right.
      2)   Thus, the wind around a low is counterclockwise.
      3)   The low pressure and its wind system are called a *cyclone*.

   c.   Figure 26 below shows winds moving parallel to isobars (called contours on upper level charts). The winds are clockwise around highs and counterclockwise around lows.

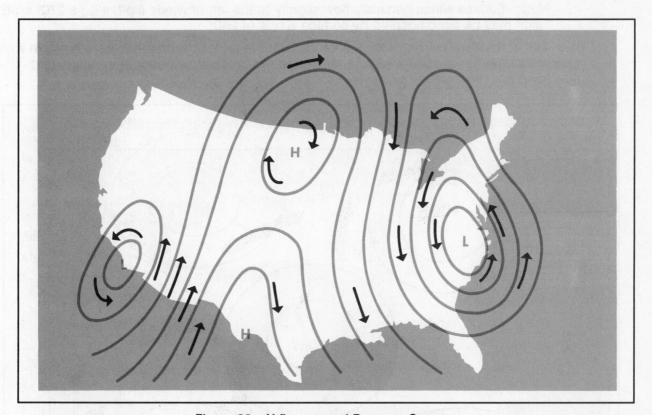

Figure 26. Airflow around Pressure Systems

5.   The high pressure belt between 30° and 60° latitude forces air outward to the north and to the south, resulting in three major wind belts, as shown in Figure 25 on page 36.

   a.   The southward moving air is deflected by the Coriolis force, becoming the well-known subtropical northeast trade winds (see Part II, Chapter 15, page 163).

      1)   These winds carry tropical storms from east to west, e.g., hurricanes from the ocean off North Africa west into the Caribbean.

   b.   The northbound air becomes entrained into the mid-latitude storms.

      1)   High-level winds are deflected to the right and are known as the prevailing westerlies.

      2)   These westerlies drive mid-latitude storms generally from west to east.

   c.   Polar easterlies dominate low-level circulation north of about 60° latitude where few major storm systems develop.

F.   Friction

1.   Friction between the wind and the terrain surface slows the wind.

    a.   The rougher the terrain, the greater the frictional force
    b.   The stronger the wind speed, the greater the friction
    c.   Frictional force always acts in opposition to the wind direction.

2.   When the wind is within approximately 3,000 ft. of the surface, friction reduces the wind speed, which in turn reduces the Coriolis force.

    a.   Recall that in the absence of surface friction, the Coriolis force turns the wind 90° to the right to parallel the isobars.

    b.   Surface friction slows the wind, which reduces the Coriolis force, causing the wind to turn to the left, across the isobars at an angle, toward the area of lower pressure.

    c.   Note: Surface winds generally flow slightly to the left of winds aloft, e.g., a 270° wind aloft may be accompanied by surface winds of 240°.

3.   Thus, in the Northern Hemisphere, friction causes the surface winds that are moving in a clockwise direction around a high pressure area to cross the isobars at an angle and flow toward an area of low pressure, as shown in Figure 28 below.

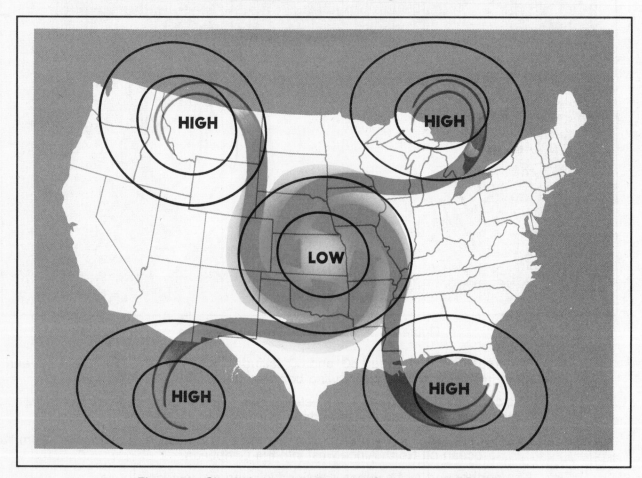

Figure 28. Circulation around Pressure Systems at the Surface

4.   The angle of surface wind to isobars is about 10° over water, increasing with roughness of terrain.

    a.   In mountainous regions, one often has difficulty relating surface wind to pressure gradient because of immense friction and also because local terrain effects on pressure could cause the angle to be 40° or more.

    b.   The average angle across the isobars is about 30°, for all surfaces.

    c.   The speed of the wind will also affect this angle.

        1)   Normally, the angle is smaller for high winds (i.e., due to a stronger Coriolis force) and large for slower winds (i.e., weaker Coriolis force).

G. Jet Stream

1.   A discussion of the general circulation is incomplete when it does not mention the **jet stream**.

    a.   Winds on the average increase with height throughout the troposphere, culminating in a maximum near the level of the tropopause.

    b.   These maximum winds tend to be further concentrated in narrow bands.

2.   A jet stream, then, is a narrow band of strong winds meandering through the atmosphere at a level near the tropopause.

    a.   Since it is of interest primarily to high-level flight, further discussion of the jet stream is reserved for Part II, Chapter 13, High Altitude Weather, beginning on page 144.

H. **Local and Small-Scale Winds**

1. The previous discussion has dealt only with the general circulation and major wind systems. Local terrain features such as mountains and shore lines influence local winds and weather.

2. **Mountain and Valley Winds**

a. In the daytime, air next to a mountain slope is heated by contact with the ground as the ground receives radiation from the sun.

1) This air usually becomes warmer than air at the same altitude but farther from the slope.

2) Colder, denser air in the surroundings settles downward and forces the warmer air near the ground up the mountain slope.

a) This wind is a "valley wind," so called because the air is flowing up out of the valley.

b. At night, the air in contact with the mountain slope is cooled by terrestrial radiation and becomes heavier than the surrounding air.

1) It sinks along the slope, producing the "mountain wind" which flows like water down the mountain slope.

c. Mountain winds are usually stronger than valley winds, especially in winter.

1) The mountain wind often continues down the more gentle slopes of canyons and valleys, and in such cases takes the name "drainage wind."

2) It can become quite strong over some terrain conditions and in extreme cases can become hazardous when flowing through canyon restrictions as discussed in Part I, Chapter 9, Turbulence, beginning on page 93.

3. **Katabatic Wind**

a. A **katabatic wind** is any wind blowing down an incline when the incline is influential in causing the wind.

1) Thus, the mountain wind is a katabatic wind.

b. Any katabatic wind originates because cold, heavy air spills down sloping terrain displacing warmer, less dense air ahead of it.

1) Air is heated and dried as it flows down slope.
2) Sometimes the descending air becomes warmer than the air it replaces.

c. Many katabatic winds recurring in local areas have been given colorful names to highlight their dramatic local effect.

1) Some of these include:

a) The Bora, a cold northerly wind blowing from the Alps to the Mediterranean coast;

b) The Chinook, a warm wind down the east slope of the Rocky Mountains often reaching hundreds of miles into the high plains;

c) The Taku, a cold wind in Alaska blowing off the Taku glacier; and

d) The Santa Ana, a warm wind descending from the Sierras into the Santa Ana Valley of California.

4.   Land and Sea Breezes

   a.   As frequently stated earlier, land surfaces warm and cool more rapidly than do water surfaces.

   b.   Land is warmer than the sea during the day.

      1)   Wind blows from the cool water to warm land -- the "sea breeze," so called because it blows from the sea.

      2)   At night, the wind reverses, blows from cool land to warmer water, and creates a "land breeze."

   c.   Figure 30 below diagrams land and sea breezes.

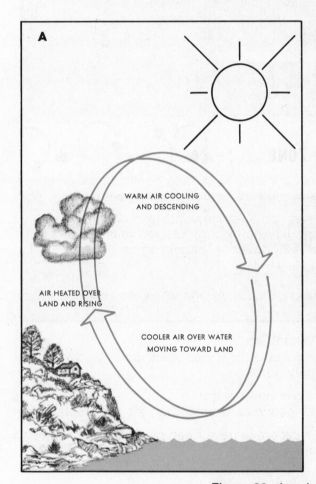 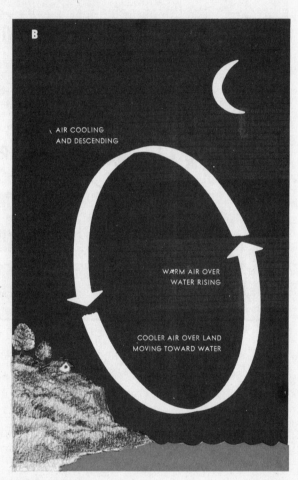

Figure 30. Land and Sea Breezes

   d.   Land and sea breezes develop only when the overall pressure gradient is weak.

      1)   Wind occurring as a result of a strong pressure gradient mixes the air so rapidly that local temperature and pressure gradients do not develop along the shore line.

I.   **Wind Shear**

1.   Rubbing two objects against each other creates friction.

 a.   If the objects are solid, no exchange of mass occurs between the two.

 b.   However, if the objects are fluid currents, friction creates eddies along a common shallow mixing zone, and a mass transfer takes place in the shallow mixing layer.

  1)   This zone of induced eddies and mixing is a shear zone.

2.   Figure 31 below shows two adjacent currents of air and their accompanying shear zone.

3.   Part I, Chapter 9, Turbulence, beginning on page 93, relates wind shear to turbulence.

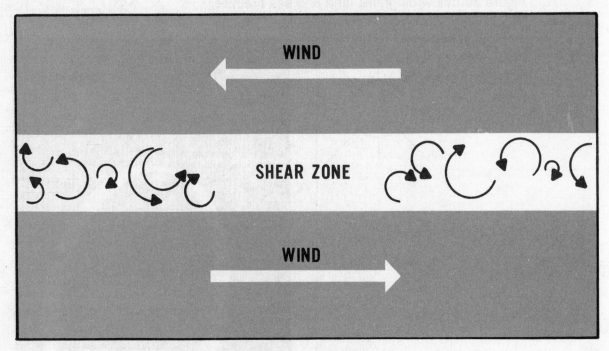

Figure 31. Wind Shear

J. **In Closing**

1. The horizontal air flow in a convective current is wind.

a. Wind speed is proportional to the spacing of isobars or contours on a weather map.

1) However, with the same spacing, wind speed at the surface will be less than aloft because of surface friction.

b. You can also determine wind direction from a weather map.

1) If you visualize standing with the wind aloft at your back (Northern Hemisphere), low pressure should be on your left and high pressure on your right.

a) If you visualize standing with the surface wind to your back, then turn clockwise about 30°, low pressure should be on your left and high pressure on your right.

2) On a surface map, wind will cross the isobar at an angle toward lower pressure; on an upper air chart, it will be nearly parallel to the contour.

2. As the pressure gradient force drives the wind from high- to low-pressure areas, the Coriolis force deflects the wind to the right (in the Northern Hemisphere), resulting in the general circulation around high- and low-pressure systems.

a. In the Northern Hemisphere, wind blows counterclockwise around a low and clockwise around a high.

1) At the surface where winds cross the isobars at an angle, you can see a transport of air from high to low pressure.

2) Although winds are virtually parallel to contours on an upper air chart, there still is a slow transport of air from high to low pressure.

3. At the surface when air converges into a low, it cannot go outward against the pressure gradient, nor can it go downward into the ground; it must go upward.

a. Thus, a low or trough is an area of rising air.

1) Rising air is conducive to cloudiness and precipitation; thus, we have the general association of low pressure and bad weather.

b. By similar reasoning, air moving out of a high or ridge depletes the quantity of air. Highs and ridges, therefore, are areas of descending air.

1) Descending air favors dissipation of cloudiness; hence the association of high pressure and good weather.

4. Three exceptions to the low pressure -- bad weather, high pressure -- good weather rule:

a. Many times weather is more closely associated with an upper air pattern than with features shown by the surface map. Although features on the two charts are related, they seldom are identical.

1) A weak surface system often loses its identity in the upper air pattern, while another system may be more evident on the upper air chart than on the surface map.

2) Widespread cloudiness and precipitation often develop in advance of an upper trough or low.

a) A line of showers and thunderstorms is not uncommon with a trough aloft even though the surface pressure pattern shows little or no cause for the development.

b. Downward motion in a high or ridge places a "cap" on convection, preventing any upward motion.

   1) Air may become stagnant in a high, trap moisture and contamination in low levels, and restrict ceiling and visibility.

      a) Low stratus, fog, haze, and smoke are not uncommon in high pressure areas.

   2) However, a high or ridge aloft with moderate surface winds most often produces good flying weather.

c. A dry, sunny region becomes quite warm from intense surface heating, thus generating a surface low pressure area, called a thermal low.

   1) The warm air is carried to high levels by convection, but cloudiness is scant because of lack of moisture.

      a) Since in warm air, pressure decreases slowly with altitude, the warm surface low is not evident at upper levels.

   2) The thermal low is relatively shallow with weak pressure gradients and no well-defined cyclonic circulation.

      a) It generally supports good flying weather.

      b) However, during the heat of the day, one must be alert for high density altitude and convective turbulence.

5. Due to surface friction, highs and lows tend to *lean* from the surface into the upper atmosphere, i.e., the upper portion moves faster than the lower portion, which is slowed by friction with the surface.

   a. Winds aloft often flow across the associated surface systems.

   b. Upper winds tend to steer surface systems in the general direction of the upper wind flow.

   c. An intense, cold, low pressure vortex *leans less* than does a weaker system.

      1) The intense low becomes oriented almost vertically and is clearly evident on both surface and upper air charts.

         a) Upper winds encircle the surface low and do not blow across it.

         b) Thus, the storm moves very slowly and usually causes an extensive and persistent area of clouds, precipitation, strong winds, and generally adverse flying weather.

6. There are three major wind belts in the Northern Hemisphere:

   a. Subtropical northeasterly trade winds, which carry tropical storms from east to west

   b. Midlatitude westerlies, which drive midlatitude storms generally from west to east (which encompasses the United States)

   c. Polar easterlies, which primarily contribute to the development of midlatitude storms

7. Local terrain features influence local winds and weather, e.g.,

   a. Mountain and valley winds
   b. Katabatic wind
   c. Land and sea breezes

END OF CHAPTER

CHAPTER FIVE
MOISTURE, CLOUD FORMATION, AND PRECIPITATION

Please take a few minutes to study each of the concepts listed above and anticipate/imagine what they are and how they relate to the other listed concepts.

A. **Water Vapor**

1. Water evaporates into the air and becomes an ever-present but variable part of the atmosphere.

2. Water vapor is invisible, just as oxygen and other gases are invisible.

    a. We can readily measure water vapor and express the results in different ways. Two commonly used terms are:

        1) Relative humidity, and
        2) Dew point.

3. **Relative humidity** relates the actual water vapor present to that which could be present.

    a. Relative humidity is routinely expressed as a percentage.

    b. Temperature largely determines the maximum amount of water vapor a parcel of air can hold.

        1) As Figure 32 below shows, warm air can hold more water vapor than cool air.

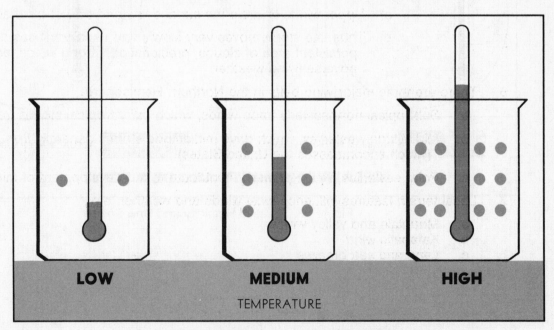

Figure 32. Air Temperature and Water Vapor

c.   Relative humidity expresses the degree of saturation. Air with 100% relative humidity is saturated; at less than 100%, unsaturated.

1)   If a given volume of air is cooled to some specific temperature, it can hold no more water vapor than is actually present; relative humidity becomes 100%, and saturation occurs.

4.   **Dew point** is the temperature to which air must be cooled to become saturated by the water vapor already present in the air.

a.   Aviation weather reports normally include the air temperature and dew point.

b.   Comparing the dew point to the air temperature reveals how close the air is to saturation.

c.   Figure 33 below relates water vapor, temperature, and relative humidity.

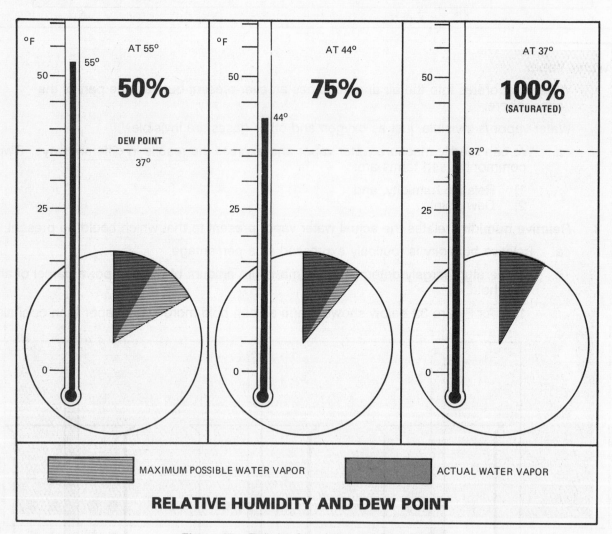

Figure 33. Relative Humidity and Dew Point

5.   The difference between air temperature and dew point is commonly known as the **temperature-dew point spread**.

    a.   As this spread becomes smaller, relative humidity increases, and it reaches 100% when temperature and dew point are the same.

    b.   Surface temperature-dew point spread is important in anticipating fog, but has little bearing on precipitation.

        1)   To support precipitation, air must be saturated through thick layers aloft.

    c.   Sometimes the spread at ground level may be quite large, yet at higher altitudes the air is saturated and clouds form.

        1)   Some rain may reach the ground, or it may evaporate as it falls into the drier air.

            a)   Figure 34 below is a photograph of "virga" -- streamers of precipitation trailing beneath clouds but evaporating before reaching the ground.

Figure 34. Virga

B. **Change of State**

1. Evaporation, condensation, freezing, melting, and sublimation are changes of state in water.

    a. Evaporation is the changing of liquid water to invisible water vapor.

        1) Condensation is the reverse process, i.e., water vapor into water.

    b. Freezing is the changing of liquid water into solid water, or ice.

        1) Melting is the reverse process, i.e., ice into water.

    c. Sublimation is the changing of ice directly to water vapor, or water vapor to ice, bypassing the liquid state in each process.

        1) Snow or ice crystals result from the sublimation of water vapor directly to the solid state.

2. **Latent Heat**

    a. The exchange of heat energy is required to change a substance from one state to another.

        1) This energy is called "latent heat" and is significant, as we will learn in later chapters.

        2) Figure 35 below diagrams the heat exchanges between the different states.

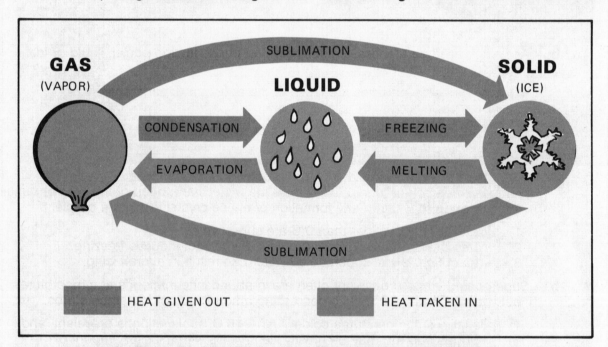

Figure 35. Heat Transactions When Water Changes State

    b. Evaporation of water, for example, requires heat energy that may come from the water itself, or some other source such as the surrounding air.

        1) This energy is known as the "latent heat of vaporization," and its transference to the water vapor cools the source it comes from.

            a) Evaporation is thus a cooling process.

            b) When you step out of the shower, the evaporation of water from your body removes heat from your skin, making you feel cold.

2) The heat energy is stored or hidden in the water vapor, thus the term "latent (hidden) heat."

    a) It is considered hidden because the temperature of the evaporated water remains the same.

3) When the water vapor condenses back into liquid water, this energy is released to the atmosphere as the "latent heat of condensation."

    a) Thus, condensation is a warming process.

c. Melting and freezing involve the exchange of "latent heat of fusion" in a similar manner.

1) The latent heat of fusion is much less than that of condensation and evaporation; however, each plays an important role in aviation weather.

3. **Condensation Nuclei.** As air becomes saturated, water vapor begins to condense on the nearest available surface.

a. The atmosphere is never completely clean; an abundance of microscopic solid particles suspended in the air are condensation surfaces.

1) These particles, such as salt, dust, and combustion by-products are condensation nuclei.

2) Some condensation nuclei have an affinity for water and can induce condensation or sublimation even when air is almost, but not completely, saturated.

b. As water vapor condenses or sublimates on condensation nuclei, liquid or ice particles begin to grow.

1) Whether the particles are liquid or ice does not depend entirely on temperature.

    a) Liquid water may be present at temperatures well below freezing.

4. **Supercooled Water**

a. Freezing is complex, and liquid water droplets often condense or persist at temperatures colder than 0°C because the molecular motion of the droplet remains large enough to weaken any formation of an ice crystal within the droplet.

1) Water droplets colder than 0°C are supercooled.
2) When they strike an exposed object, the impact induces freezing.
3) Impact freezing of supercooled water can result in aircraft icing.

b. Supercooled water drops very often are in abundance in clouds at temperatures between 0°C and −15°C, with decreasing amounts at colder temperatures.

1) Usually, at temperatures colder than −15°C, sublimation is prevalent, and clouds and fog may be mostly ice crystals with a lesser amount of supercooled water.

2) Strong vertical currents may carry supercooled water to great heights where temperatures are much colder than −15°C.

    a) Supercooled water has been observed at temperatures colder than −40°C.

5.   **Dew and Frost**

    a.   During clear nights with little or no wind, vegetation often cools by radiation to a temperature at or below the dew point of the adjacent air.

        1)   Moisture then collects on the leaves just as it does on a pitcher of ice water in a warm room.

        2)   Heavy dew often collects on grass and plants while none collects on pavements or large solid objects.

            a)   These larger objects absorb abundant heat during the day, lose it slowly during the night, and cool below the dew point only in extreme cases.

    b.   Frost forms in much the same way as dew. The difference is that the dew point of surrounding air must be colder than freezing.

        1)   Water vapor then sublimates directly as ice crystals or frost rather than condensing as dew.

        2)   Sometimes dew forms and later freezes; however, frozen dew is easily distinguished from frost.

            a)   Frozen dew is hard and transparent while frost is white and opaque.

C.   **Cloud Formation**

1.   Normally, air must become saturated for condensation or sublimation to occur.

    a.   Saturation may result from cooling temperature, increasing dew point, or both.

        1)   Cooling is far more predominant.

2.   Three basic processes may cool air to saturation.

    a.   Air moving over a colder surface,
    b.   Stagnant air overlying a cooling surface,
    c.   Expansional (adiabatic) cooling in upward-moving air.

        1)   Expansional cooling is the major cause of cloud formation.
        2)   See Part I, Chapter 6, Stable and Unstable Air, beginning on page 55.

3.   A cloud is a visible collection of minute water or ice particles suspended in air.

    a.   If the cloud is near the ground, it is called fog.

    b.   When entire layers of air cool to saturation, fog or sheet-like clouds result.

    c.   Saturation of a localized updraft produces a towering cloud.

    d.   A cloud may be composed entirely of liquid water, ice crystals, or a mixture of the two.

D. **Precipitation**

1. Precipitation is an all-inclusive term denoting drizzle, rain, snow, ice pellets, hail, and ice crystals.

    a. Precipitation occurs when these particles grow in size and weight until the atmosphere no longer can suspend them and they fall.

2. These particles grow primarily in two ways.

    a. Once a water droplet or ice crystal forms, it continues to grow by added condensation or sublimation directly onto the particle.

        1) This is the slower of the two methods and usually results in drizzle or very light rain or snow.

    b. Cloud particles collide and merge into a larger drop in the more rapid growth process. This process produces larger precipitation particles and does so more rapidly than the simple condensation growth process.

        1) Upward currents enhance the growth rate and also support larger drops, as shown in Figure 36 below.

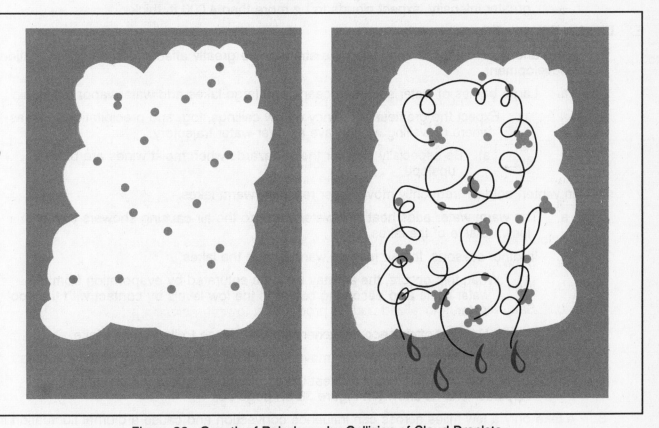

Figure 36. Growth of Raindrops by Collision of Cloud Droplets

        2) Precipitation formed by merging drops with mild upward currents can produce light to moderate rain and snow.

        3) Strong upward currents support the largest drops and build clouds to great heights.

            a) They can produce heavy rain, heavy snow, and hail.

3. **Liquid, Freezing, and Frozen Precipitation**

    a. Precipitation that forms and remains in a liquid state is either rain or drizzle.

    b. Sublimation forms snowflakes, and they reach the ground as snow if temperatures remain below freezing.

    c. Precipitation can change its state as the temperature of its environment changes.

        1) Falling snow may melt in warmer layers of air at lower altitudes to form rain.

        2) Rain falling through colder air may become supercooled, freezing on impact as freezing rain; or it may freeze during its descent, falling as ice pellets.

            a) Ice pellets always indicate freezing rain at higher altitude.

    d. Sometimes strong upward currents sustain large supercooled water drops until some freeze; subsequently, other drops freeze to them forming hailstones.

4. To produce significant precipitation, clouds usually must be 4,000 ft. thick or more.

    a. The thicker the clouds, the heavier the precipitation is likely to be.

    b. When arriving at or departing from an airport reporting precipitation of light or greater intensity, expect clouds to be more than 4,000 ft. thick.

E. **Land and Water Effects**

1. Land and water surfaces underlying the atmosphere greatly affect cloud and precipitation development.

    a. Large bodies of water such as oceans and large lakes add water vapor to the air.

        1) Expect the greatest frequency of low ceilings, fog, and precipitation in areas where prevailing winds have an over-water trajectory.

            a) Be especially alert for these hazards when moist winds are blowing upslope.

2. In winter, cold air frequently moves over relatively warm lakes.

    a. The warm water adds heat and water vapor to the air causing showers to the leeward side of the lakes.

    b. In other seasons, the air may be warmer than the lakes.

        1) When this occurs, the air may become saturated by evaporation from the water while also becoming cooler in the low levels by contact with the cool water.

            a) Fog often becomes extensive and dense to the lee of a lake.

    c. Figure 37 on page 53 illustrates movement of air over both warm and cold lakes.

    d. Strong cold winds across the Great Lakes often carry precipitation to the Appalachians as shown in Figure 38 on page 53.

3. A lake only a few miles across can influence convection and cause a diurnal fluctuation in cloudiness.

    a. During the day, cool air over the lake blows toward the land, and convective clouds form over the land.

    b. At night, the pattern reverses; clouds tend to form over the lake as cool air from the land flows over the lake, creating convective clouds over the water.

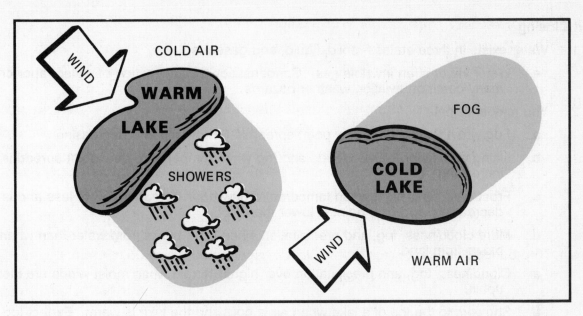

Figure 37. Lake Effects

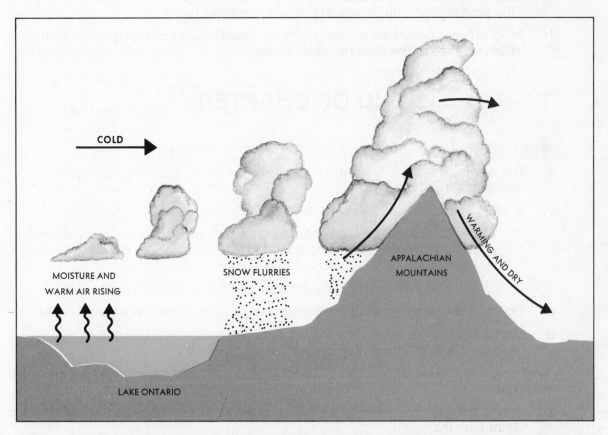

Figure 38. Strong Cold Winds across the Great Lakes Absorb Water Vapor and May Carry
Showers as Far Eastward as the Appalachians

F. **In Closing**

1. Water exists in three states -- solid, liquid, and gaseous.

    a. Water vapor is an invisible gas. Condensation or sublimation of water vapor creates many common aviation weather hazards.

2. You may anticipate

    a. Fog when temperature-dew point spread is 5°F or less and decreasing.

    b. Lifting or clearing of low clouds and fog when temperature-dew point spread is increasing.

    c. Frost on a clear night when temperature-dew point spread is 5°F or less and is decreasing, and dew point is lower than 32°F.

    d. More cloudiness, fog, and precipitation when wind blows from water than when it blows from land.

    e. Cloudiness, fog, and precipitation over higher terrain when moist winds are blowing uphill.

    f. Showers to the lee of a lake when air is cold and the lake is warm. Expect fog to the lee of the lake when the air is warm and the lake is cold.

    g. Clouds at least 4,000 ft. thick when significant precipitation is reported. The heavier the precipitation, the thicker the clouds are likely to be.

    h. Icing on your aircraft when flying through liquid clouds or precipitation with temperatures at the freezing level or lower.

END OF CHAPTER

CHAPTER SIX
STABLE AND UNSTABLE AIR

> Please take a few minutes to study each of the concepts listed above and anticipate/imagine what they are and how they relate to the other listed concepts.

A. Definitions

1. A *stable* atmosphere resists any upward or downward displacement.

2. An *unstable* atmosphere allows an upward or downward disturbance to grow into a vertical, or convective, current.

B. Adiabatic Cooling and Heating

1. Any time air moves upward, it expands because of decreasing atmospheric pressure.

   a. Conversely, downward moving air is compressed by increasing pressure.
   b. But as pressure and volume change, temperature also changes.

2. When air expands, it cools, and when compressed, it warms.

   a. These changes are **adiabatic**, i.e., no heat is removed from or added to the air.

   b. We frequently use the terms **expansional**, or **adiabatic**, **cooling** and **compressional**, or **adiabatic**, **heating**.

   c. The adiabatic rate of change of temperature is constant in unsaturated air but varies with the amount of moisture in saturated air.

3. **Unsaturated air** moving upward and downward cools and warms at about 3°C per 1,000 ft. (5.4°F per 1,000 ft.).

   a. This rate is the dry adiabatic rate of temperature change and is independent of the temperature of the mass of air through which the vertical movements occur.

   b. Figure 41 illustrates a "Chinook Wind," which is a warm wind blowing down the side of a mountain -- an excellent example of dry adiabatic warming.

      1) As the cold, relatively dry air moves, it sinks down the side of the mountain.

      2) As it is compressed by the increasing atmospheric pressure, it becomes warmer through adiabatic heating.

4. When **saturated air** moves upward and cools, condensation occurs.

   a. Latent heat released through condensation (see Part I, Chapter 5, Moisture, Cloud Formation, and Precipitation, beginning on page 45) partially offsets the expansional cooling.

      1) Thus, the saturated adiabatic rate of cooling is slower than the dry rate.

      2) The saturated cooling rate depends on saturation temperature (dew point) of the air, i.e., the relative humidity.

   b. Because warmer air can hold more water before becoming saturated, more latent heat is released through condensation in saturated warm air than in cold.

      1) Thus, the saturated adiabatic rate of cooling is less in warm air than in cold.

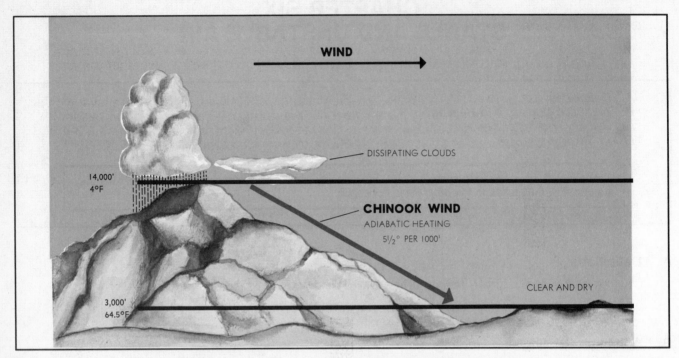

Figure 41. Adiabatic Warming of Downward Moving Air

c.   When saturated air moves downward, it heats at the same rate as it cools on ascent *provided* liquid water evaporates rapidly enough to maintain saturation.

    1)   Minute water droplets in the air evaporate rapidly enough.

    2)   Larger drops evaporate more slowly and complicate the moist adiabatic process in downward moving air.

5.   **Adiabatic Cooling and Vertical Air Movement**

a.   At this point we should clarify the terms "ambient, or existing, lapse rate" and "adiabatic rates of cooling."

    1)   As you move upward through the atmosphere, the temperature of the air around you generally changes with altitude.

        a)   This change is the ambient (existing) lapse rate.

        b)   In the standard atmosphere, the lapse rate is 2°C per 1,000 ft. (see Part I, Chapter 2, Temperature, beginning on page 7).

    2)   As a parcel of air is forced upward, it expands, and therefore cools, at a given rate.

        a)   This is the adiabatic rate of cooling, or adiabatic lapse rate.

        b)   This rate depends on the amount of moisture present in the parcel of air that is forced upward.

b.   If a parcel of air is forced upward into the atmosphere, we must consider two possibilities:

    1)   As the sample cools, it may become colder than the surrounding air, or
    2)   Even though it cools, the air may remain warmer than the surrounding air.

c.   If the upward moving air becomes colder than the surrounding air, the adiabatic lapse rate is greater than the ambient (existing) lapse rate, and the parcel of air sinks.

    1)   If it remains warmer, the adiabatic lapse rate is less than the ambient (existing) lapse rate, and the parcel of air continues to rise as a convective current.

C. Stability and Instability

1.   EXAMPLE: The difference between the ambient (existing) lapse rate of a given mass of air and the adiabatic rates of cooling in upward moving air determines whether the air is stable or unstable.

      a.   In Figure 42 below we have, for three situations, filled a balloon at sea level with air at 31°C -- the same as the ambient temperature. We have carried the balloon to 5,000 ft. In each situation, the air in the balloon expanded and cooled at the dry adiabatic rate of 3°C for each 1,000 ft. to a temperature of 16°C at 5,000 ft.

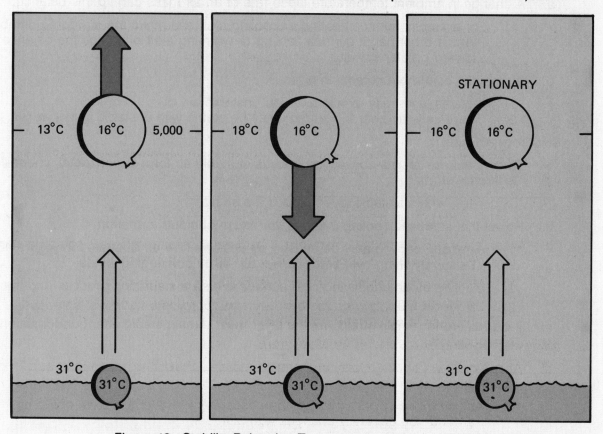

Figure 42. Stability Related to Temperatures Aloft and Adiabatic Cooling

1)   In the first situation (left), air inside the balloon, even though cooling adiabatically, remains warmer than surrounding air.

      a)   Vertical motion is favored.

      b)   The colder, more dense surrounding air forces the balloon further upward.

      c)   This air is unstable, and a convective current develops.

2)   In the second situation (center), the air aloft is warmer.

      a)   Air inside the balloon, cooling adiabatically, now becomes colder than the surrounding air.

      b)   The balloon sinks under its own weight, returning to its original position when the lifting force is removed.

      c)   The air is stable, and spontaneous convection is impossible.

3)   In the third situation, temperature of air inside the balloon is the same as that of surrounding air. The balloon will remain at rest.

      a)   This condition is neutrally stable; that is, the air is neither stable nor unstable.

4)   Note that, in all three situations, temperature of air in the expanding balloon cooled at a fixed rate.

   a)   The differences in the three conditions depend, therefore, on the temperature differences between the surface and 5,000 ft., that is, on the ambient (existing) lapse rates.

2.   Stability runs the gamut from absolutely stable to absolutely unstable, and the atmosphere usually is in a delicate balance somewhere in between.

   a.   A change in ambient temperature lapse rate of an air mass can tip this balance.

      1)   For example, surface heating or cooling aloft can make the air more unstable; on the other hand, surface cooling or warming aloft often tips the balance toward greater stability.

   b.   Air may be stable or unstable in layers.

      1)   A stable layer may overlie and cap unstable air; or
      2)   Conversely, air near the surface may be stable with unstable layers above.

3.   **Stratiform Clouds**

   a.   Since stable air resists convection, clouds in stable air form in horizontal, sheet-like layers or "strata."

      1)   Thus, within a stable layer, clouds are *stratiform*.

   b.   Recall that adiabatic cooling is the major cause of cloud formation.

      1)   Adiabatic cooling may be caused by upslope flow as illustrated in Figure 43 below, by lifting over cold, denser air, or by converging winds.

      2)   Cooling by an underlying cold surface is also a stabilizing process and may produce fog.

   c.   If clouds are to remain stratiform, the layer must remain stable after condensation has occurred.

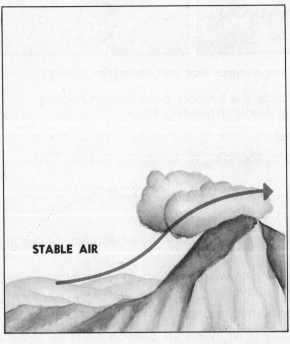
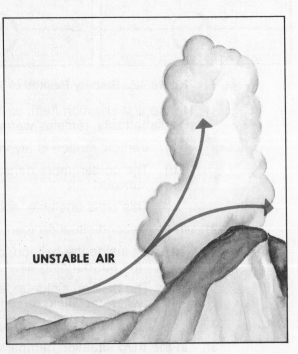

Figure 43. Stable and Unstable Air Forced Upward

4. Cumuliform Clouds

    a.   Unstable air favors convection. A cumulus (heaping) cloud forms in a convective updraft and builds upward, also shown in Figure 43 on the previous page.

        1)   Thus, within an *unstable* layer, clouds are *cumuliform*, and the vertical extent of the cloud depends on the depth of the unstable layer.

    b.   Initial lifting to trigger a cumuliform cloud may be the same as that for lifting stable air.

        1)   In addition, convection may be set off by surface heating (see Part I, Chapter 4, Wind, beginning on page 29).

    c.   Air may be unstable or slightly stable before condensation occurs, but for convective cumuliform clouds to develop, it must be unstable after saturation.

        1)   Cooling in the updraft is now at the slower moist adiabatic rate because of the release of latent heat of condensation.

            a)   Temperature in the saturated updraft is warmer than ambient temperature, and convection is spontaneous.

        2)   Updrafts accelerate until temperature within the cloud cools below the ambient temperature.

            a)   This condition occurs where the unstable layer is capped by a stable layer often marked by a temperature inversion.

        3)   Vertical heights range from the shallow fair weather cumulus to the giant thunderstorm cumulonimbus -- the ultimate in atmospheric instability capped by the tropopause.

    d.   You can estimate the heights of cumuliform cloud bases using surface temperature-dew point spread.

        1)   Unsaturated air in a convective current cools at about 5.4°F (3°C) per 1,000 ft.; dew point decreases at about 1°F (5/9°C).

            a)   Thus, in a convective current, temperature and dew point converge at about 4.4°F (2.5°C) per 1,000 ft. as illustrated in Figure 44 on page 60.

            b)   The point at which they converge is the base of the clouds.

        2)   We can get a quick *estimate* of a convective cloud base in thousands of feet by rounding these values and dividing into the spread.

            a)   When using Fahrenheit, divide by 4; when using Celsius, divide by 2.

            b)   This method of estimating is reliable only with instability clouds and during the warmer part of the day.

        3)   EXAMPLE: If the surface temperature is 88°F and the dew point is 48°F, the spread is 40°F. Dividing by 4 yields 10. Thus, the base of the cloud is 10,000 ft.

    e.   When unstable air lies above stable air, convective currents aloft sometimes form middle and high level cumuliform clouds.

        1)   In relatively shallow layers, they occur as altocumulus and ice crystal cirrocumulus clouds.

        2)   Altocumulus castellanus clouds develop in deeper midlevel unstable layers.

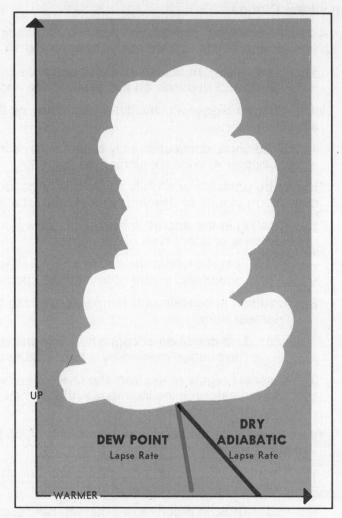

DEW POINT
Lapse Rate

DRY ADIABATIC
Lapse Rate

UP

WARMER

Figure 44. Cloud Base Determination

5. **Merging Stratiform and Cumuliform**

   a.   A layer of stratiform clouds may sometimes form in a mildly stable layer while a few ambitious convective clouds penetrate the layer, thus merging stratiform with cumuliform.

   b.   Convective clouds may be almost or entirely embedded in a massive stratiform layer and pose an unseen threat to instrument flight.

D. **In Closing**

1. The usual convection in unstable air gives a bumpy ride; only at times is it violent enough to be hazardous.

    a. In stable air, flying is usually smooth but sometimes can be plagued by low ceiling and visibility.

    b. In preflight planning, you need to take into account stability or instability and any associated hazards.

2. Certain observations you can make on your own:

    a. Thunderstorms are sure signs of violently unstable air. Give these storms a wide berth.

    b. Showers and clouds towering upward indicate strong updrafts and rough (turbulent) air. Stay clear of these clouds.

    c. Fair weather cumulus clouds often indicate bumpy turbulence beneath and in the clouds. The cloud tops indicate the approximate upper limit of convection; flight above is usually smooth.

    d. Dust devils are a sign of dry, unstable air, usually to considerable height. Your ride may be fairly rough unless you can get above the instability.

    e. Stratiform clouds indicate stable air. Flight generally will be smooth, but low ceiling and visibility might require IFR.

    f. Restricted visibility at or near the surface over large areas usually indicates stable air. Expect a smooth ride, but poor visibility may require IFR.

    g. Thunderstorms may be embedded in stratiform clouds posing an unseen threat to instrument flight.

    h. Even in clear weather, you have some clues to stability, viz.:

        1) When temperature decreases uniformly and rapidly as you climb (approaching 3°C per 1,000 ft.), you have an indication of unstable air.

        2) If temperature remains unchanged or decreases only slightly with altitude, the air tends to be stable.

        3) If the temperature increases with altitude through a layer -- an inversion -- the layer is stable and convection is suppressed. Air may be unstable beneath the inversion.

        4) When air near the surface is warm and moist, suspect instability. Surface heating, cooling aloft, converging or upslope winds, or an invading mass of colder air may lead to instability and cumuliform clouds.

END OF CHAPTER

CHAPTER SEVEN
CLOUDS

A. **Introduction** -- To you as a pilot, clouds are your weather "signposts in the sky."

    1.   They give you an indication of air motion, stability, and moisture.

    2.   Clouds help you visualize weather conditions and potential weather hazards you might encounter in flight.

    3.   The photographs on pages 64 through 71 illustrate some of the basic cloud types discussed below.

        a.   The caption with each photograph describes the cloud type and its significance to flight.

        b.   Study the descriptions and potential hazards posed by each type.

B. **Classification** -- Clouds are classified according to the way they are formed.

    1.   Clouds formed by vertical currents in unstable air are *cumulus*, meaning *accumulation* or *heap*.

        a.   They are characterized by their lumpy, billowy appearance.

    2.   Clouds formed by the cooling of a stable layer are *stratus*, meaning *stratified* or *layered*.

        a.   They are characterized by their uniform, sheet-like appearance.

    3.   In addition to the above, the prefix *nimbo* or the suffix *nimbus* means raincloud.

        a.   Thus, stratified clouds from which rain is falling are *nimbostratus*.

        b.   A heavy, swelling cumulus type cloud which produces precipitation is a *cumulonimbus*.

    4.   Clouds broken into fragments are often identified by adding the suffix *fractus*.

        a.   EXAMPLE: Fragmentary cumulus is *cumulus fractus*.

C. **Identification** -- For identification purposes, you must also be concerned with the four families of clouds.

    1.   The **high cloud** family is cirriform and includes cirrus, cirrocumulus, and cirrostratus.

        a.   They are composed almost entirely of ice crystals.

        b.   The height of the bases of these clouds ranges from about 16,500 to 45,000 ft. in middle latitudes.

        c.   Figures 45 through 47 on pages 64 and 65 are photographs of high clouds.

2.   In the **middle cloud** family are the altostratus, altocumulus, and nimbostratus clouds.

    a.   These clouds are primarily water, much of which may be supercooled.

    b.   The height of the bases of these clouds ranges from about 6,500 to 23,000 ft. in middle latitudes.

    c.   Figures 48 through 52 on pages 66 through 69 are photographs of middle clouds.

3.   In the **low cloud** family are the stratus, stratocumulus, and fair weather cumulus clouds.

    a.   Low clouds are almost entirely water, but at times the water may be supercooled.

        1)   Low clouds at subfreezing temperatures can also contain snow and ice particles.

    b.   The bases of these clouds range from near the surface to about 6,500 ft. in middle latitudes.

    c.   Figures 53 through 55 on pages 69 and 70 are photographs of low clouds.

4.   The **clouds with extensive vertical development** family includes towering cumulus and cumulonimbus.

    a.   These clouds usually contain supercooled water above the freezing level.

        1)   But when a cumulus grows to great heights, water in the upper part of the cloud freezes into ice crystals, forming a cumulonimbus.

    b.   The heights of cumuliform cloud bases range from 1,000 ft. or less to above 10,000 ft.

    c.   Figures 56 and 57 on page 71 are photographs of clouds with extensive vertical development.

FIGURE 45.   CIRRUS. Cirrus are thin, feather-like ice crystal clouds in patches or narrow bands. Larger ice crystals often trail downward in well-defined wisps called "mares' tails." Wispy, cirrus-like, these contain no significant icing or turbulence. Dense, banded cirrus, which often are turbulent, are discussed in chapter 13.

FIGURE 46.   CIRROCUMULUS. Cirrocumulus are thin clouds, the individual elements appearing as small white flakes or patches of cotton. May contain highly supercooled water droplets. Some turbulence and icing.

FIGURE 47.   CIRROSTRATUS. Cirrostratus is a thin whitish cloud layer appearing like a sheet or veil. Cloud elements are diffuse, sometimes partially striated or fibrous. Due to their ice crystal makeup, these clouds are associated with halos— large luminous circles surrounding the sun or moon. No turbulence and little if any icing. The greatest problem flying in cirriform clouds is restriction to visibility. They can make the strict use of instruments mandatory.

FIGURE 48.   ALTOCUMULUS. Altocumulus are composed of white or gray colored layers or patches of solid cloud. The cloud elements may have a waved or roll-like appearance. Some turbulence and small amounts of icing.

FIGURE 49.   ALTOSTRATUS. Altostratus is a bluish veil or layer of clouds. It is often associated with altocumulus and sometimes gradually merges into cirrostratus. The sun may be dimly visible through it. Little or no turbulence with moderate amounts of ice.

FIGURE 50. ALTOCUMULUS CASTELLANUS. Altocumulus castellanus are middle level convective clouds. They are characterized by their billowing tops and comparatively high bases. They are a good indication of mid-level instability. Rough turbulence with some icing.

FIGURE 51.   STANDING LENTICULAR ALTOCUMULUS CLOUDS. Standing lenticular altocumulus clouds are formed on the crests of waves created by barriers in the wind flow. The clouds show little movement, hence the name *standing*. Wind, however, can be quite strong blowing through such clouds. They are characterized by their smooth, polished edges. The presence of these clouds is a good indication of very strong turbulence and should be avoided. Chapter 9, "Turbulence," further explains the significance of this cloud.

FIGURE 52. NIMBOSTRATUS. Nimbostratus is a gray or dark massive cloud layer, diffused by more or less continuous rain, snow, or ice pellets. This type is classified as a middle cloud although it may merge into very low stratus or stratocumulus. Very little turbulence, but can pose a serious icing problem if temperatures are near or below freezing.

FIGURE 53. STRATUS. Stratus is a gray, uniform, sheet-like cloud with relatively low bases. When associated with fog or precipitation, the combination can become troublesome for visual flying. Little or no turbulence, but temperatures near or below freezing can create hazardous icing conditions.

FIGURE 54. STRATOCUMULUS. Stratocumulus bases are globular masses or rolls unlike the flat, sometimes indefinite, bases of stratus. They usually form at the top of a layer mixed by moderate surface winds. Sometimes, they form from the breaking up of stratus or the spreading out of cumulus. Some turbulence, and possible icing at subfreezing temperatures. Ceiling and visibility usually better than with low stratus.

FIGURE 55. CUMULUS. Fair weather cumulus clouds form in convective currents and are characterized by relatively flat bases and dome-shaped tops. Fair weather cumulus do not show extensive vertical development and do not produce precipitation. More often, fair weather cumulus indicates a shallow layer of instability. Some turbulence and no significant icing.

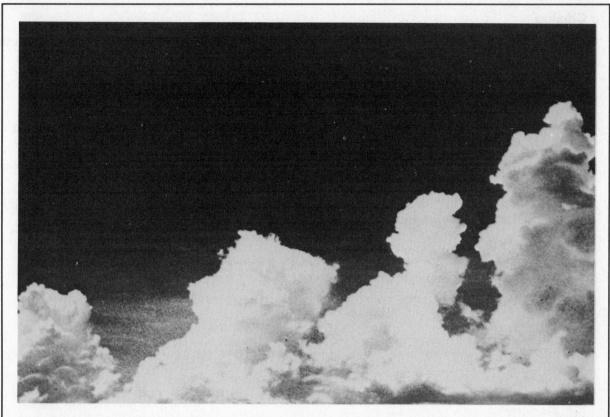

FIGURE 56. TOWERING CUMULUS. Towering cumulus signifies a relatively deep layer of unstable air. It shows considerable vertical development and has billowing *cauliflower* tops. Showers can result from these clouds. Very strong turbulence; some clear icing above the freezing level.

FIGURE 57. CUMULONIMBUS. Cumulonimbus are the ultimate manifestation of instability. They are vertically developed clouds of large dimensions with dense *boiling* tops often crowned with thick veils of dense cirrus (the anvil). Nearly the entire spectrum of flying hazards are contained in these clouds including violent turbulence. They should be avoided at all times! This cloud is the thunderstorm cloud and is discussed in detail in chapter 11, "Thunderstorms."

D. **In Closing**

1. Clouds give an indication of weather conditions and weather hazards you may encounter. Remember there are four families of clouds:

    a. The high cloud family, including cirrus, cirrocumulus, and cirrostratus

    b. The middle cloud family, including altostratus, altocumulus, and nimbostratus

    c. The low cloud family, including stratus, stratocumulus, and fair weather cumulus clouds

    d. The clouds with extensive vertical development family, including towering cumulus and cumulonimbus

2. Study the photographs and descriptions on the preceding pages so you will be familiar with each cloud type and its potential hazards.

END OF CHAPTER

CHAPTER EIGHT
AIR MASSES AND FRONTS

A. Air Masses

1. When a body of air comes to rest or moves slowly over an extensive area that has fairly uniform properties of temperature and moisture, the body of air takes on those properties.

   a. Thus, the air over the area becomes a kind of entity as illustrated in Figure 58 below and has a fairly uniform horizontal distribution of its properties.

   b. The area over which the air mass acquires its identifying distribution of moisture and temperature is known as its **source region**.

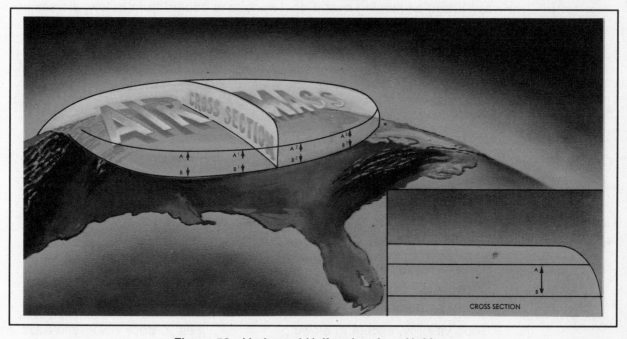

Figure 58. Horizontal Uniformity of an Air Mass

2.   Source regions are many and varied, but the best source regions for air masses are large snow or ice-covered polar regions, cold northern oceans, tropical oceans, and large desert areas.

   a.   Mid-latitudes are poor source regions because air movement is so constantly varied that air masses have little opportunity to stagnate and take on the properties of the underlying region.

3.   Just as an air mass took on the properties of its source region, it will tend to take on properties of a new underlying surface when it moves away from its source region, thus becoming modified.

   a.   The degree of modification depends on the speed with which the air mass moves, the nature of the region over which it moves, and the temperature difference between the new surface and the air mass.

   b.   Some ways air masses are modified are

      1)   Warming from below.

         a)   Cool air moving over a warm surface is heated from below, generating instability and increasing the possibility of showers.

      2)   Cooling from below.

         a)   Warm air moving over a cool surface is cooled from below, increasing stability. If air is cooled to its dew point, stratus and/or fog forms.

      3)   Addition of water vapor.

         a)   Evaporation from water surfaces and falling precipitation adds water vapor to the air. When the water is warmer than the air, evaporation can raise the dew point sufficiently to saturate the air and form stratus or fog.

      4)   Subtraction of water vapor.

         a)   Water vapor is removed by condensation and precipitation.

4.   The **stability** of an air mass determines its typical weather characteristics.

   a.   Characteristics typical of an unstable and a stable air mass are as follows:

Unstable Air	Stable Air
Cumuliform clouds	Stratiform clouds and fog
Showery precipitation	Continuous precipitation
Rough air (turbulence)	Smooth air
Good visibility, except in blowing obstructions	Fair to poor visibility in haze and smoke

B. Fronts

1. As air masses move out of their source regions, they come in contact with other air masses of different properties. The zone between two different air masses is a *frontal zone*, or *front*.

    a. Across this zone, temperature, humidity, and wind often change rapidly over short distances.

2. **Discontinuities**. When you pass through a front, the change from the properties of one air mass to those of the other is sometimes quite abrupt. Abrupt changes indicate a narrow frontal zone. At other times, the change of properties is very gradual, indicating a broad and diffuse frontal zone.

    a. **Temperature** is one of the most easily recognized discontinuities across a front.

        1) At the surface, the passage of a front usually causes a noticeable temperature change.

        2) When flying through a front, you note a significant change in temperature, especially at low altitudes.

        3) Remember that the temperature change, even when gradual, is faster and more pronounced than a change during a flight wholly within one air mass.

            a) Thus, for safety, obtain a new altimeter setting after flying through a front.

    b. **Dew point** and temperature-dew point spread usually differ across a front.

        1) This difference helps identify the front and may give a clue to changes in cloudiness and/or fog.

    c. **Wind** always changes across a front.

        1) Wind discontinuity may be in direction, in speed, or in both.
        2) Be alert for a wind shift when flying in the vicinity of a frontal surface.
        3) The relatively sudden change in wind also creates wind shear.

    d. **Pressure**. A front lies in a pressure trough, and pressure generally is higher in the cold air.

        1) Thus, when you cross a front directly into colder air, pressure usually rises abruptly.

        2) When you approach a front toward warm air, pressure generally falls until you cross the front and then remains steady or falls slightly in the warm air.

        3) However, pressure patterns vary widely across fronts, and your course may not be directly across a front.

        4) The important thing to remember is that when crossing a front, you will encounter a difference in the rate of pressure change; be especially alert in keeping your altimeter setting current.

3.   The three principal types of fronts are the **cold front**, the **warm front**, and the **stationary front**.

a.   The leading edge of an advancing cold air mass is a **cold front**.

1)   At the surface, cold air is overtaking and replacing warmer air.

2)   Cold fronts move at about the speed of the wind component perpendicular to the front just above the frictional layer.

3)   Figure 59 on page 77 shows the vertical cross section of a cold front and the symbol depicting it on a surface weather chart.

a)   The vertical cross section in the top illustration shows the frontal slope.

i)   The frontal slope is steep near the leading edge as cold air replaces warm air.

ii)   The solid heavy arrow shows movement of the front.

iii)   Warm air may descend over the front as indicated by the dashed arrows, but more commonly, the cold air forces warm air upward over the frontal surface as shown by the solid arrows.

b)   The symbol in the bottom illustration is a line with barbs pointing in the direction of movement.

i)   If a map is in color, a blue line represents the cold front.

4)   A shallow cold air mass or a slow moving cold front may have a frontal slope more like a warm front shown in Figure 60 on page 79.

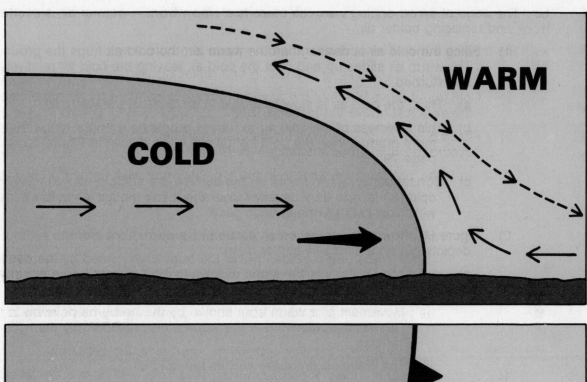

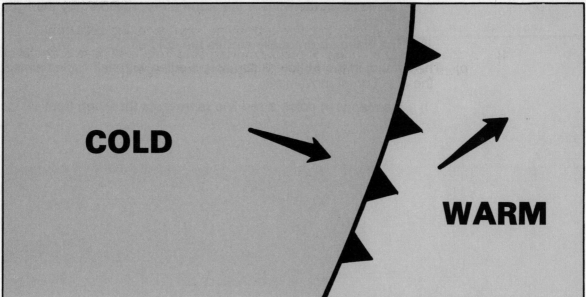

Figure 59. Cross Section and Weather Map Symbol of a Cold Front

b.   The edge of an advancing warm air mass is a **warm front** -- warmer air is overtaking and replacing colder air.

    1)   Since the cold air is denser than the warm air, the cold air hugs the ground. The warm air slides up and over the cold air leaving the cold air relatively undisturbed.

        a)   Thus, the cold air is slow to retreat in advance of the warm air.

        b)   This slowness of the cold air to retreat produces a frontal slope that is more gradual than the cold frontal slope, as shown in Figure 60 on page 79.

        c)   Consequently, warm fronts on the surface are seldom as well marked as cold fronts, and they usually move about half as fast when the general wind flow is the same in each case.

    2)   Figure 60 shows the vertical cross section of a warm front and the symbol depicting it on a surface weather chart.

        a)   In the top illustration, the slope of a warm front generally is more shallow than slope of a cold front.

            i)   Movement of a warm front shown by the heavy black arrow is slower than the wind in the warm air represented by the light solid arrows.

            ii)   The warm air gradually erodes the cold air.

        b)   The symbol in the bottom illustration is a line with half circles aimed in the direction of movement.

            i)   If a map is in color, a red line represents the warm front.

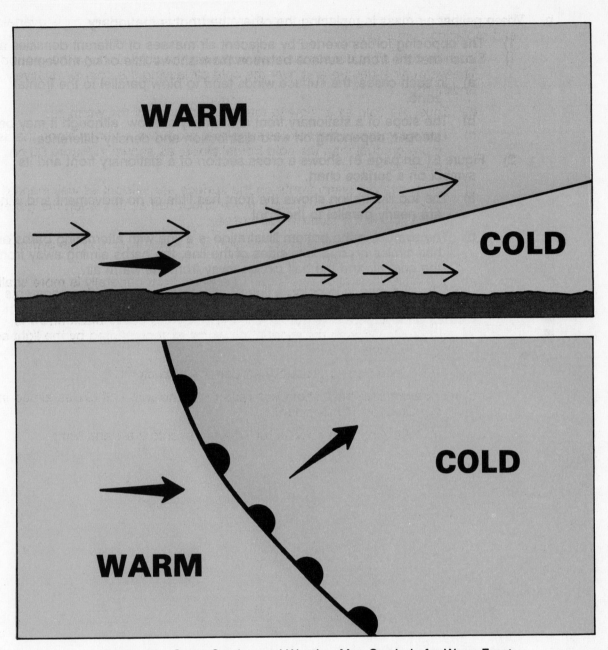

Figure 60. Cross Section and Weather Map Symbol of a Warm Front

   c.   When neither air mass is replacing the other, the front is **stationary**.

      1)   The opposing forces exerted by adjacent air masses of different densities are such that the frontal surface between them shows little or no movement.

         a)   In such cases, the surface winds tend to blow parallel to the frontal zone.

         b)   The slope of a stationary front is normally shallow, although it may be steeper, depending on wind distribution and density difference.

      2)   Figure 61 on page 81 shows a cross section of a stationary front and its symbol on a surface chart.

         a)   The top illustration shows the front has little or no movement and winds are nearly parallel to the front.

         b)   The symbol in the bottom illustration is a line with alternating barbs and half circles on opposite sides of the line, the barbs aiming away from the cold air and the half circles away from the warm air.

            i)   If a map is in color, a line of alternating red and blue segments represents the stationary front.

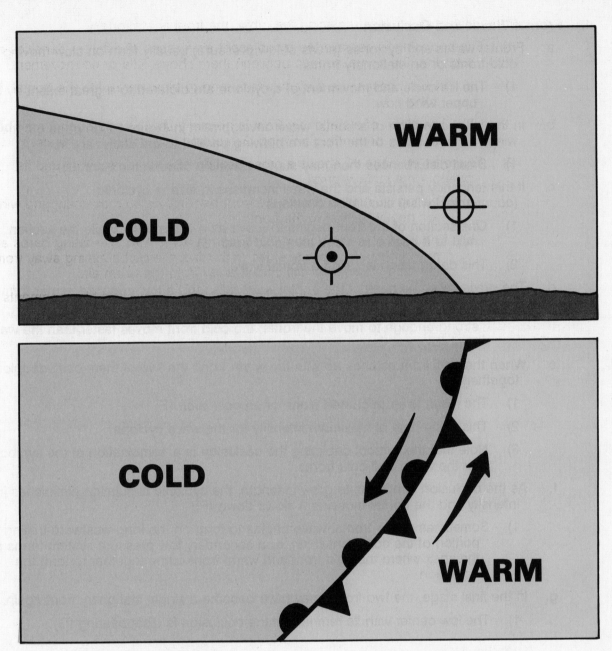

Figure 61. Cross Section and Weather Map Symbol of a Stationary Front

4. **Frontal Waves and Occlusion**

a. Frontal waves and cyclones (areas of low pressure) usually form on slow-moving cold fronts or on stationary fronts.

   1) The life cycle and movement of a cyclone are dictated to a great extent by the upper wind flow.

b. In the initial condition of a frontal wave development in Figure 62 on page 83, the winds on both sides of the front are blowing parallel to the stationary front (A).

   1) Small disturbances then may start a wavelike bend in the front (B).

c. If this tendency persists and the wave increases in size, a cyclonic (counterclockwise) circulation develops.

   1) One section of the front begins to move as a warm front, while the section next to it begins to move as a cold front (C).

   2) This deformation is called a *frontal wave*.

d. The pressure at the peak of the frontal wave falls, and a low-pressure center forms.

   1) The cyclonic circulation becomes stronger, and the surface winds are now strong enough to move the fronts; the cold front moves faster than the warm front (D).

e. When the cold front catches up with the warm front, the two of them *occlude* (close together).

   1) The result is an **occluded front**, or an *occlusion* (E).

   2) This is the time of maximum intensity for the wave cyclone.

   3) Note that the symbol depicting the occlusion is a combination of the symbols for the warm and cold fronts.

f. As the occlusion continues to grow in length, the cyclonic circulation diminishes in intensity and the frontal movement slows down (F).

   1) Sometimes a new frontal wave begins to form on the long westward-trailing portion of the cold front (F,G), or a secondary low pressure system forms at the apex where the cold front and warm front come together to form the occlusion.

g. In the final stage, the two fronts may have become a single stationary front again.

   1) The low center with its remnant of the occlusion is disappearing (G).

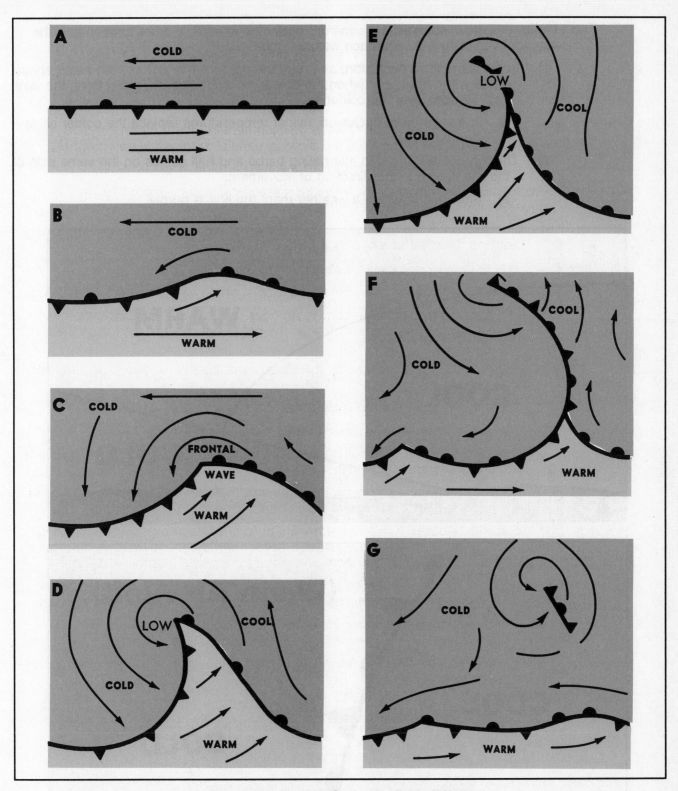

Figure 62. The Life Cycle of a Frontal Wave

h.   Figure 63 below indicates a warm-front occlusion in vertical cross section and the symbol depicting it on a surface weather chart.

1)   In the warm front occlusion, air under the cold front is not as cold as air ahead of the warm front; and when the cold front overtakes the warm front, the less cold air rides over the colder air.

a)   In a warm front occlusion, milder temperatures replace the colder air at the surface.

2)   The symbol is a line with alternating barbs and half circles on the same side of the line aiming in the direction of movement.

a)   Shown in color on a weather map, the line is purple.

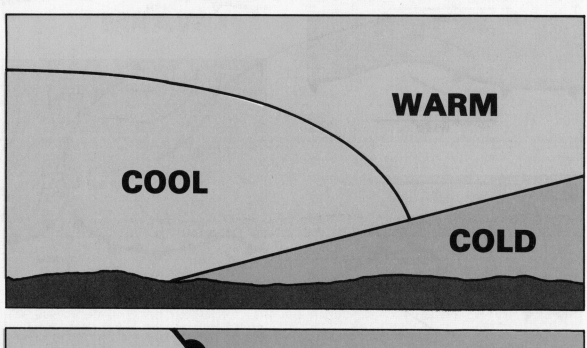

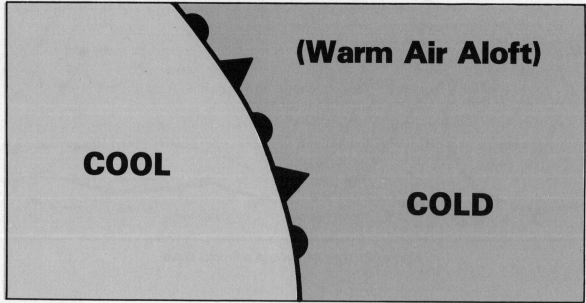

Figure 63. Cross Section and Weather Map Symbol for a Warm-Front Occlusion

i.   Figure 64 below indicates a cold-front occlusion in vertical cross section.

      1)   In the cold-front occlusion, the coldest air is under the cold front.

      2)   When it overtakes the warm front, it lifts the warm front aloft; and cold air replaces cool air at the surface.

      3)   The weather map symbol is the same as used for a warm-front occlusion.

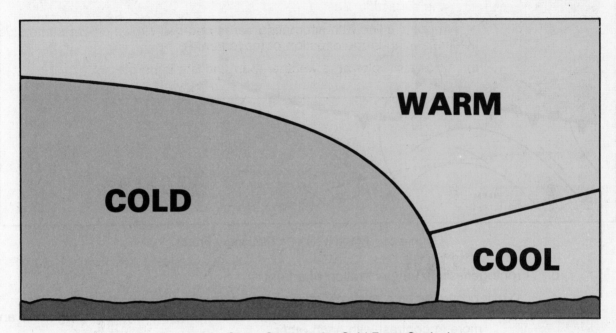

Figure 64. Cross Section of a Cold-Front Occlusion

5. Nonfrontal Lows

a.   As we have learned, fronts lie in troughs, or elongated areas of low pressure.

      1)   These frontal troughs mark the boundaries between air masses of different properties.

      2)   Low-pressure areas lying solely in a homogeneous air mass are called nonfrontal lows.

b.   Nonfrontal lows are infrequent east of the Rocky Mountains in mid-latitudes but do occur occasionally during the warmer months.

      1)   Small nonfrontal lows over the western mountains are common as is the semistationary thermal low in the extreme southwestern United States.

c.   Tropical lows are also nonfrontal.

6.   **Frontolysis and Frontogenesis**

a.   As adjacent air masses modify and as temperature and pressure differences equalize across a front, the front dissipates.

1)   This process, frontolysis, is illustrated in Figure 65 below.

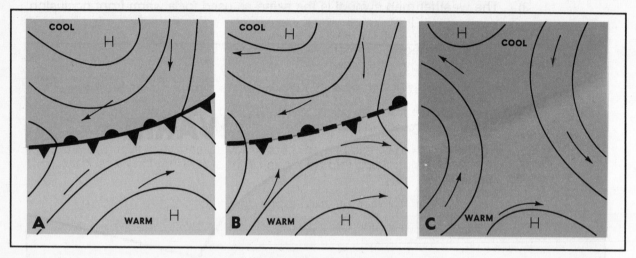

Figure 65. Frontolysis of a Stationary Front

b.   Frontogenesis is the generation of a front.

1)   It occurs when a relatively sharp zone of transition develops over an area between two air masses which have densities gradually becoming more and more in contrast with each other.

a)   The necessary wind flow pattern develops at the same time.

2)   Figure 66 below shows an example of frontogenesis with the symbol.

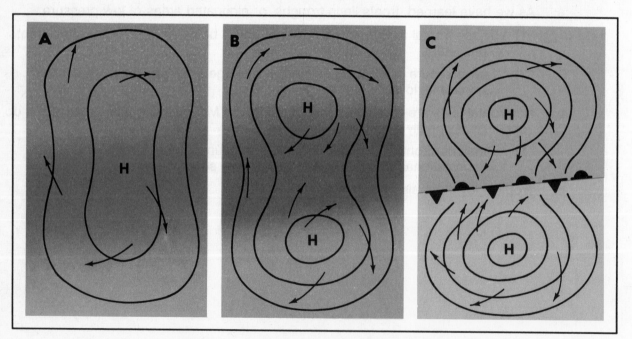

Figure 66. Frontogenesis of a Stationary Front

7.   **Frontal Weather**

   a.   In fronts, flying weather varies from virtually clear skies to extreme hazards including hail, turbulence, icing, low clouds, and poor visibility.

   1)   Weather occurring with a front depends on

   a)   The amount of moisture available,
   b)   The degree of stability of the air that is forced upward,
   c)   The slope of the front,
   d)   The speed of frontal movement, and
   e)   The upper wind flow.

   b.   Sufficient moisture must be available for clouds to form, or there will be no clouds.

   1)   As an inactive front (i.e., no precipitation) comes into an area of moisture, clouds and precipitation may develop rapidly.

   2)   A good example of this is a cold front moving eastward from the dry slopes of the Rocky Mountains into a tongue of moist air from the Gulf of Mexico over the Plains States.

   a)   Thunderstorms may build rapidly and catch a pilot unaware.

   c.   The degree of stability of the lifted air determines whether cloudiness will be predominately stratiform or cumuliform.

   1)   If the warm air overriding the front is stable, stratiform clouds develop.

   a)   Precipitation from stratiform clouds is usually steady, as illustrated in Figure 67 below, and there is little or no turbulence.

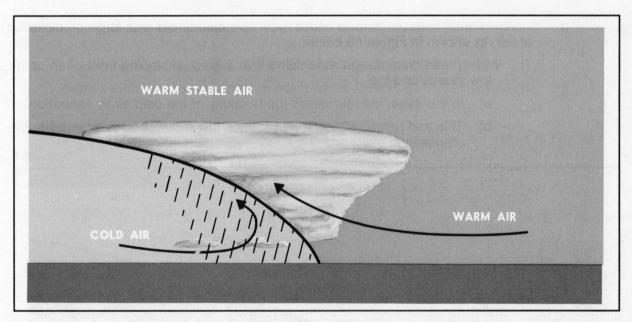

Figure 67. A Cold Front Underrunning Warm, Moist, Stable Air

2)   If the warm air is unstable, cumuliform clouds develop.

  a)   Precipitation from cumuliform clouds is showery, as in Figure 68 below, and the clouds are turbulent.

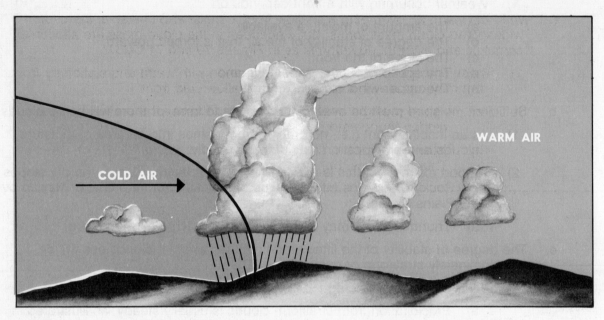

Figure 68. A Cold Front Underrunning Warm, Moist, Unstable Air

d.   Shallow frontal surfaces tend to cause extensive cloudiness with large precipitation areas, as shown in Figure 69 below.

1)   Widespread precipitation associated with a gradual sloping front often causes low stratus and fog.

  a)   In this case, the rain raises the humidity of the cold air to saturation.

  b)   This and related effects may produce low ceiling and poor visibility over thousands of square miles.

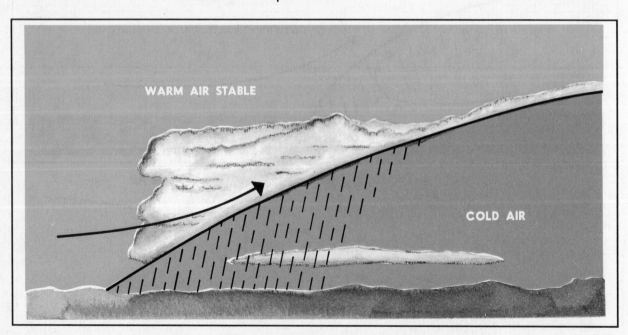

Figure 69. A Warm Front with Overrunning Moist, Stable Air

2)   If temperature of the cold air near the surface is below freezing but the warmer air aloft is above freezing, precipitation falls as freezing rain or ice pellets.

a)   However, if temperature of the warmer air aloft is well below freezing, precipitation forms as snow.

e.   When the warm air overriding a shallow front is moist and unstable, the usual widespread cloud mass forms; but embedded in the cloud mass are altocumulus, cumulus, and even thunderstorms as in Figures 70 and 71 below.

1)   These embedded storms are more common with warm and stationary fronts but may occur with a slow moving, shallow cold front.

a)   A good preflight briefing helps you to foresee the presence of these hidden thunderstorms.

b)   Radar also helps in this situation and is discussed in Part I, Chapter 11, Thunderstorms, beginning on page 115.

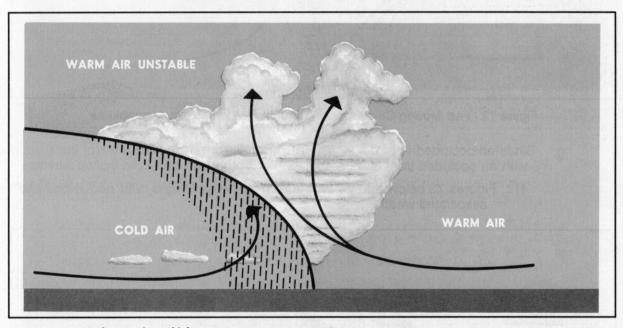

Figure 70. A Slow-Moving Cold Front Underrunning Warm, Moist, Unstable Air

Figure 71. Warm Front with Overrunning Warm, Moist, Unstable Air

f.   A fast moving, steep cold front forces upward motion of the warm air along its leading edge.

1)   If the warm air is moist, precipitation occurs immediately along the surface position of the front as shown in Figure 72 below.

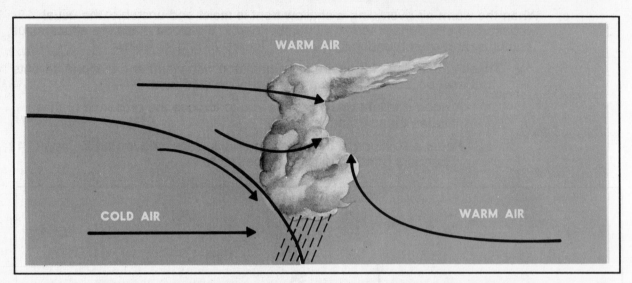

Figure 72. Fast Moving Cold Front Underrunning Warm, Moist, Unstable Air

g.   Since an occluded front develops when a cold front overtakes a warm front, weather with an occluded front is a combination of both warm and cold frontal weather.

1)   Figures 73 below and 74 on page 91 show warm and cold occlusions and associated weather.

Figure 73. Warm Front Occlusion Lifting Warm, Moist, Unstable Air

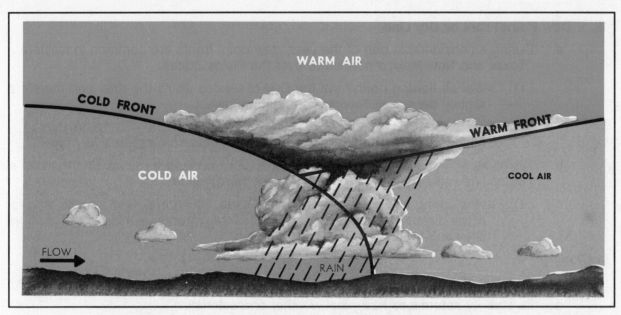

Figure 74. Cold Front Occlusion Lifting Warm, Moist, Stable Air

h.   A front may have little or no cloudiness associated with it.

1)   *Dry fronts* occur when the warm air aloft is flowing down the frontal slope or the air is so dry that any cloudiness that occurs is at high levels.

i.   The upper wind flow dictates to a great extent the amount of cloudiness and rain accompanying a frontal system as well as movement of the front itself.

1)   When winds aloft blow across a front, it tends to move with the wind.

a)   When winds aloft parallel a front, the front moves slowly, if at all.

2)   A deep, slow moving trough aloft forms extensive cloudiness and precipitation, while a rapid moving minor trough more often restricts weather to a rather narrow band.

a)   However, the latter often breeds severe, fast moving, turbulent spring weather.

8.   **Instability Line**

a.   An instability line is a narrow, nonfrontal line or band of convective activity.

1)   If the activity consists of fully developed thunderstorms, the line is a *squall line* (see Part I, Chapter 11, Thunderstorms, beginning on page 115).

b.   Instability lines form in moist, unstable air.

1)   An instability line may develop far from any front.

2)   More often, it develops ahead of a cold front, and sometimes a series of these lines move out ahead of the front.

3)   A favored location for instability lines which frequently erupt into severe thunderstorms is a dew point front or dry line.

9. **Dew Point Front or Dry Line**

   a. During a considerable part of the year, dew point fronts are common in western Texas and New Mexico northward over the Plains States.

      1) Moist air flowing north from the Gulf of Mexico abuts the drier and therefore slightly denser air flowing from the southwest.

      2) Except for moisture differences, there is seldom any significant air mass contrast across this front, and thus, it is commonly called a "dry line."

   b. Nighttime and early morning fog and low-level clouds often prevail on the moist side of the line while generally clear skies mark the dry side.

      1) In the spring and early summer over Texas, Oklahoma, and Kansas, and for some distance eastward, the dry line is a favored spawning area for squall lines and tornadoes.

C. **In Closing**

   1. An air mass is a body of air with fairly uniform properties of temperature and moisture.

      a. Its stability determines its typical weather characteristics.

   2. A front is the zone between two different air masses.

      a. There are three principal types of fronts: cold, warm, and stationary.

   3. Frontal waves usually form on slow moving cold fronts or on stationary fronts.

      a. One section of the front begins to move as a warm front, and the section beside it begins to move as a cold front.

      b. The cold front moves faster than the warm front, and the two close together to form an occlusion.

   4. Nonfrontal lows are low-pressure areas lying solely in a homogeneous air mass.
   5. Frontolysis is the dissipation of a front.
   6. Frontogenesis is the development of a front.
   7. In fronts, flying weather varies from virtually clear skies to extremely hazardous conditions.

      a. Surface weather charts pictorially portray fronts and, in conjunction with other forecast charts and special analyses, help you in determining expected weather conditions along your proposed route.

         1) Knowing the locations of fronts and associated weather helps you determine whether you can proceed as planned.

         2) Often you can change your route to avoid adverse weather.

      b. Frontal weather may change rapidly.

         1) For example, there may be only cloudiness associated with a cold front over northern Illinois during the morning but with a strong squall line forecast by afternoon.

      c. A mental picture of what is happening and what is forecast should greatly help you in avoiding adverse weather conditions.

         1) If unexpected adverse weather develops en route, your mental picture helps you in planning the best diversion.

         2) *Always obtain a good preflight weather briefing.*

END OF CHAPTER

CHAPTER NINE
TURBULENCE

> Please take a few minutes to study each of the concepts listed above and anticipate/imagine what they are and how they relate to the other listed concepts.

A. **Introduction** -- A turbulent atmosphere is one in which air currents vary greatly over short distances.

   1. These currents may range from rather mild eddies to strong currents of relatively large dimensions.

   2. As an aircraft moves through these currents, it undergoes changing accelerations which jostle it from its smooth flight path.

      a. This jostling is turbulence.

         1) Turbulence ranges from bumpiness which can annoy crew and passengers to severe jolts which can structurally damage the aircraft or injure passengers.

      b. Aircraft reaction to turbulence varies with the difference in wind speed in adjacent currents, size of the aircraft, wing loading, airspeed, and aircraft attitude.

         1) When an aircraft travels rapidly from one current to another, it undergoes abrupt changes in acceleration.

         2) Obviously, if the aircraft were to move more slowly, the changes in acceleration would be less.

         3) The first rule in flying through turbulence, then, is to reduce airspeed.

            a) Your aircraft manual most likely lists recommended airspeed for penetrating turbulence.

   3. The main causes of turbulence are

      a. Convective currents,
      b. Obstructions to wind flow, and
      c. Wind shear.

   4. Turbulence also occurs in the wake of moving aircraft whenever the airfoils exert lift, i.e., wake turbulence.

   5. Any combination of causes may occur at one time.

B. **Convective Currents**

   1. Convective currents are a common cause of turbulence, especially at low altitudes. These currents are localized vertical air movements, both ascending and descending.

      a. For every rising current, there is a compensating downward current.

      b. The downward currents frequently occur over broader areas than do the upward currents. Therefore, they have a slower vertical speed than do the rising currents.

2.   Convective currents are most active on warm summer afternoons when winds are light.

    a.   Heated air at the surface creates a shallow, unstable layer because the warm air is forced upward.

        1)   Convection increases in strength and to greater heights as surface heating increases.

    b.   Barren surfaces such as sandy or rocky wastelands and plowed fields become hotter than open water or ground covered by vegetation.

        a)   Thus, air at and near the surface heats unevenly.

        1)   Because of this uneven heating, the strength of convective currents can vary considerably within short distances.

3.   When cold air moves over a warm surface, it becomes unstable in lower levels.

    a.   Convective currents may extend several thousand feet above the surface, resulting in rough, choppy turbulence when you are flying through the cold air.

        1)   This condition often occurs in any season after the passage of a cold front.

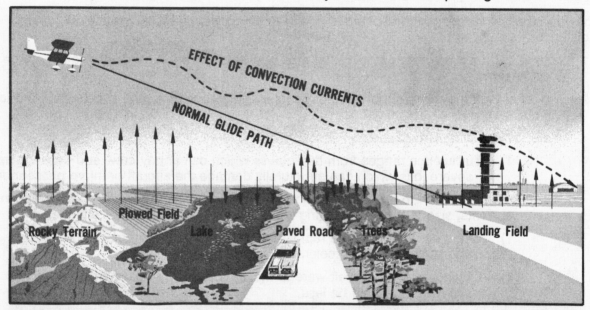

Figure 76. Effect of Convective Currents on Final Approach

b.   Figure 76 on page 94 illustrates the effect of low-level convective turbulence on an aircraft approaching a landing field.

1)   Predominantly upward currents (top) tend to cause the aircraft to overshoot the intended touchdown point.

2)   Predominantly downward currents (bottom) tend to cause the aircraft to undershoot the intended touchdown point.

4.   Turbulence on approach can cause abrupt changes in airspeed and may even result in a stall at a dangerously low altitude.

a.   To guard against this danger, increase airspeed slightly over normal approach speed.

1)   This procedure may appear to conflict with the rule of reducing airspeed for turbulence penetration, but remember, the approach speed for your aircraft is well below the recommended turbulence penetration speed.

5.   As air moves upward, it cools by expansion.

a.   A convective current continues upward until it reaches a level where its temperature cools to that of the surrounding air.

1)   If it cools to saturation, a cloud forms.

b.   Billowy fair weather cumulus clouds, usually seen on sunny afternoons, are signposts in the sky indicating convective turbulence.

1)   The cloud top usually marks the approximate upper limit of the convective current.

2)   You can expect to encounter turbulence beneath or in the clouds. While above the clouds, air generally is smooth.

3)   You will find flight more comfortable above the cumulus as illustrated in Figure 77 below.

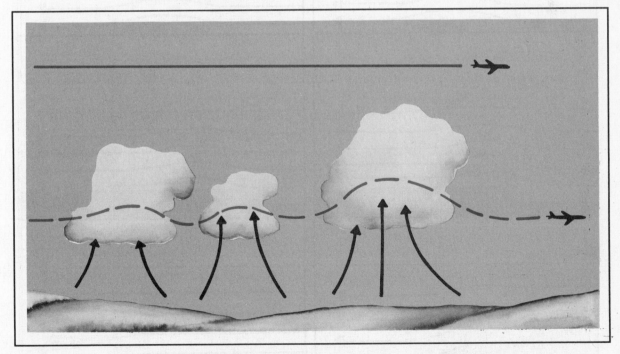

Figure 77. Avoiding Turbulence by Flying above Convective Clouds

    c. When convection extends to greater heights, it develops larger towering cumulus clouds and cumulonimbus with anvil-like tops.

        1) The cumulonimbus (i.e., thunderstorm) provides visual warning of violent convective turbulence. See Part I, Chapter 11, Thunderstorms, beginning on page 115.

    d. You should also know that when air is too dry for cumulus to form, convective currents still can be active.

        1) There is little indication of their presence until you encounter turbulence.

        2) This is sometimes referred to as clear air turbulence (CAT).

C. Obstructions to Wind Flow

    1. Obstructions such as buildings, trees, and rough or mountainous terrain disrupt smooth wind flow into a complex snarl of eddies as diagramed in Figure 78 below.

       a. An aircraft flying through these eddies experiences turbulence.

       b. This turbulence is classified as "mechanical" since it results from mechanical disruption of the ambient wind flow.

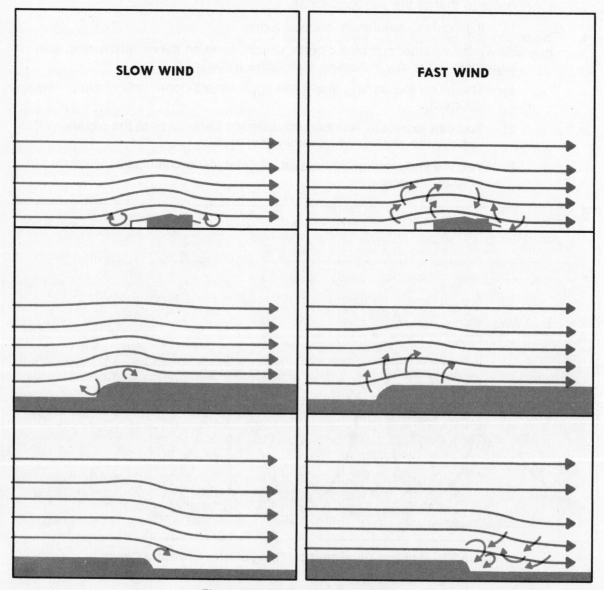

Figure 78. Mechanical Turbulence

2.   The degree of mechanical turbulence depends on wind speed and roughness of the obstructions.

   a.   The higher the speed and/or the rougher the surface, the greater is the turbulence.

   b.   The wind carries the turbulent eddies downstream -- how far depends on wind speed and stability of the air.

      1)   Unstable air allows larger eddies to form than those that form in stable air.

      2)   However, the instability breaks up the eddies quickly, whereas in stable air they dissipate slowly.

3.   Mechanical turbulence can cause cloudiness near the top of the mechanically disturbed layer, just as convective turbulence can.

   a.   The type of cloudiness tells you whether it is from mechanical or convective mixing.

      1)   Mechanical mixing produces stratocumulus clouds in rows or bands while convective clouds form a random pattern.

      2)   The cloud rows developed by mechanical mixing may be parallel to or perpendicular to the wind depending on meteorological factors which we do not discuss here.

4.   The airport area is especially vulnerable to mechanical turbulence which invariably causes gusty surface winds.

   a.   When an aircraft is in a low-level approach or a climb, airspeed fluctuates in the gusts, and the aircraft may even stall.

      1)   During extremely gusty conditions, maintain a margin of airspeed above normal approach or climb speed to allow for changes in airspeed.

   b.   When landing with a gusty crosswind, be alert for control problems in mechanical turbulence caused by airport structures upwind.

      1)   Surface gusts can also create taxi problems.

5.   Mechanical turbulence can affect low-level cross-country flight almost anywhere.

   a.   When flying over rolling hills, you may experience mechanical turbulence.

      1)   Generally, such turbulence is not hazardous, but it may be annoying or uncomfortable.

      2)   A climb to higher altitude should reduce the turbulence.

   b.   When flying over rugged hills or mountains, however, you may have some real turbulence problems.

      1)   When wind speed across a mountain exceeds about 40 kt., you can anticipate turbulence. Where and to what extent depends largely on air stability.

      2)   If the air crossing the mountains is unstable, turbulence on the windward side is almost certain.

         a)   If sufficient moisture is present, convective clouds form, intensifying the turbulence.

         b)   Convective clouds over a mountain or along a ridge are a sure sign of unstable air and turbulence on the windward side and over the mountain crest.

3) As the unstable air crosses the barrier, it spills down the leeward slope often as a violent downdraft.

   a) Sometimes the downward speed exceeds the maximum climb rate for your aircraft and may drive the craft into the mountainside as shown in Figure 80 below.

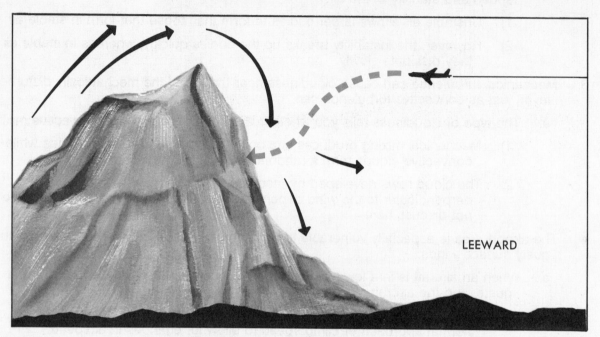

LEEWARD

Figure 80. Wind Flow in Mountain Areas

4) In the process of crossing the mountains, mixing reduces the instability to some extent.

   a) Thus, hazardous turbulence in unstable air generally does not extend a great distance downwind from the barrier.

6. **Mountain Wave**

   a. When stable air crosses a mountain barrier, the turbulent situation is somewhat reversed.

      1) Air flowing up the windward side is relatively smooth.
      2) Wind flow across the barrier is laminar -- that is, it tends to flow in layers.

   b. The barrier may set up waves in these layers much as waves develop on a disturbed water surface.

      1) The waves remain nearly stationary while the wind blows rapidly through them.

      2) The wave pattern, diagramed in Figure 81 on the opposite page, is a "standing" or "mountain" wave, so named because it remains essentially stationary and is associated with the mountains.

   c. Wave crests extend well above the highest mountains, sometimes into the lower stratosphere.

      1) Under each wave crest is a rotary circulation also diagramed in Figure 81.

         a) The "rotor" forms below the elevation of the mountain peaks.

         b) One of the most dangerous features of a mountain wave is the turbulent areas in and below rotor clouds.

      2) Updrafts and downdrafts in the waves can also create violent turbulence.

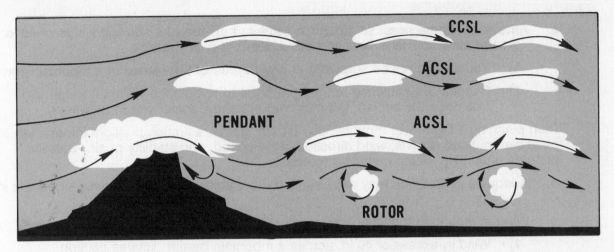

Figure 81. Cross Section of a Mountain Wave

d.   Figure 81 above further illustrates clouds often associated with a mountain wave.

1)   When moisture is sufficient to produce clouds on the windward side, they are stratified.

2)   Crests of the standing waves may be marked by stationary, lens-shaped clouds known as "standing lenticular" clouds. (See Figure 51 on page 68.)

a)   They form in the updraft and dissipate in the downdraft, so they do not move as the wind blows through them.

3)   The rotor may also be marked by a "rotor" cloud.

a)   Figure 83 below is a photograph of a series of rotor clouds, each under the crest of a wave.

Figure 83. Standing Wave Rotor Clouds

e.   Always anticipate possible mountain wave turbulence when strong winds of 40 kt. or greater blow across a mountain or ridge and the air is stable.

   1)   You should not be surprised at any degree of turbulence in a mountain wave.

   2)   Reports of turbulence range from none to turbulence violent enough to damage the aircraft, but most reports show something in between.

7.   When planning a flight over mountainous terrain, gather as much preflight information as possible on cloud reports, wind direction, wind speed, and stability of air. Satellite pictures often help locate mountain waves.

   a.   Adequate information may not always be available, so remain alert for signposts in the sky.

   b.   Wind at mountain top level in excess of 25 kt. suggests some turbulence.

   1)   Wind in excess of 40 kt. across a mountain barrier dictates caution.

   c.   Stratified clouds mean stable air.

   1)   Standing lenticular and/or rotor clouds suggest a mountain wave.

   a)   Expect turbulence many miles to the lee of mountains and relatively smooth flight on the windward side.

   d.   Convective clouds on the windward side of mountains mean unstable air.

   1)   Expect turbulence in close proximity to and on either side of the mountain.

   e.   When approaching mountains from the leeward side during strong winds, begin your climb well away from the mountains -- 100 miles in a mountain wave and 30 to 50 miles otherwise.

   1)   Climb to an altitude 3,000 to 5,000 ft. above mountain tops before attempting to cross.

   2)   It is recommended that you approach a ridge at a 45° angle to enable a rapid retreat to calmer air.

   a)   If unable to make good on your first attempt and you have higher altitude capabilities, you may choose to back off and make another attempt at higher altitude.

   b)   Sometimes you may have to choose between turning back or detouring the area.

   f.   Flying through mountain passes and valleys is not a safe procedure in high winds.

   1)   The mountains funnel the wind into passes and valleys, thus increasing wind speed and intensifying turbulence.

   2)   If winds at mountain top level are strong, go high or go around.

   g.   Surface wind may be relatively calm in a valley surrounded by mountains when wind aloft is strong.

   1)   If taking off in the valley, climb above mountain top level before leaving the valley.

   2)   Maintain lateral clearance from the mountains sufficient to allow recovery if caught in a downdraft.

D. Wind Shear

1.   Wind shear generates eddies between two wind currents of differing velocities.

   a.   The differences may be in wind speed, wind direction, or in both.

   1)   Wind shear may be associated with either a wind shift or a wind speed gradient at any level in the atmosphere.

b. Three conditions are of special interest.

1) Wind shear with a low-level temperature inversion,

2) Wind shear in a frontal zone, and

3) Clear air turbulence (CAT) at high levels associated with a jet stream or strong circulation.

a) High-level CAT is discussed in detail in Part II, Chapter 13, High Altitude Weather, beginning on page 144.

2. Wind shear with a low-level temperature inversion

a. A temperature inversion forms near the surface on a clear night with calm or light surface wind.

b. Wind just above the inversion may be relatively strong.

1) As illustrated in Figure 86 below, a wind shear zone develops between the calm and the stronger winds above.

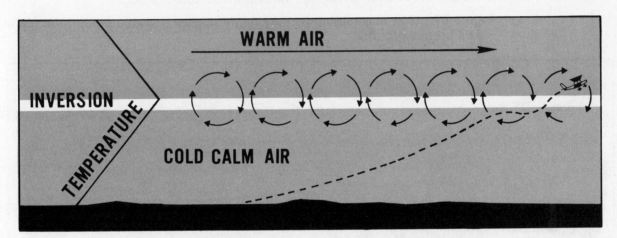

Figure 86. Wind Shear in a Low-level Temperature Inversion

c. Eddies in the shear zone cause airspeed fluctuations as an aircraft climbs or descends through the inversion.

1) An aircraft most likely is either climbing from takeoff or approaching to land when passing through the inversion; therefore, airspeed is relatively slow -- perhaps only a few knots greater than stall speed.

2) The fluctuation in airspeed can induce a stall precariously close to the ground.

d. Since surface wind is calm or very light, takeoff or landing can be in any direction.

1) Takeoff may be in the direction of the wind above the inversion.

a) If so, the aircraft encounters a sudden tailwind and a corresponding loss of airspeed when climbing through the inversion. Stall is possible.

2) If the landing approach is into the wind above the inversion, the headwind is suddenly lost when descending through the inversion. Again, a sudden loss in airspeed may induce a stall.

e. When taking off or landing in calm wind under clear skies within a few hours before or after sunrise, be prepared for a temperature inversion near the ground.

1) You can be relatively certain of a shear zone in the inversion if you know the wind at 2,000 to 4,000 ft. is 25 kt. or more.

2) Allow a margin of airspeed above normal climb or approach speed to alleviate danger of stall in event of turbulence or sudden change in wind velocity.

3.   Wind shear in a frontal zone

   a.   Wind changes abruptly in the frontal zone and can induce wind shear turbulence.

   b.   The degree of turbulence depends on the magnitude of the wind shear.

   c.   When turbulence is expected in a frontal zone, follow turbulence penetration procedures recommended in your aircraft manual.

E. **Wake Turbulence**

1.   An aircraft generates lift due, in part, to relatively high pressure beneath its wings, and relatively low pressure above.

   a.   This causes air spillage at the wingtips from the underside, up and over the wings.

   b.   Thus, whenever the wings are producing lift, they generate rotary motions or vortices off the wingtips.

2.   When the landing gear bears the entire weight of the aircraft, no wingtip vortices develop.

   a.   But the instant the aircraft rotates on takeoff, these vortices begin.

      1)   Figure 87 below illustrates how they might appear if visible behind the plane as it breaks ground.

Figure 87. Wingtip Vortices

   b.   These vortices continue throughout the flight until the craft again settles firmly on its landing gear (i.e., no lift is produced).

3.   These vortices spread downward and outward from the flight path. They also drift with the wind. Avoid flying through these vortices.

   a.   The strength of the vortices is proportional to the weight of the aircraft as well as other factors.

      1)   Therefore, wake turbulence is more intense behind large, transport category aircraft than behind small aircraft.

      2)   Generally, it is a problem only when following the larger aircraft.

4.   The turbulence persists several minutes and may linger after the aircraft is out of sight.

   a.   Most jets when taking off lift the nose wheel about midpoint in the takeoff roll; therefore, vortices begin at approximately the middle of the takeoff roll.

   b.   Vortices behind propeller aircraft begin only a short distance behind liftoff.

   c.   Following a landing of either type of aircraft, vortices end at approximately the point where the nose wheel touches down.

5.   When using the same runway as a heavier aircraft:

a.   If landing behind another aircraft, keep your approach above its approach and keep your touchdown beyond the point where its nose wheel touched the runway, as in Figure 88 (A);

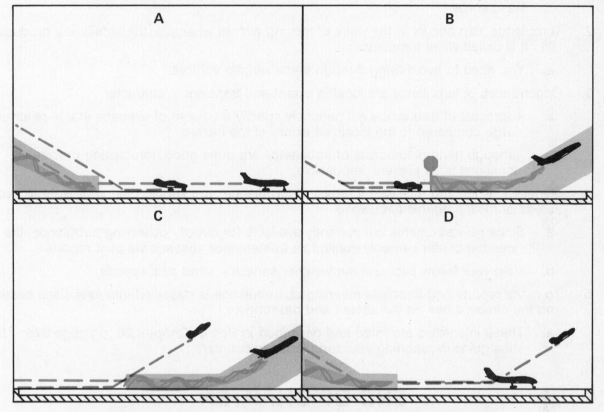

Figure 88. Planning Landing or Takeoff to Avoid Wake Turbulence

b.   If landing behind a departing aircraft, land only if you can complete your landing roll before reaching the midpoint of its takeoff roll, as in Figure 88 (B).

c.   If departing behind another departing aircraft, take off only if you can become airborne before reaching the midpoint of its takeoff roll and only if you can climb fast enough to stay above its flight path, as in Figure 88 (C); and

d.   Do not depart behind a landing aircraft unless you can taxi onto the runway beyond the point at which its nose wheel touched down and have sufficient runway left for safe takeoff, as in Figure 88 (D).

e.   If parallel runways are available and the heavier aircraft takes off with a crosswind on the downwind runway, you may safely use the upwind runway.

f.   Never land or take off downwind from the heavier aircraft.

g.   When using a runway crossing the heavier aircraft's runway, you may safely use the upwind portion of your runway.

h.   You may cross behind a departing aircraft behind the midpoint of its takeoff roll.

i.   If none of these procedures is possible, wait 5 minutes or so for the vortices to dissipate or to blow off the runway.

6.   The problem of wake turbulence is more operational than meteorological.

F. **In Closing**

1. The main causes of turbulence are

    a. Convective currents,
    b. Obstructions to wind flow, and
    c. Wind shear.

2. Turbulence also occurs in the wake of moving aircraft whenever the airfoils are producing lift. It is called wake turbulence.

    a. You need to avoid flying through these wingtip vortices.

3. Occurrences of turbulence are local in extent and transient in character.

    a. A forecast of turbulence will generally specify a volume of airspace that is relatively large compared to the localized extent of the hazard.

    b. Although general forecasts of turbulence are quite good, forecasting precise locations is, at present, impossible.

4. Generally, when you receive a forecast of turbulence, you should plan your flight to avoid areas of *most probable turbulence*.

    a. Since no instruments are currently available for directly observing turbulence, the weather briefer can only confirm its existence or absence via pilot reports.

    b. Help your fellow pilot and the weather service -- send pilot reports.

5. To make reports and forecasts meaningful, turbulence is classified into intensities based on the effects it has on the aircraft and passengers.

    a. These intensities are listed and described in Part III, Chapter 26, on page 375. Use this guide in reporting your turbulence encounters.

END OF CHAPTER

CHAPTER TEN
ICING

> Please take a few minutes to study each of the concepts listed above and anticipate/imagine what they are and how they relate to the other listed concepts.

A. **Introduction** -- Aircraft icing is one of the major weather hazards to aviation. It is a cumulative hazard.

    1. Icing reduces aircraft efficiency by increasing weight, reducing lift, decreasing thrust, and increasing drag.

        a. As shown in Figure 89 below, each effect tends either to slow the aircraft or to force it downward.

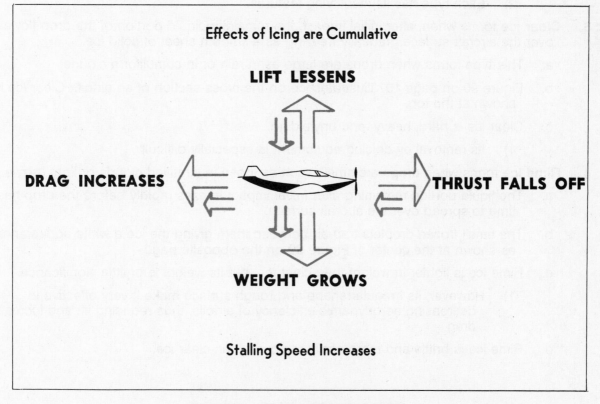

Figure 89. Effects of Structural Icing

2.   Icing also seriously impairs aircraft engine performance.

3.   Other icing effects include false indications on flight instruments, loss of radio communications, and loss of operation of control surfaces, brakes, and landing gear.

4.   In this chapter we discuss the principles of structural, induction system, and instrument icing and relate icing to cloud types and other factors.

    a.   Although ground icing and frost are structural icing, we discuss them separately because of their different effect on an aircraft.

B.   Structural Icing

1.   Two conditions are necessary for structural icing in flight:

    a.   The aircraft must be flying through visible moisture such as rain droplets or clouds.

    b.   The temperature at the point where the moisture strikes the aircraft must be 0°C or colder.

        1)   Note that aerodynamic cooling can lower temperature of an airfoil to 0°C even though the ambient temperature is a few degrees warmer.

2.   Supercooled water increases the rate of icing and is essential to rapid accumulation.

    a.   Supercooled water is in an unstable liquid state. When an aircraft strikes a supercooled drop, part of the drop freezes instantaneously.

    b.   The latent heat of fusion released by the freezing portion raises the temperature of the remaining portion to the melting point.

        1)   Aerodynamic effects may cause the remaining portion to freeze.

    c.   The way in which the remaining portion freezes determines the type of icing.

        1)   The types of structural icing are clear, rime, and a mixture of the two.
        2)   Each type has its identifying features.

3.   **Clear ice** forms when, after initial impact, the remaining liquid portion of the drop flows out over the aircraft surface gradually freezing as a smooth sheet of solid ice.

    a.   This type forms when drops are large as in rain or in cumuliform clouds.

    b.   Figure 90 on page 107 illustrates ice on the cross-section of an airfoil. Clear ice is shown at the top.

    c.   Clear ice is hard, heavy, and unyielding.

        1)   Its removal by deicing equipment is especially difficult.

4.   **Rime ice** forms when drops are small, such as those in stratified clouds or light drizzle.

    a.   The liquid portion remaining after initial impact freezes rapidly before the drop has time to spread over the aircraft surface.

    b.   The small frozen droplets trap air between them giving the ice a white appearance as shown at the center of Figure 90 on the opposite page.

    c.   Rime ice is lighter in weight than clear ice, but its weight is of little significance.

        1)   However, its irregular shape and rough surface make it very effective in decreasing aerodynamic efficiency of airfoils, thus reducing lift and increasing drag.

    d.   Rime ice is brittle and more easily removed than clear ice.

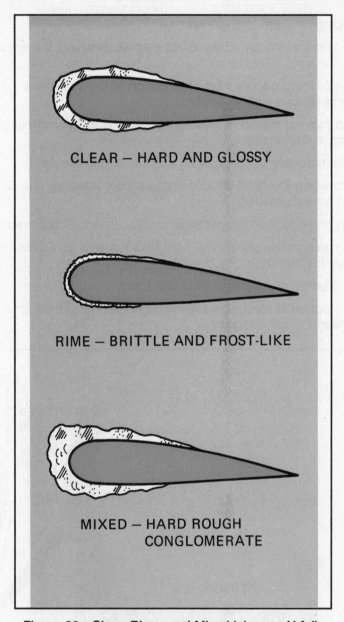

Figure 90. Clear, Rime, and Mixed Icing on Airfoils

5.   **Mixed ice** forms when drops vary in size or when liquid drops are intermingled with snow or ice particles. It can form rapidly.

    a.   Ice particles become imbedded in clear ice, building a very rough accumulation sometimes in a mushroom shape on leading edges as shown at the bottom of Figure 90 above.

6.   The FAA, National Weather Service, the military aviation weather services, and aircraft operating organizations have classified aircraft structural icing into intensity categories.

    a.   Part III, Chapter 26, Tables and Conversion Graphs, beginning on page 373, contains a table listing these intensities.

        1)   The table is your guide in estimating how ice of a specific intensity will affect your aircraft.

        2)   Use the table also in reporting ice when you encounter it.

C. **Induction System Icing**

1. Ice frequently forms in the air intake of an engine, robbing the engine of air to support combustion.

    a. This type of icing occurs with both piston and jet engines.
    b. Carburetor icing is one example.

2. The downward moving piston in a piston engine or the compressor in a jet engine forms a partial vacuum in the intake.

    a. Adiabatic expansion in the partial vacuum cools the air.

    b. Ice forms when the temperature drops below freezing and sufficient moisture is present for sublimation.

    c. In piston engines, fuel evaporation produces additional cooling.

3. Induction icing always lowers engine performance and can even reduce intake flow below that necessary for the engine to operate.

    a. Figure 95 below illustrates carburetor icing.

4. Induction icing potential varies greatly among different aircraft and occurs under a wide range of meteorological conditions.

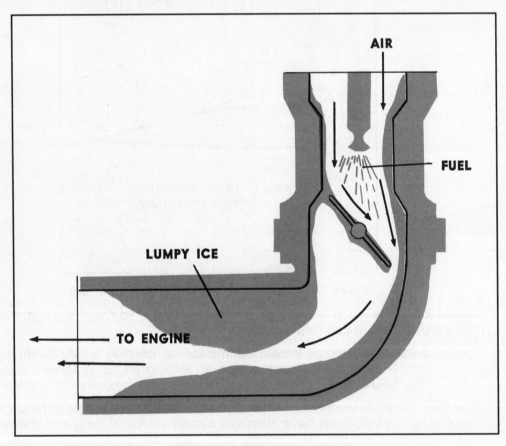

Figure 95. Carburetor Icing

D. Instrument Icing

1. Icing of the pitot tube as seen in Figure 96 below reduces ram air pressure to the airspeed indicator and renders the instrument unreliable.

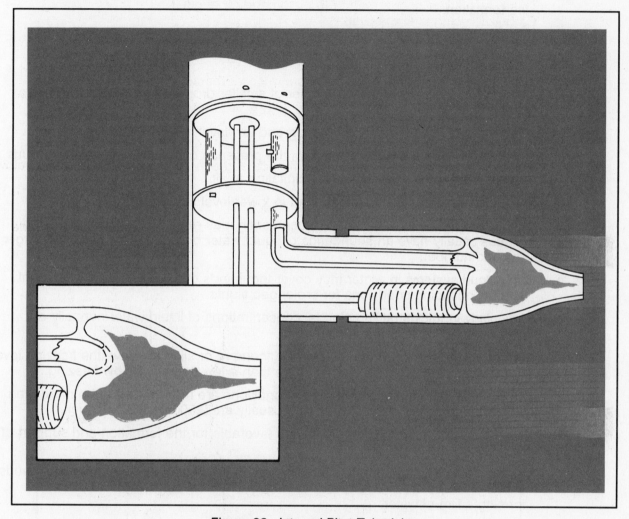

Figure 96. Internal Pitot Tube Icing

2. Icing of the static pressure port reduces reliability of all instruments on the pitot-static system -- the airspeed and vertical speed indicators and the altimeter.

3. Ice forming on the radio antenna distorts its shape, increases drag, and imposes vibrations that may result in failure in the communications system of the aircraft.

    a. The severity of this icing depends upon the shape, location, and orientation of the antenna.

E. **Icing and Cloud Types**

1. All clouds at subfreezing temperatures have icing potential.

    a. However, drop size and distribution and aerodynamic effects of the aircraft influence ice formation.

    b. Ice may not form even though the potential exists.

2. The condition most favorable for very hazardous icing is the presence of many large, supercooled water drops.

    a. Conversely, an equal or lesser number of smaller droplets favors a slower rate of icing.

3. Small water droplets occur most often in fog and low-level clouds.

    a. Drizzle or very light rain is evidence of the presence of small drops in such clouds.

        1) In many cases there is no precipitation at all.

    b. The most common type of icing found in lower-level stratus clouds is rime.

4. Thick extensive stratified clouds that produce continuous rain such as altostratus and nimbostratus usually have an abundance of liquid water because of the relatively larger drop size and number.

    a. Such cloud systems in winter may cover thousands of square miles and present very serious icing conditions for prolonged flights.

    b. Particularly in thick stratified clouds, concentrations of liquid water normally are greater with warmer temperatures.

        1) Thus, heaviest icing usually will be found at or slightly above the freezing level where temperature is never more than a few degrees below freezing.

    c. In layer type clouds, continuous icing conditions are rarely found to be more than 5,000 ft. above the freezing level, and usually are 2,000 or 3,000 ft. thick.

5. The upward currents in cumuliform clouds are favorable for the formation and support of many large water drops.

    a. When an aircraft enters the heavy water concentrations found in cumuliform clouds, the large drops break and spread rapidly over the leading edge of the airfoil, forming a film of water.

        1) If temperatures are freezing or colder, the water freezes quickly to form a solid sheet of clear ice.

    b. You should avoid cumuliform clouds when possible.

    c. The updrafts in cumuliform clouds lift large amounts of liquid water far above the freezing level.

        1) On rare occasions icing has been encountered in thunderstorm clouds at altitudes of 30,000 to 40,000 ft. where the free air temperature was colder than −40°C.

    d. While the vertical extent of critical icing potential cannot be specified in cumuliform clouds, their individual cell-like distribution usually limits the horizontal extent of icing conditions.

        1) An exception, of course, may be found in a prolonged flight through a broad zone of thunderstorms or heavy showers.

F. **Other Factors in Icing**

1. **Fronts**. A condition favorable for rapid accumulation of clear icing is freezing rain below a frontal surface.

a. Rain forms above the frontal surface at temperatures warmer than freezing.

1) Subsequently, it falls through air at temperatures below freezing and becomes supercooled.

2) The supercooled drops freeze on impact with an aircraft surface.

b. Figure 98 below diagrams this type of icing. It may occur with either a warm front (top) or a cold front.

1) The icing can be critical because of the large amount of supercooled water.

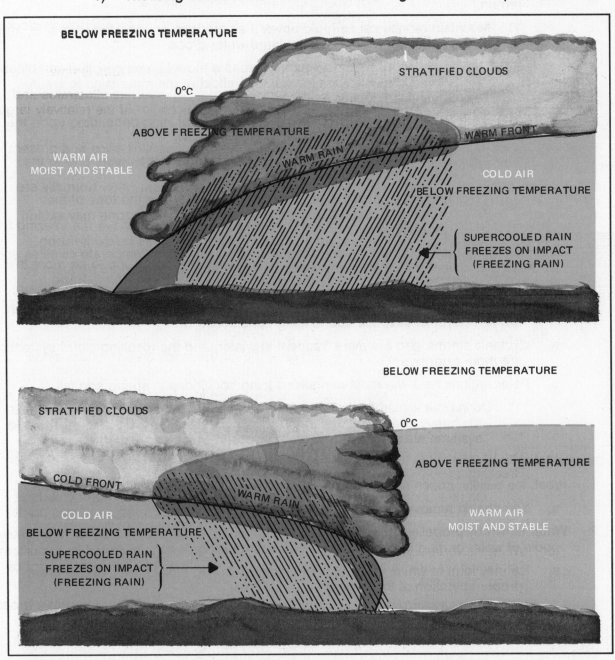

Figure 98. Freezing Rain with a Warm Front and a Cold Front

      c.   Icing can also become serious in cumulonimbus clouds along a surface cold front, along a squall line, or embedded in the cloud shield of a warm front.

  2.   **Terrain**. Air blowing upslope is cooled adiabatically. When the air is cooled below the freezing point, any water it contains becomes supercooled.

      a.   In stable air blowing up a gradual slope, the cloud drops generally remain comparatively small since larger drops fall out as rain.

          1)   Ice accumulation is rather slow and you should have ample time to get out of it before the accumulation becomes extremely dangerous.

      b.   When air is unstable, convective clouds develop a more serious hazard.

      c.   Icing is more probable and more hazardous in mountainous regions than over other terrain.

          1)   Mountain ranges cause rapid upward air motions on the windward side, and these vertical currents support large water drops.

          2)   The movement of a frontal system across a mountain range often combines the normal frontal lift with the upslope effect of the mountains to create extremely hazardous icing zones.

          3)   Each mountainous region has preferred areas of icing depending upon the orientation of mountain ranges to the wind flow.

              a)   The most dangerous icing takes place above the crests and to the windward side of the ridges.

              b)   This zone usually extends about 5,000 ft. above the tops of the mountains, but when clouds are cumuliform, the zone may extend much higher.

  3.   **Seasons**. Icing may occur during any season of the year, but in temperate climates such as those that cover most of the contiguous United States, icing is more frequent in winter.

      a.   The freezing level is nearer the ground in winter than in summer leaving a smaller low-level layer of airspace free of icing conditions.

      b.   Cyclonic storms also are more frequent in winter, and the resulting cloud systems are more extensive.

      c.   Polar regions have the most dangerous icing conditions in spring and fall.

          1)   During the winter the air is normally too cold in the polar regions to contain the heavy concentrations of moisture necessary for icing, and most cloud systems are stratiform and are composed of ice crystals.

G. Ground Icing

  1.   Frost, ice pellets, frozen rain, or snow may accumulate on parked aircraft.

      a.   You should remove all ice prior to takeoff.

  2.   Water blown by propellers or splashed by wheels of an airplane as it taxis or runs through pools of water or mud may result in serious aircraft icing.

      a.   Ice may form in wheel wells, brake mechanisms, flap hinges, etc., and prevent proper operation of these parts.

  3.   Ice on runways and taxiways can create traction and braking problems.

H. **Frost**

    1.   Frost is a hazard to flying.

        a.   Pilots must remove all frost from airfoils prior to takeoff.

    2.   Frost forms near the surface primarily in clear, stable air and with light winds -- conditions which in all other respects make weather ideal for flying.

        a.   Because of this, the real hazard is often minimized.

    3.   Thin metal airfoils are especially vulnerable surfaces on which frost will form.

    4.   Frost does not change the basic aerodynamic shape of the wing, but the roughness of its surface spoils the smooth flow of air thus causing a slowing of the airflow.

        a.   This slowing of the air causes early air flow separation over the affected airfoil resulting in a loss of lift.

        b.   A heavy coat of hard frost will cause a 5% to 10% increase in stall speed.

        c.   Even a small amount of frost on airfoils may prevent an aircraft from becoming airborne at normal takeoff speed.

        d.   Also possible is that, once airborne, an aircraft could have insufficient margin of airspeed above stall so that moderate gusts or turning flight could produce incipient or complete stalling.

    5.   Frost formation in flight offers a more complicated problem. The extent to which it will form is still a matter of conjecture.

        a.   At most, it is comparatively rare.

I. **In Closing**

    1.   Icing reduces aircraft efficiency by increasing weight, reducing lift, decreasing thrust, and increasing drag.

        a.   Icing also seriously impairs aircraft engine performance.

    2.   Structural icing can occur as clear ice, rime ice, or a mixture of the two.

    3.   Induction system icing, e.g., carburetor icing, lowers engine performance by reducing the intake of air necessary to support combustion.

    4.   Instrument icing affects the airspeed and vertical speed indicators and the altimeter. Also, ice forming on the radio antenna may result in failure of the communication and/or radio navigation systems.

    5.   Icing is where you find it. As with turbulence, icing may be local in extent and transient in character.

    6.   Forecasters can identify regions in which icing is possible.

        a.   However, they cannot define the precise small pockets in which it occurs.

    7.   You should plan your flight to avoid those areas where icing probably will be heavier than your aircraft can handle.

        a.   Also, you must be prepared to avoid or to escape the hazard when it is encountered en route.

8. Here are a few specific points to remember:

    a. Before takeoff, check weather for possible icing areas along your planned route.

        1) Check for pilot reports, and if possible talk to other pilots who have flown along your proposed route.

    b. If your aircraft is not equipped with deicing or anti-icing equipment, avoid areas of icing.

        1) Water (clouds or precipitation) must be visible and outside air temperature must be near 0°C or colder for structural ice to form.

    c. Always remove ice or frost from airfoils before attempting takeoff.

    d. In cold weather, avoid, when possible, taxiing or taking off through mud, water, or slush.

        1) If you have taxied through any of these, make a preflight check to ensure freedom of controls.

    e. When climbing out through an ice layer, climb at an airspeed a little faster than normal to avoid a stall.

    f. Use deicing or anti-icing equipment before accumulations of ice become too great.

        1) When such equipment becomes less than totally effective, change course or altitude to get out of the icing as rapidly as possible.

    g. If your aircraft is not equipped with a pitot-static system deicer, be alert for erroneous readings from your airspeed indicator, vertical speed indicator, and altimeter.

    h. In stratiform clouds, you can probably alleviate icing by changing to an altitude with above-freezing temperatures or to one colder than −10°C.

        1) An altitude change also may take you out of clouds.
        2) Rime icing in stratiform clouds can be very extensive horizontally.

    i. In frontal freezing rain, you may be able to climb or descend to a layer warmer than freezing.

        1) Temperature is always warmer than freezing at some higher altitude.

        2) If you are going to climb, move quickly; procrastination may leave you with too much ice.

            a) If you are going to descend, you must know the temperature and terrain below.

    j. Avoid cumuliform clouds if at all possible. Clear ice may be encountered anywhere above the freezing level.

        1) Most rapid accumulations are usually at temperatures from 0°C to −15°C.

    k. Avoid abrupt maneuvers when your aircraft is heavily coated with ice because the aircraft has lost some of its aerodynamic efficiency.

    l. When "iced up," fly your landing approach with power.

9. Help your fellow pilots and the weather service by sending pilot reports when you encounter icing or when icing is forecast but none is encountered.

END OF CHAPTER

CHAPTER ELEVEN
THUNDERSTORMS

> Please take a few minutes to study each of the concepts listed above and anticipate/imagine what they are and how they relate to the other listed concepts.

A. Introduction

1. This chapter looks at where and when thunderstorms occur, explains what creates a storm, and looks inside the storm at what goes on and what it can do to an aircraft.

2. In some tropical regions, thunderstorms occur year-round.

   a. In midlatitudes, they develop most frequently in spring, summer, and fall.
   b. Arctic regions occasionally experience thunderstorms during summer.

3. Figures 100 through 104 depict various thunderstorm activity. Note the frequency shown by location (south-central and southeastern states) and by season (summer).

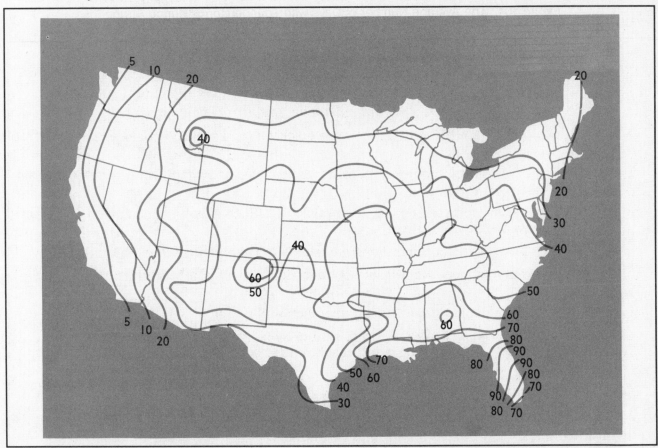

Figure 100. The Average Number of Thunderstorms Each Year

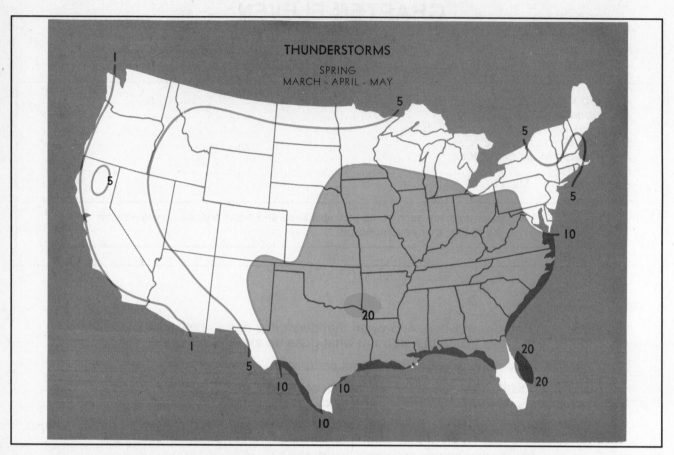

Figure 101. The Average Number of Days with Thunderstorms during Spring

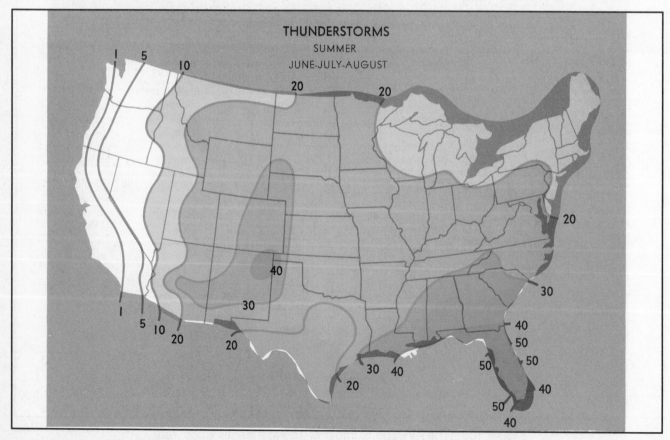

Figure 102. The Average Number of Days with Thunderstorms during Summer

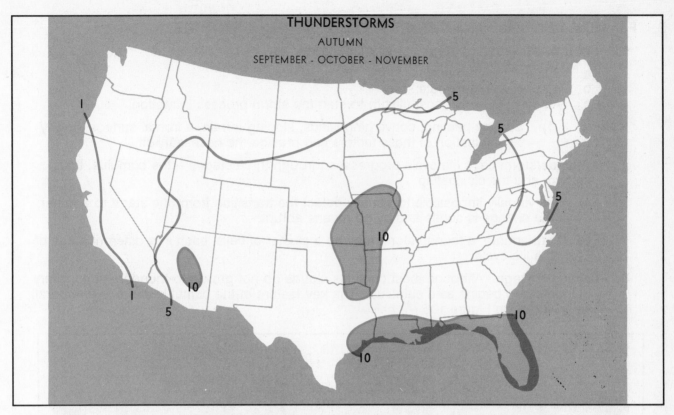

Figure 103. The Average Number of Days with Thunderstorms during Fall

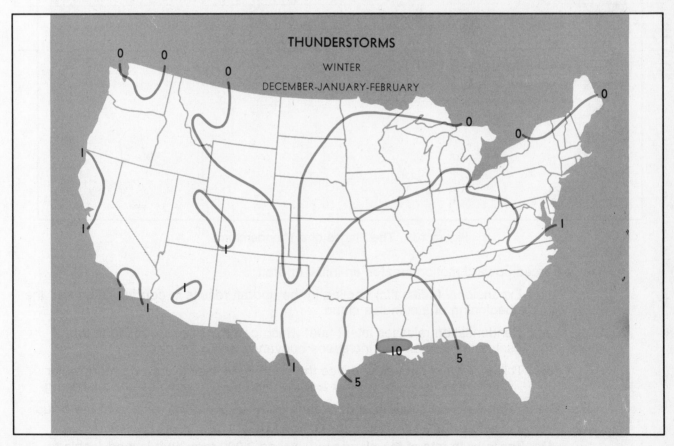

Figure 104. The Average Number of Days with Thunderstorms during Winter

B.   **Formation of Thunderstorms**

1.   For a thunderstorm to form, the air must have

   a.   Sufficient water vapor.
   b.   An unstable lapse rate.
   c.   An initial upward boost (lifting) to start the storm process in motion.

      1)   Surface heating, converging winds, sloping terrain, a frontal surface, or any combination of these factors can provide the necessary lift.

2.   A thunderstorm cell's life cycle progresses through three stages -- the cumulus, the mature, and the dissipating.

   a.   It is virtually impossible to visually detect the transition from one stage to another; the change is subtle and by no means abrupt.

   b.   Furthermore, a thunderstorm may be a cluster of cells, each in a different stage of the life cycle.

3.   **Cumulus Stage.** Although most cumulus clouds do not grow into thunderstorms, every thunderstorm begins as a cumulus. The key feature of the cumulus stage is an updraft, as illustrated in Figure 105 (A) below.

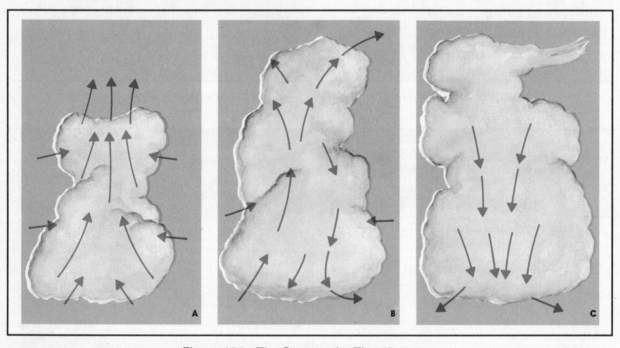

Figure 105. The Stages of a Thunderstorm

   a.   Forced upward motion creates an initial updraft.

      1)   Expansional (adiabatic) cooling in the updraft results in condensation and the beginning of a cumulus cloud.

      2)   Condensation releases latent heat which partially offsets cooling in the saturated updraft and increases buoyancy within the cloud.

      3)   This increased buoyancy drives the updraft still faster, drawing more water vapor into the cloud, and for a while, the updraft becomes self-sustaining.

   b.   The updraft varies in strength and extends from very near the surface to the cloud top.

      1)   The growth rate of the cloud may exceed 3,000 fpm, so it is inadvisable to attempt to climb over rapidly building cumulus clouds.

      c.   Early during the cumulus stage, water droplets are quite small but grow to raindrop size as the cloud grows.

           1)   The upwelling air carries the liquid water above the freezing level, creating an icing hazard.

           2)   As the raindrops grow still heavier, they fall.

      d.   The cold rain drags air with it, creating a cold downdraft coexisting with the updraft; the cell has reached the mature stage.

4.   **Mature Stage**. Precipitation beginning to fall from the cloud base is a sign that a downdraft has developed and that the cell has entered the mature stage.

      a.   Cold rain in the downdraft retards compressional (adiabatic) heating, and the downdraft remains cooler than surrounding air.

           1)   Thus, its downward speed is accelerated and can exceed 2,500 fpm.

           2)   The downrushing air spreads outward at the surface, as shown in Figure 105 (B) on page 118, producing strong, gusty surface winds, a sharp temperature drop, and a rapid rise in pressure.

                a)   The surface wind surge is a "plow wind" and its leading edge is the "first gust."

      b.   Meanwhile, updrafts reach a maximum with speeds sometimes exceeding 6,000 fpm.

           1)   Updrafts and downdrafts in close proximity create strong vertical shear and a very turbulent environment.

      c.   All thunderstorm hazards (discussed later in this chapter) reach their greatest intensity during the mature stage.

           1)   Duration of the mature stage is closely related to severity of the thunderstorm.

5.   **Dissipating Stage**. Downdrafts characterize the dissipating stage of the thunderstorm cell, as shown in Figure 105 (C) on page 118, and the storm dies rapidly.

      a.   When rain has ended and downdrafts have abated, the dissipating stage is complete.

      b.   When all cells of the thunderstorm have completed this stage, only harmless cloud remnants remain.

6.   Individual thunderstorms can measure from less than 5 mi. to more than 30 mi. in diameter.

      a.   Cloud bases range from a few hundred feet in very moist climates to 10,000 ft. or higher in drier regions.

           1)   Tops generally range from 25,000 to 45,000 ft. but occasionally extend above 65,000 ft.

C. **Types of Thunderstorms**

1.   **Air mass thunderstorms** most often result from surface heating.

      a.   When the storm reaches the mature stage, rain falls through or immediately beside the updraft.

           1)   Falling precipitation induces frictional drag, retards the updraft and reverses it to a downdraft. Thus, the storm is self-destructive.

           2)   The downdraft and cool precipitation cool the lower portion of the storm and the underlying surface.

                a)   This cooling cuts off the inflow of water vapor, causing the storm to run out of energy and die.

           3)   A self-destructive cell usually has a life cycle of 20 min. to 1½ hr.

b.   Since air mass thunderstorms generally result from surface heating, they reach maximum intensity and frequency over land during middle and late afternoon.

1)   Offshore, they reach a maximum during late hours of darkness when land temperature is coolest and cool air off the land flows over the relatively warm water.

2.   **Steady state thunderstorms** are usually associated with weather systems.

a.   Fronts, converging winds, and troughs aloft induce upward motion, spawning thunderstorms which often form into squall lines (see page 122).

1)   Afternoon heating intensifies these storms.

b.   In a steady state storm, precipitation falls outside the updraft, as shown in Figure 106 below, allowing the updraft to continue unabated.

1)   Thus, the mature stage updrafts become stronger and last much longer than in air mass storms -- hence, the name, "steady state."

2)   A steady state cell may persist for several hours.

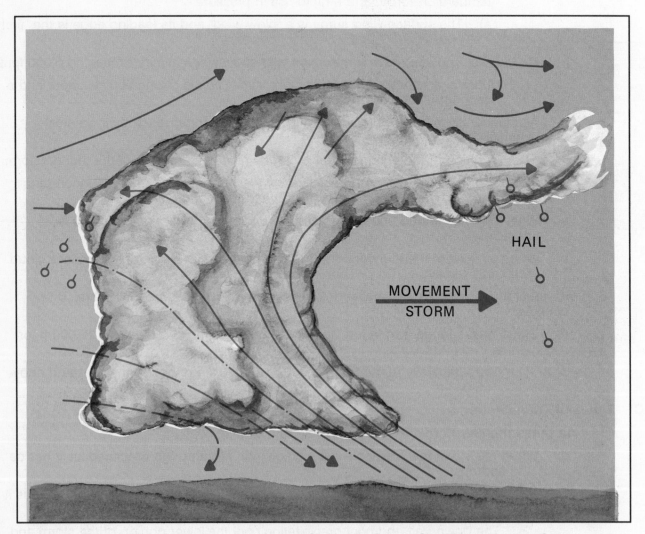

Figure 106. Schematic of the Mature Stage of a Steady State Thunderstorm Cell

D. Hazards

1. Tornadoes

    a. The most violent thunderstorms draw air into their cloud bases with great vigor.

        1) If the incoming air has any initial rotating motion, it often forms an extremely concentrated vortex from the surface well into the cloud.

            a) Meteorologists have estimated that wind in such a vortex can exceed 200 kt.; pressure inside the vortex is quite low.

        2) The strong winds gather dust and debris, and the low pressure generates a funnel-shaped cloud extending downward from the cumulonimbus base.

            a) If the cloud does not reach the surface, it is a *funnel cloud*.
            b) If it touches a land surface, it is a *tornado*, Figure 107 below.
            c) If it touches water, it is a *water spout*, Figure 108 below.

Figure 107. A Tornado

Figure 108. A Waterspout

    b. Tornadoes may occur with isolated thunderstorms, but form more frequently with steady state thunderstorms associated with cold fronts or squall lines.

        1) Reports or forecasts of tornadoes indicate that atmospheric conditions are favorable for violent turbulence.

    c. Families of tornadoes have been observed as appendages of the main cloud extending several miles outward from the area of lightning and precipitation.

        1) Thus, any cloud connected to a severe thunderstorm carries a threat of violence.

d.   Frequently, cumulonimbus mamma clouds (see Figure 110) occur in connection with violent thunderstorms and tornadoes.

1)   The cloud displays rounded, irregular pockets or festoons from its base and is a sign of violent turbulence and extreme instability.

Figure 110. Cumulonimbus Mamma Clouds

2)   Surface aviation reports specifically mention this and other especially hazardous clouds.

e.   Tornadoes occur most frequently in the Great Plains states east of the Rocky Mountains.

1)   As shown in Figure 111 on page 123, however, they have occurred in every state.

f.   An aircraft entering a tornado vortex is almost certain to suffer structural damage.

1)   Since the vortex extends well into the cloud, any pilot inadvertently caught on instruments in a severe thunderstorm could encounter a hidden vortex.

2.   **Squall Lines**. A squall line is a non-frontal, narrow band of active thunderstorms.

a.   Often it develops ahead of a cold front in moist, unstable air, but it may also develop in unstable air far removed from any front.

b.   The line may be too long to easily detour and too wide and severe to penetrate.

c.   It often contains severe steady-state thunderstorms and presents the single most intense weather hazard to aircraft.

d.   A squall line usually forms rapidly, generally reaching maximum intensity during the late afternoon and the first few hours of darkness.

e.   Figure 112 on page 123 is a photograph of an advancing squall line.

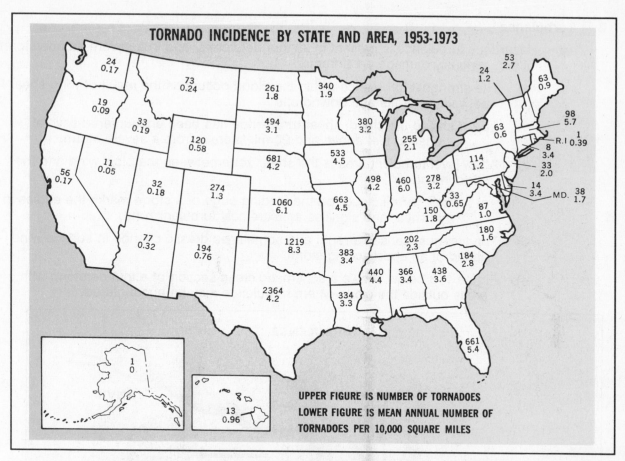

Figure 111. Tornado Incidence by State and Area (1953-1973)

Figure 112. Squall Line Thunderstorms

3. **Turbulence**

    a. Hazardous turbulence is present in *all* thunderstorms, and in a severe thunderstorm it can seriously damage an airframe.

        1) The strongest turbulence within the cloud occurs as the result of wind shear between updrafts and downdrafts.

            a) Outside the cloud, shear turbulence has been encountered several thousand feet above and 20 mi. laterally from a severe storm.

        2) A low-level turbulent area is the shear zone between the plow wind and the surrounding air.

            a) Often, a "roll cloud" on the leading edge of a storm marks the eddies in this shear and signifies an extremely turbulent zone.

        3) The first gust causes a rapid and sometimes drastic change in surface wind ahead of an approaching storm.

        4) Figure 113 below shows a schematic cross section of a thunderstorm with areas outside the cloud where turbulence may be encountered.

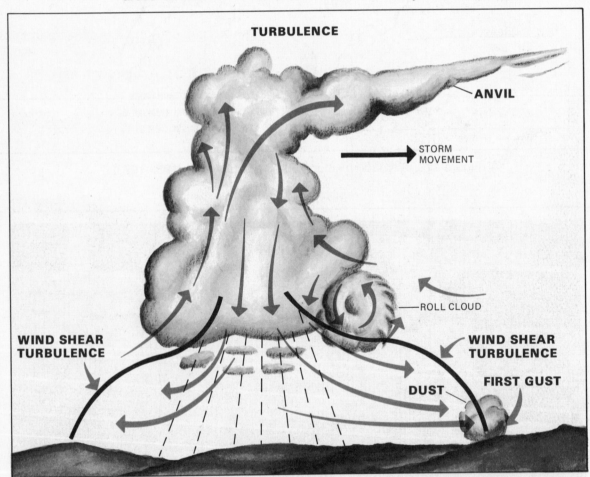

Figure 113. Schematic Cross Section of a Thunderstorm

    b. It is almost impossible to hold a constant altitude in a thunderstorm, and maneuvering in an attempt to do so greatly increases stresses on the aircraft.

        1) Stresses will be least if the aircraft is held in a constant *attitude* and allowed to "ride the waves."

4. **Hail**

    a. Hail competes with turbulence as the greatest thunderstorm hazard to aircraft.

        1) Supercooled drops above the freezing level begin to freeze.

            a) Once a drop has frozen, other drops attach and freeze to it, so the hailstone grows, sometimes into a huge iceball.

        2) Large hail occurs with severe thunderstorms usually built to great heights.

            a) Eventually the hailstones fall, possibly some distance from the storm core.

            b) Hail has been observed in clear air several miles from the parent thunderstorm.

    b. As hailstones fall below the freezing level, they begin to melt, and precipitation may reach the ground as either hail or rain.

        1) Rain at the surface does not mean the absence of hail aloft.

        2) You should anticipate possible hail with *any* thunderstorm, especially beneath the anvil of a large cumulonimbus.

    c. Hailstones larger than ½ in. in diameter can significantly damage an aircraft in a few seconds.

        1) Figure 114 below is a photograph of an aircraft flown through hail.

Figure 114. Hail Damage to an Aircraft

5. **Icing**

    a. Updrafts in a thunderstorm support abundant liquid water.

        1) When carried above the freezing level, the water becomes supercooled.

        2) When temperature in the upward current cools to about −15°C, much of the remaining water vapor sublimates as ice crystals.

            a) Above this level, the amount of supercooled water decreases.

    b. Supercooled water freezes on impact with an aircraft (see Part I, Chapter 10, Icing, beginning on page 105).

        1) Clear icing can occur at any altitude above the freezing level, but at high levels, icing may be rime or mixed rime and clear.

        2) The abundance of supercooled water makes clear icing very rapid between 0°C and −15°C, and encounters can be frequent in a cluster of cells.

6. **Low Ceiling and Visibility**

    a. Visibility generally is near zero within a thunderstorm cloud.

    b. Ceiling and visibility also can become restricted in precipitation and dust between the cloud base and the ground.

        1) The restrictions create the same problem as all ceiling and visibility restrictions.

    c. The hazards are increased many times when associated with the other thunderstorm hazards of turbulence, hail, and lightning, which make precision instrument flying virtually impossible.

7. **Effect on Altimeters**

    a. Pressure usually falls rapidly with the approach of a thunderstorm, then rises sharply with the onset of the first gust and arrival of the cold downdraft and heavy rain showers, falling back to normal as the storm moves on.

        1) This cycle of pressure change may occur in as little as 15 min.

    b. If the altimeter setting is not corrected, the indicated altitude may be in error by over 100 ft.

8. **Thunderstorm Electricity**

    a. Electricity generated by thunderstorms is rarely a great hazard to aircraft, but it may cause damage and is annoying to flight crews.

        1) Lightning is the most spectacular of the electrical discharges.

    b. **Lightning**

        1) A lightning strike can puncture the skin of an aircraft and can damage communication and electronic navigational equipment.

            a) Lightning has been suspected of igniting fuel vapors causing explosion; however, serious accidents due to lightning strikes are extremely rare.

            b) Nearby lightning can blind the pilot rendering him momentarily unable to navigate either by instrument or by visual reference.

            c) Nearby lightning can also induce permanent errors in the magnetic compass.

            d) Lightning discharges, even distant ones, can disrupt radio communications on low and medium frequencies.

    2) A few pointers on lightning:

        a) The more frequent the lightning, the more severe the thunderstorm.

        b) Increasing frequency of lightning indicates a growing thunderstorm.

        c) Decreasing lightning indicates a storm nearing the dissipating stage.

        d) At night, frequent distant flashes playing along a large sector of the horizon suggest a probable squall line.

  c. **Precipitation Static**

    1) A steady, high level of noise in radio receivers is called precipitation static and is caused by intense corona discharges from sharp metallic points and edges of flying aircraft.

        a) It is encountered often in the vicinity of thunderstorms.

        b) When an aircraft flies through clouds, precipitation, or a concentration of solid particles (ice, sand, dust, etc.), it accumulates a charge of static electricity.

        c) The electricity discharges onto a nearby surface or into the air, causing a noisy disturbance at lower frequencies.

    2) The corona discharge is weakly luminous and may be seen at night. Although it has a rather eerie appearance, it is harmless.

        a) It was named "St. Elmo's Fire" by Mediterranean sailors, who saw the brushy discharge at the top of ships' masts.

E. **Thunderstorms and Radar**

  1. Weather radar detects water droplets of precipitation size.

    a. The greater the number of drops, the stronger the echo; similarly, the larger the drops, the stronger the echo

    b. Drop size determines echo intensity to a much greater extent than does drop number.

  2. Meteorologists have shown that drop size is almost directly proportional to rainfall rate and the greatest rainfall rate is in thunderstorms.

    a. Thus, thunderstorms yield the strongest echoes.

    b. Hailstones usually are covered with a film of water and, therefore, act as huge water droplets giving the strongest echo of all types of precipitation.

    c. Showers show less intense echoes, and gentle rain and snow return the weakest of all echoes.

  3. Since the strongest echoes identify thunderstorms, they also mark the areas of greatest hazards.

    a. Realize, however, that severe turbulence associated with a thunderstorm can also occur outside and below the cell, and thus not appear on the radar.

    b. Radar information can be valuable both from ground-based radar for preflight planning and from airborne radar for severe weather avoidance.

  4. Thunderstorms build and dissipate rapidly, and they also may move rapidly.

    a. **Do not attempt to preflight plan a course between echoes.**

    b. The best use of ground radar information is to isolate general areas and coverage of echoes.

      1) You must avoid individual storms from in-flight observations either by visual sighting or by airborne radar.

5.   Airborne weather avoidance radar is, as its name implies, for avoiding severe weather, not for penetrating it.

    a.   Whether to fly into an area of radar echoes depends on echo intensity, spacing between the echoes, and your capabilities and those of your aircraft.

    b.   Remember that weather radar detects only precipitation drops; it does not detect minute cloud droplets.

    c.   **The radar scope provides no assurance of avoiding instrument weather due to clouds and fog.**

        1)   Your scope may be clear between intense echoes; this clear area does not necessarily mean you can fly between the storms and maintain visual sighting of them.

6.   The most intense echoes are severe thunderstorms.

    a.   Remember that hail may fall several miles from the cloud, and hazardous turbulence may extend as much as 20 mi. from the cloud.

        1)   Avoid the most intense echoes by at least 20 mi.; that is, echoes should be separated by at least 40 mi. before you fly between them.

        2)   As echoes diminish in intensity, you can reduce the distance by which you avoid them.

    b.   Figure 116 below illustrates use of airborne radar in avoiding thunderstorms.

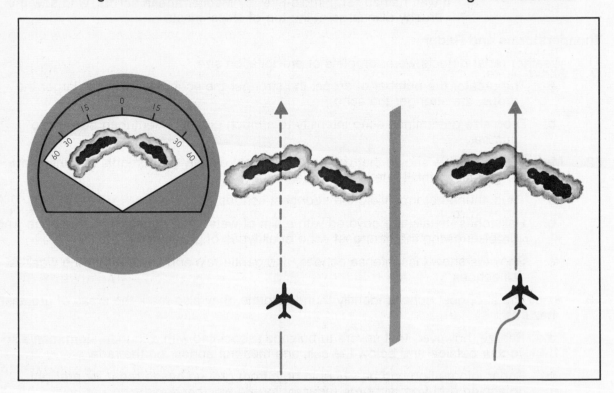

Figure 116. Use of Airborne Radar to Avoid Heavy Precipitation and Turbulence

7. A **stormscope** senses and displays electrical discharges, as opposed to precipitation.

   a. By definition, a thunderstorm has lightning, which is a discharge of static electricity.

       1) Thus, an indication of discharge on the stormscope implies turbulence.

   b. However, a clear display only indicates a lack of electrical discharge.

       1) Convective activity, hazardous precipitation, and other thunderstorm hazards may still be present.

   c. Experts agree that a combination of stormscope and radar is the best thunderstorm detection system.

F. **In Closing**

   1. Thunderstorms progress through three stages: cumulus, mature, and dissipating.

   2. Air mass thunderstorms generally result from surface heating, reaching maximum intensity and frequency over land during middle and late afternoon.

   3. Steady state thunderstorms, usually associated with weather systems, may persist for several hours.

   4. Hazards associated with thunderstorms are tornadoes, squall lines, turbulence, icing, hail, low ceiling and visibility, altimeter inaccuracy due to pressure changes, lightning, and precipitation static.

   5. The strongest echoes on weather radar identify thunderstorms, and the most intense echoes are severe thunderstorms.

   6. Weather radar detects precipitation, whereas a stormscope senses and displays electrical discharges.

   7. Above all, remember this: **never regard any thunderstorm as "light"** even when radar observers report the echoes are of light intensity.

      a. **Avoiding thunderstorms is the best policy.**

   8. The following are some DOs and DON'Ts of thunderstorm *avoidance*:

      a. Don't land or take off in the face of an approaching thunderstorm. A sudden wind shift or low level turbulence could cause loss of control.

      b. Don't attempt to fly under a thunderstorm even if you can see through to the other side. Turbulence under the storm could be disastrous.

      c. Do circumnavigate the entire area if the area has 6/10 thunderstorm coverage.

      d. Don't fly without airborne radar into a cloud mass containing scattered embedded thunderstorms. Scattered thunderstorms not embedded usually can be visually circumnavigated.

      e. Don't trust the visual appearance to be a reliable indicator of the turbulence associated with a thunderstorm.

      f. Do avoid by at least 20 NM any thunderstorm identified as severe or giving an intense radar echo. This is especially true under the anvil of a large cumulonimbus.

      g. Do clear the top of a known or suspected severe thunderstorm by at least 1,000 ft. altitude for each 10 kt. of wind speed at the cloud top. This rule would exceed the altitude capability of most aircraft.

      h. Do remember that vivid and frequent lightning indicates the probability of a severe thunderstorm.

      i. Do regard as severe any thunderstorm with tops 35,000 ft. or higher, whether the top is visually sighted or determined by radar.

9. If you **cannot** avoid penetrating a thunderstorm, following are some DOs **before** entering the storm:

    a. Tighten your safety belt, put on your shoulder harness if you have one, and secure all loose objects.

    b. Plan your course to take you through the storm in a minimum time and *hold* it.

    c. To avoid the most critical icing, establish a penetration altitude below the freezing level or above the level of −15°C.

    d. Turn on pitot heat and carburetor or jet inlet heat. Icing can be rapid at any altitude and cause almost instantaneous power failure or loss of airspeed indication.

    e. Establish power settings for reduced turbulence penetration airspeed recommended in your aircraft manual. Reduced airspeed lessens the structural stresses on the aircraft.

    f. Turn up cockpit lights to highest intensity to lessen danger of temporary blindness from lightning.

    g. If using automatic pilot, disengage altitude hold mode and speed hold mode. The automatic altitude and speed controls will increase maneuvers of the aircraft, thus increasing structural stresses.

    h. If using airborne radar, tilt your antenna up and down occasionally.

        1) Tilting it up may detect a hail shaft that will reach a point on your course by the time you do.

        2) Tilting it down may detect a growing thunderstorm cell that may reach your altitude.

10. The following are some DOs and DON'Ts **during** thunderstorm penetration:

    a. Do keep your eyes on your instruments. Looking outside the cockpit can increase danger of temporary blindness from lightning.

    b. Don't change power settings; maintain settings for reduced airspeed.

    c. Do maintain a constant *attitude*; let the aircraft "ride the waves." Maneuvers in trying to maintain constant altitude increase stresses on the aircraft.

    d. Don't turn back once you are in a thunderstorm. A straight course through the storm most likely will get you out of the hazards most quickly. In addition, turning maneuvers increase stresses on the aircraft.

G. Addendum: Microbursts

1. Microbursts are small-scale intense downdrafts which, on reaching the surface, spread outward in all directions from the downdraft center. This causes the presence of both vertical and horizontal wind shears that can be extremely hazardous to all types and categories of aircraft, especially at low altitudes.

2. Parent clouds producing microburst activity can be any of the low or middle layer convective cloud types.

    a. Microbursts commonly occur within the heavy rain portion of thunderstorms, but also in much weaker, benign-appearing convective cells that have little or no precipitation reaching the ground.

3.   The life cycle of a microburst as it descends in a convective rain shaft is illustrated below.

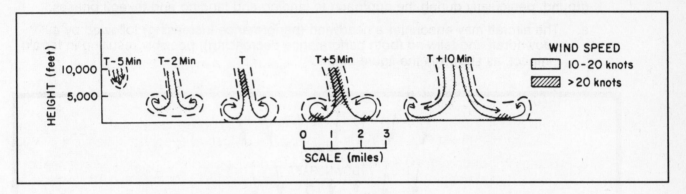

   a.   T is the time the microburst strikes the ground.

4.   Characteristics of microbursts include:

   a.   Size -- The microburst downdraft is typically less than 1 mi. in diameter as it
        descends from the cloud base to about 1,000-3,000 ft. above the ground.

        1)   In the transition zone near the ground, the downdraft changes to a horizontal
             outflow that can extend to approximately 2½ mi. in diameter.

   b.   Intensity -- The downdrafts can be as strong as 6,000 fpm.

        1)   Horizontal winds near the surface can be as strong as 45 kt. resulting in a
             90-kt. shear (headwind to tailwind change for a traversing aircraft) across the
             microburst.

        2)   These strong horizontal winds occur within a few hundred feet of the ground.

   c.   Visual signs -- Microbursts can be found almost anywhere there is convective
        activity.

        1)   They may be embedded in heavy rain associated with a thunderstorm or in
             light rain in benign-appearing virga.

        2)   When there is little or no precipitation at the surface accompanying the
             microburst, a ring of blowing dust may be the only visual clue of its
             existence.

   d.   Duration -- An individual microburst will seldom last longer than 15 min. from the
        time it strikes the ground until dissipation.

        1)   An important consideration for pilots is that the microburst intensifies for about
             5 min. after it strikes the ground, with the maximum intensity winds lasting
             approximately 2 to 4 min.

        2)   Once microburst activity starts, multiple microbursts in the same general area
             are not uncommon and should be expected.

        3)   Sometimes microbursts are concentrated into a line structure, and under these
             conditions, activity may continue for as long as an hour.

5.   Microburst wind shear may create a severe hazard for aircraft within 1,000 ft. of the ground, particularly during the approach to landing and landing and takeoff phases.

a.   The aircraft may encounter a headwind (performance increasing) followed by a downdraft and tailwind (both performance decreasing), possibly resulting in terrain impact, as shown in the figure below.

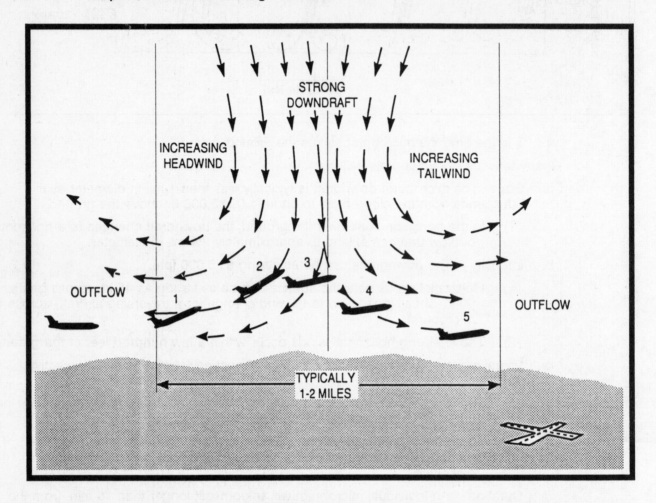

6.   Flight in the vicinity of suspected or reported microburst activity should always be avoided.

END OF CHAPTER

CHAPTER TWELVE
COMMON IFR PRODUCERS

Please take a few minutes to study each of the concepts listed above and anticipate/imagine what they are and how they relate to the other listed concepts.

A. Fog

1. Fog is a surface-based cloud composed of either water droplets or ice crystals.

   a. Fog is the most frequent cause of reducing surface visibility to below 3 SM, and is one of the most common and persistent weather hazards encountered in aviation.

      1) The rapidity with which fog can form makes it especially hazardous.

         a) It is not unusual for visibility to drop from VFR to less than 1 SM in a few minutes.

      2) It is primarily a hazard during takeoff and landing, but it is also important to VFR pilots who must maintain visual reference to the ground.

   b. A small temperature-dew point spread is essential for fog to form.

      1) Therefore, fog is prevalent in coastal areas where moisture is abundant.

         a) However, fog can occur anywhere.

      2) Abundant condensation nuclei enhance the formation of fog.

         a) Thus, fog is prevalent in industrial areas where by-products of combustion provide a high concentration of these nuclei.

      3) Fog occurs most frequently in the colder months, but the season and frequency of occurrence vary from one area to another.

   c. Fog is classified by the way it forms. Formation may involve more than one process. Fog may form

      1) By cooling air to its dew point.
      2) By adding moisture to air near the ground.

2.   **Radiation fog** is relatively shallow fog. It may be dense enough to hide the entire sky or may conceal only part of the sky.

    a.   "Ground fog" is a form of radiation fog.

        1)   As viewed in flight, dense radiation fog may obliterate the entire surface below you.

            a)   A less dense fog may permit your observation of a small portion of the surface directly below you.

            b)   Tall objects such as buildings, hills, and towers may protrude upward through ground fog giving you fixed references for VFR flight.

            c)   Figure 117 below illustrates ground fog as seen from the air.

Figure 117. Ground Fog as Seen from the Air

    b.   Conditions favorable for radiation fog are clear sky, little or no wind, and a small temperature-dew point spread (high relative humidity).

        1)   The fog forms almost exclusively at night or near daybreak.

        2)   Terrestrial radiation cools the ground; in turn, the cool ground cools the air in contact with it.

            a)   When the air is cooled to its dew point, fog forms.

        3)   When rain soaks the ground, followed by clearing skies, radiation fog is not uncommon the following morning.

    c.   Radiation fog is restricted to land because water surfaces cool little from nighttime radiation.

        1)   It is shallow when wind is calm.

        2)   Winds up to about 5 kt. mix the air slightly and tend to deepen the fog by spreading the cooling through a deeper layer.

            a)   Stronger winds disperse the fog or mix the air through a still deeper layer with stratus clouds forming at the tip of the mixing layer.

    d.   Ground fog usually "burns off" rather rapidly after sunrise.

        1)   Other radiation fog generally clears before noon unless clouds move in over the fog.

3. **Advection fog** forms when moist air moves over colder ground or water.

    a. It is most common along coastal areas but often develops deep in continental areas.

        1) At sea it is called "sea fog."

    b. Advection fog deepens as wind speed increases up to about 15 kt.

        1) Wind much stronger than 15 kt. lifts the fog into a layer of low stratus or stratocumulus.

    c. The west coast of the United States is quite vulnerable to advection fog.

        1) This fog frequently forms offshore as a result of cold water as shown in Figure 118 below and then is carried inland by the wind.

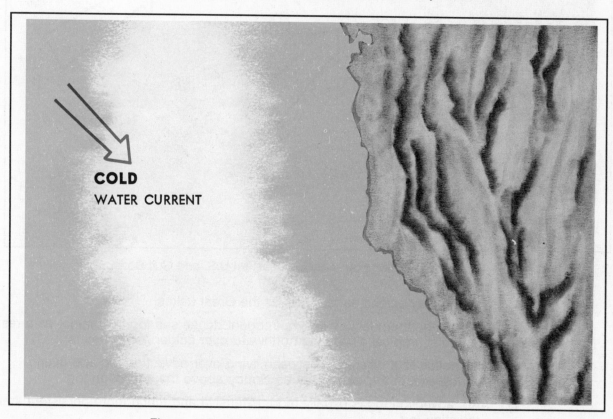

COLD
WATER CURRENT

Figure 118. Advection Fog off the Coast of California

d.   During the winter, advection fog over the central and eastern United States results when moist air from the Gulf of Mexico spreads northward over cold ground as shown in Figure 119 below.

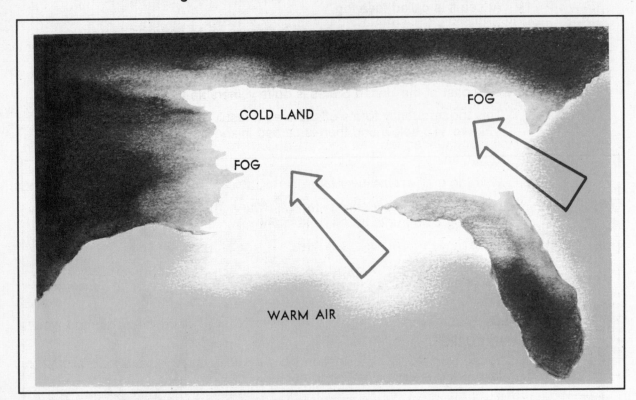

Figure 119. Advection Fog over Southeastern U.S. and Gulf Coast

1)   The fog may extend as far north as the Great Lakes.

e.   Water areas in northern latitudes have frequent dense sea fog in summer as a result of warm, moist, tropical air flowing northward over colder Arctic waters.

f.   A pilot will notice little difference between flying over advection fog and over radiation fog except that skies may be cloudy above the advection fog.

1)   Also, advection fog is usually more extensive and much more persistent than radiation fog.

2)   Advection fog can move in rapidly regardless of the time of day or night.

4. **Upslope fog** forms as a result of moist, stable air being cooled adiabatically as it moves up sloping terrain.

   a. Once the upslope wind ceases, the fog dissipates.

   b. Unlike radiation fog, it can form under cloudy skies.

   c. Upslope fog is common along the eastern slopes of the Rockies and somewhat less frequent east of the Appalachians.

   d. Upslope fog often is quite dense and extends to high altitudes.

5. **Precipitation-induced fog** forms when relatively warm rain or drizzle falls through cool air, and evaporation from the precipitation saturates the cool air.

   a. Precipitation-induced fog can become quite dense and continue for an extended period of time.

      1) This fog may extend over large areas, completely suspending air operations.

      2) It is most commonly associated with warm fronts, but can occur with slow moving cold fronts and with stationary fronts.

   b. Fog induced by precipitation is in itself hazardous, as is any fog.

      1) It is especially critical, however, because it occurs in the proximity of precipitation and other possible hazards such as icing, turbulence, and thunderstorms.

6. **Ice fog** occurs in cold weather when the temperature is well below freezing and water vapor sublimates directly as ice crystals.

   a. Conditions favorable for its formation are the same as for radiation fog except for cold temperature, usually −25°F or colder.

      1) It occurs mostly in the Arctic regions, but is not unknown in middle latitudes during the cold season.

   b. Ice fog can be blinding to someone flying in the direction of the Sun.

B. **Low Stratus Clouds**

1. Stratus clouds, like fog, are composed of extremely small water droplets or ice crystals suspended in air.

2. Stratus and fog frequently exist together.

   a. In many cases there is no real line of distinction between fog and stratus; rather, one gradually merges into the other.

3. Flight visibility may approach zero in stratus clouds.

4. Stratus tends to be lowest during night and early morning, lifting or dissipating due to solar heating during the late morning or afternoon.

   a. Low stratus clouds often occur when moist air mixes with a colder air mass or in any situation where temperature-dew point spread is small.

C. **Haze and Smoke**

    1.   Haze is a concentration of salt particles or other dry particles not readily classified as dust or other phenomena.

        a.   It occurs in stable air, is usually only a few thousand feet thick, but sometimes may extend as high as 15,000 ft.

        b.   Haze layers often have definite tops above which horizontal visibility is good.

            1)   However, downward visibility from above a haze layer is poor, especially on a slant.

        c.   Visibility in haze varies greatly depending upon whether the pilot is facing the Sun.

            1)   Landing an aircraft into the Sun is often hazardous when haze is present.

    2.   Smoke concentrations form primarily in industrial areas when air is stable.

        a.   It is most prevalent at night or early morning under a temperature inversion but it can persist throughout the day.

            1)   Figure 120 below illustrates smoke trapped under a temperature inversion.

Figure 120. Smoke Trapped in Stagnant Air under an Inversion

    3.   When skies are clear above haze or smoke, visibility generally improves during the day; however, the improvement is slower than the clearing of fog.

        a.   Fog evaporates, but haze or smoke must be dispersed by movement of air.

            1)   Haze or smoke may be blown away; or heating during the day may cause convective mixing that spreads the smoke or haze to a higher altitude, decreasing the concentration near the surface.

        b.   At night or in the early morning, radiation fog or stratus clouds often combine with haze or smoke.

            1)   The fog and stratus may clear rather rapidly during the day but the haze and smoke will linger.

            2)   A heavy cloud cover above haze or smoke may block sunlight, preventing dissipation; visibility will improve little, if any, during the day.

D. **Blowing Restrictions to Visibility**

   1.   Strong wind lifts blowing dust in both stable and unstable air.

      a.   When air is unstable, dust is lifted to great heights (as much as 15,000 ft.) and may be spread over wide areas by upper winds.

         1)   Visibility is restricted both at the surface and aloft.

      b.   When air is stable, dust does not extend to as great a height as in unstable air and usually is not as widespread.

      c.   Dust, once airborne, may remain suspended and restrict visibility for several hours after the wind subsides.

   2.   Blowing sand is more local than blowing dust; the sand is seldom lifted above 50 ft. However, visibilities within it may be near zero.

      a.   Blowing sand may occur in any dry area where loose sand is exposed to strong wind.

   3.   Blowing snow can cause visibility at ground level to be near zero and the sky may become obscured when the particles are raised to great heights.

E. **Precipitation**

   1.   Rain, drizzle, and snow are the forms of precipitation which most commonly present ceiling and/or visibility problems.

   2.   Drizzle or snow restricts visibility to a greater degree than rain.

      a.   Drizzle falls in stable air and, therefore, often accompanies fog, haze, or smoke, frequently resulting in extremely poor visibility.

      b.   Visibility may be reduced to zero in heavy snow.

   3.   Rain seldom reduces surface visibility below 1 SM except in brief, heavy showers, but it does limit cockpit visibility.

      a.   When rain streams over the aircraft windshield, freezes on it, or fogs over the inside surface, the pilot's visibility to the outside is greatly reduced.

F. **Obscured or Partially Obscured Sky**

   1.   To be classified as obscuring phenomena, smoke, haze, fog, precipitation, or other visibility restricting phenomena must extend upward from the surface.

      a.   When the sky is totally hidden by the surface-based phenomena, the ceiling is the vertical visibility from the ground upward into the obscuration.

      b.   If clouds or part of the sky can be seen above the obscuring phenomena, the condition is defined as a partial obscuration.

         1)   A partial obscuration does not define a ceiling.
         2)   However, a cloud layer above a partial obscuration may constitute a ceiling.

   2.   An obscured ceiling differs from a cloud ceiling.

      a.   With a cloud ceiling you normally can see the ground and runway once you descend below the cloud base.

      b.   With an obscured ceiling, the obscuring phenomena restrict visibility between your altitude and the ground, and you have restricted slant visibility.

1)   Thus, you cannot always clearly see the runway or approach lights even after penetrating the level of the obscuration ceiling as shown in Figure 122 below.

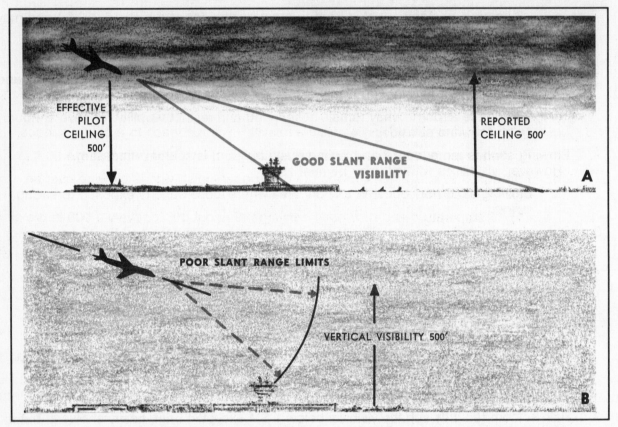

Figure 122. Difference in Visibility When a Ceiling Is Caused by a Layer Aloft (A) and by a Surface-Based Obscuration (B)

3.   Partial obscurations also present a visibility problem for the pilot on approach to land but usually to a lesser degree than the total obscuration.

    a.   However, be especially aware of erratic visibility reduction in the partial obscuration.

        1)   Visibility along the runway or on the approach can instantaneously become zero.

        2)   This abrupt and unexpected reduction in visibility can be extremely hazardous, especially on touchdown.

G. How to Avoid IFR Weather

1.   In your preflight preparation, be aware of phenomena that may produce IFR or marginal VFR flight conditions.

    a.   Current charts and special analyses along with forecast and prognostic charts are your best sources of information.

    b.   No weather observation is more current or more accurate than the one you make through your cockpit window.

    c.   Your understanding of IFR producers will help you make better preflight and in-flight decisions.

2. Do not fly VFR in weather suitable only for IFR.

    a. If you do, you endanger not only your own life but the lives of others both in the air and on the ground.

    b. Remember, the single cause of the greatest number of general aviation fatal accidents is "continued VFR into adverse weather."

3. Be especially alert for development of:

    a. Fog the following morning when at dusk temperature-dew point spread is 15°F or less, skies are clear, and winds are light.

    b. Fog when moist air is flowing from a relatively warm surface to a colder surface.

    c. Fog when temperature-dew point spread is 5°F or less and decreasing.

    d. Fog or low stratus when a moderate or stronger moist wind is blowing over an extended upslope.

        1) Temperature and dew point converge at about 4°F for every 1,000 ft. the air is lifted.

    e. Steam fog when air is blowing from a cold surface (either land or water) over warmer water.

    f. Fog when rain or drizzle falls through cool air.

        1) This is especially prevalent during winter ahead of a warm front and behind a stationary front or stagnating cold front.

    g. Low stratus clouds whenever there is an influx of low level moisture overriding a shallow cold air mass.

    h. Low visibilities from haze and smoke when a high pressure area stagnates over an industrial area.

    i. Low visibilities due to blowing dust or sand over semiarid or arid regions when winds are strong and the atmosphere is unstable.

        1) This is especially prevalent in spring.

        2) If the dust extends upward to moderate or greater heights, it can be carried many miles beyond its source.

    j. Low visibility due to snow or drizzle.

    k. An undercast when you must make a VFR descent.

4. Expect little if any improvement in visibility when:

    a. Fog exists below heavily overcast skies.
    b. Fog occurs with rain or drizzle and precipitation is forecast to continue.
    c. Dust extends to high levels and no frontal passage or precipitation is forecast.
    d. Smoke or haze exists under heavily overcast skies.
    e. A stationary high persists over industrial areas.

H. **In Closing**

1. Fog is classified by the way it forms.

    a. Radiation fog is relatively shallow, forming at night or near daybreak when the air is cooled to its dew point.

    b. Advection fog forms when moist air moves over colder ground or water.

    c. Upslope fog forms as moist, stable air is cooled adiabatically while moving up sloping terrain.

    d. Precipitation-induced fog forms when evaporation from relatively warm rain or drizzle saturates cooler air.

    e. Ice fog occurs when the temperature is well below freezing, and water vapor sublimates directly as ice crystals.

2. Low stratus clouds frequently exist with fog, and may cause flight visibility to approach zero.

3. Haze and smoke must be dispersed by movement of air and will clear more slowly than fog or stratus clouds.

4. Strong wind lifts blowing dust as high as 15,000 ft., restricting visibility at the surface and aloft.

    a. Blowing sand and blowing snow can cause visibility to be near zero.

5. Precipitation presents ceiling and/or visibility problems.

    a. Drizzle or snow restricts visibility to a greater degree than rain.

6. Surface-based phenomena can partially or completely obscure the sky.

    a. The obscuring phenomena restrict visibility between your altitude and the ground, whereas with a cloud ceiling, you can see the ground and runway once you descend below the cloud base.

7. Refer to topic G., How to Avoid IFR Weather, on page 140. Your recognition of phenomena that may produce IFR conditions will help you make better preflight and in-flight decisions.

END OF CHAPTER
END OF PART I

PART II
AVIATION WEATHER -- OVER AND BEYOND

The following four chapters are separated into Part II to facilitate study of Part I by pilots without interest in these four special topics. Our Part I outline is 142 pages long, and our Part II outline 67 pages long.

    13. High Altitude Weather
    14. Arctic Weather
    15. Tropical Weather
    16. Soaring Weather

It is less intimidating to group the first 12 chapters as Part I and Chapters 13 through 16 as Part II. You can then study only the Part II topics that are relevant to you.

CHAPTER THIRTEEN
HIGH ALTITUDE WEATHER

Please take a few minutes to study each of the concepts listed above and anticipate/imagine what they are and how they relate to the other listed concepts.

A. The Tropopause

1. The tropopause is a thin layer forming the boundary between the troposphere and stratosphere.

    a. The height of the tropopause varies from about 65,000 ft. over the Equator to 20,000 ft. or lower over the poles.

    b. The tropopause is not continuous but generally descends step-wise from the Equator to the poles.

       1) These steps occur as "breaks."

       2) Figure 123 below is a cross section of the troposphere and lower stratosphere showing the tropopause and associated features.

          a) Note the break between the tropical and the polar tropopauses.

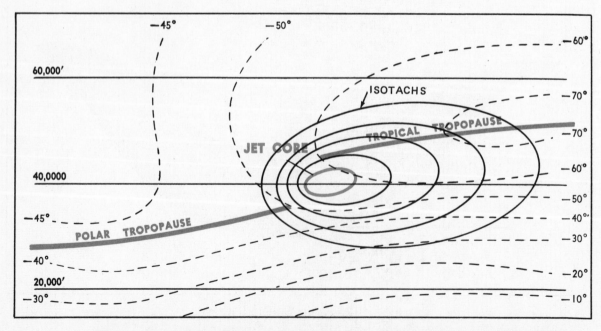

Figure 123. A Cross Section of the Upper Troposphere and Lower Stratosphere

2.   An abrupt change in temperature lapse rate characterizes the tropopause.

   a.   Note in Figure 123 how temperature above the tropical tropopause increases with height and how, over the polar tropopause, temperature remains almost constant with height.

3.   Temperature and wind vary greatly in the vicinity of the tropopause, affecting efficiency, comfort, and safety of flight.

   a.   Maximum winds generally occur at levels near the tropopause.

      1)   These strong winds create narrow zones of wind shear which often generate hazardous turbulence.

   b.   Preflight knowledge of temperature, wind, and wind shear is important to flight planning.

B. The Jet Stream

1.   The jet stream, or jet, is a narrow, shallow, meandering river of maximum winds extending around the globe in a wavelike pattern.

   a.   A second jet stream is not uncommon, and three at one time are not unknown.

      1)   A jet may be as far south as the northern Tropics.
      2)   A jet in midlatitudes generally is stronger than one in or near the Tropics.

   b.   The jet stream typically occurs in a break in the tropopause as shown in Figure 123 on page 144.

      1)   Thus, a jet stream occurs in an area of intensified temperature gradients characteristic of the break.

2.   The concentrated winds, by arbitrary definition, must be 50 kt. or greater to classify as a jet stream.

   a.   The jet maximum (concentration of greatest wind) is not constant; rather, it is broken into segments, shaped somewhat like a boomerang as diagramed in Figure 125 below.

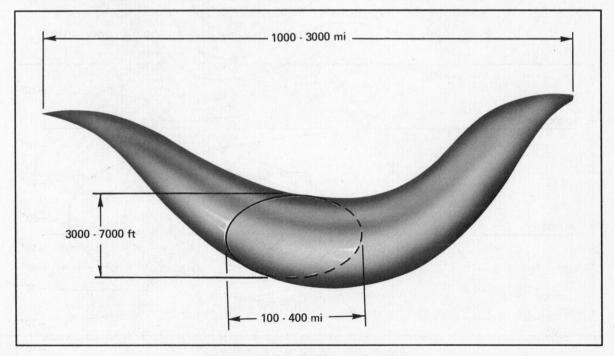

Figure 125. A Jet Stream Segment

3.   Jet stream segments move with pressure ridges and troughs in the upper atmosphere.

  a.   In general they travel faster than pressure systems, and maximum wind speed varies as the segments progress through the systems.

  b.   In midlatitude, wind speed in the jet stream averages considerably stronger in winter than in summer.

   1)   Also, the jet shifts farther south in winter than in summer.

4.   In Figure 123 on page 144, the isotachs (lines of equal wind speed) illustrate that the maximum wind speed is at the jet core and that each isotach away from the jet core the wind speed decreases.

  a.   The rate of decrease of wind speed is considerably greater on the polar side than on the equatorial side; thus, the magnitude of wind shear is greater on the polar side than on the equatorial side.

5.   Figure 126 below shows a map with two jet streams. The paths of the jets approximately conform to the shape of the pressure contours.

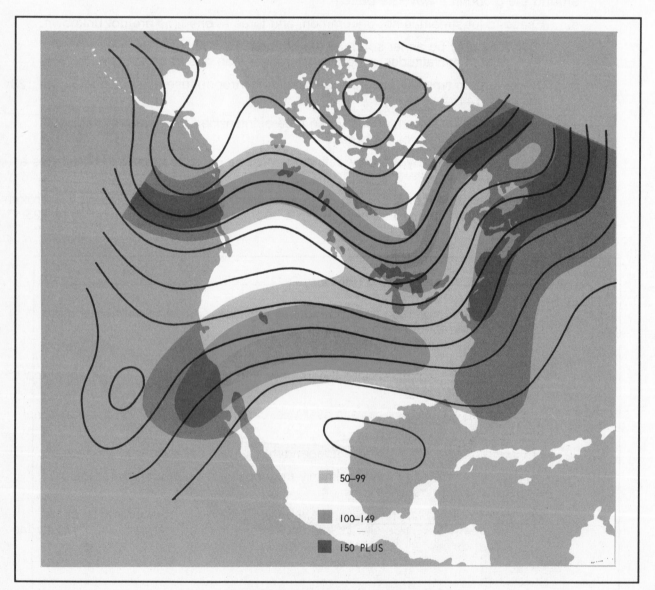

Figure 126. Multiple Jet Streams

    a.   Note how spacing of the pressure contours is closer and wind speeds higher in the vicinity of the jets than outward on either side.

        1)   Thus, horizontal wind shear is evident on both sides of the jet and is greatest near the maximum wind segments.

6.   Strong, long-trajectory jet streams usually are associated with well-developed surface lows and frontal systems beneath deep upper troughs or lows.

    a.   Cyclogenesis (i.e., development of a surface low) is usually south of the jet stream and moves nearer as the low deepens.

    b.   The occluding low moves north of the jet, and the jet crosses the frontal system near the point of occlusion (right-hand side of Figure 127).

    c.   Figure 127 below diagrams mean jet positions relative to surface systems.

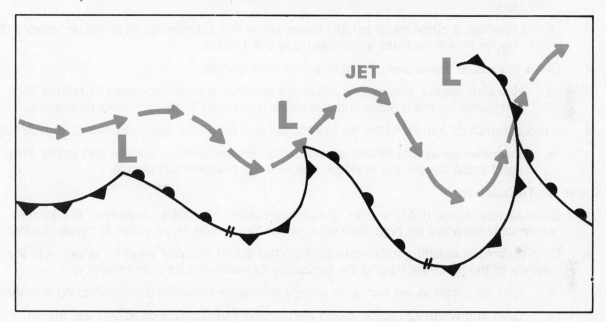

Figure 127. Mean Jet Positions Relative to Surface Systems

    d.   These long jets mark high level boundaries between warm and cold air and are favored places for cirriform cloudiness.

C. Cirrus Clouds

1.   Air travels in a corkscrew path around the jet core with upward motion on the equatorial side.

    a.   Thus, when high level moisture is available, cirriform clouds form on the equatorial side of the jet.

    b.   Jet stream cloudiness can form independently of well-defined pressure systems.

        1)   Such cloudiness ranges primarily from scattered to broken coverage in shallow layers or streaks.

        2)   The occasional fish hook and streamlined, wind-swept appearance of jet stream cloudiness always indicates very strong upper wind, usually quite far from developing or intense weather systems.

2. The densest cirriform clouds occur with well-defined systems. They appear in broad bands.

    a. Cloudiness is rather dense in an upper trough, thickens downstream, and becomes most dense at the crest of the downwind ridge.

        1) The clouds taper off after passing the ridge crest into the area of descending air.

    b. The poleward boundary of the cirrus band often is quite abrupt and frequently casts a shadow on lower clouds, especially in an occluded frontal system.

3. The upper limit of dense, banded cirrus is near the tropopause; a band may be either a single layer or multiple layers 10,000 ft. to 12,000 ft. thick.

    a. Dense, jet stream cirriform cloudiness is most prevalent along midlatitude and polar jets.

    b. However, a cirrus band usually forms along the subtropical jet in winter when a deep upper trough plunges southward into the Tropics.

4. Cirrus clouds, in themselves, have little effect on aircraft.

    a. However, dense, continuous coverage requires a pilot's constant reference to instruments; most pilots find this more tiring than flying with a visual horizon.

5. A more important aspect of the jet stream cirrus shield is its association with turbulence.

    a. Extensive cirrus cloudiness often occurs with deepening surface and upper lows; and these deepening systems produce the greatest turbulence.

D. **Clear Air Turbulence**

1. Clear air turbulence (CAT) implies turbulence devoid of clouds. However, the term is commonly reserved for high-level wind shear turbulence, even when in cirrus clouds.

2. Cold outbreaks colliding with warm air from the south intensify weather systems in the vicinity of the jet stream along the boundary between the cold and warm air.

    a. CAT develops in the turbulent energy exchange between the contrasting air masses.

    b. Cold and warm advection along with strong wind shears develop near the jet stream, especially where curvature of the jet stream sharply increases in deepening upper troughs.

    c. CAT is most pronounced in winter when temperature contrast is greatest between cold and warm air.

3. CAT is found most frequently in an upper trough on the cold (polar) side of the jet stream.

    a. Another frequent CAT location, shown in Figure 129 on page 149, is along the jet stream north and northeast of a rapidly deepening surface low.

4. Even in the absence of a well-defined jet stream, CAT is often experienced in wind shears associated with sharply curved contours of strong lows, troughs, and ridges aloft, and in areas of strong, cold or warm air advection.

    a. Also, mountain waves can create CAT.

        1) Mountain wave CAT may extend from the mountain crests to as high as 5,000 ft. above the tropopause, and can range 100 mi. or more downstream from the mountains.

5. CAT can be encountered where there seems to be no reason for its occurrence.

    a. Strong winds may carry a turbulent volume of air away from its source region.

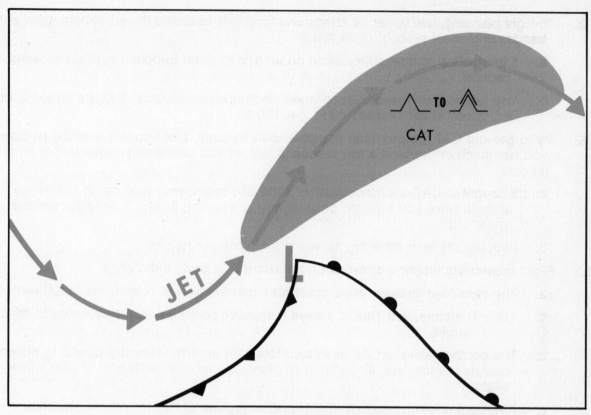

Figure 129. A Frequent CAT Location

b.   Turbulence intensity diminishes downstream, but some turbulence still may be encountered where it normally would not be expected.

c.   CAT forecast areas are sometimes elongated to indicate probable turbulence drifting downwind from the main source region.

6.   A forecast of turbulence specifies a volume of airspace which is quite small when compared to the total volume of airspace used by aviation, but is relatively large compared to the localized extent of the hazard.

a.   Since turbulence in the forecast volume is patchy, you can expect to encounter it only intermittently and possibly not at all.

b.   A flight through forecast turbulence, on the average, encounters only light and annoying turbulence 10% to 15% of the time.

1)   About 2% to 3% of the time there is a need to have all objects secured.
2)   The odds that a pilot will experience control problems are about 1 in 500.

7.   Turbulence is greatest near the jet stream windspeed maxima, usually on the polar sides, where there is a combination of strong wind shear, curvature in the flow, and cold air advection.

a.   In Figure 126 on page 146, these areas would be to the northwest of Vancouver Island, from north of the Great Lakes to east of James Bay and over the Atlantic east of Newfoundland.

1)   Also, turbulence in the form of mountain waves is probable in the vicinity of the jet stream from southern California across the Rockies into the Central Plains.

8. In flight planning, use upper air charts and forecasts to locate the jet stream, wind shears, and areas of most probable turbulence.

    a. If impractical to completely avoid an area of forecast turbulence, proceed with caution.

    b. You will do well to avoid areas where vertical shear exceeds 6 kt. per 1,000 ft. or horizontal shear exceeds 40 kt. per 150 mi.

9. If you get into CAT rougher than you care to fly through, and you are near the jet core, you can climb or descend a few thousand feet, or you can simply move farther from the jet core.

    a. If caught in CAT not associated with the jet stream, your best bet is to change altitude since you have no positive way of knowing in which direction the strongest shear lies.

    b. Pilot reports from other flights, when available, are helpful.

10. Flight maneuvers increase stresses on the aircraft, as does turbulence.

    a. The increased stresses are cumulative when the aircraft maneuvers in turbulence.

        1) Therefore, you should always maneuver gently when in turbulence to minimize stress.

    b. The patchy nature of CAT makes current pilot reports extremely helpful to observers, briefers, forecasters, air traffic controllers, and most importantly, to your fellow pilots.

    c. Always, if possible, make in-flight weather reports of CAT or other turbulence encounters.

        1) Negative reports also help when no CAT is experienced where it normally might be expected.

E. **Condensation Trails**

    1. A condensation trail, popularly contracted to "contrail," is generally defined as a cloud-like streamer which is frequently generated in the wake of aircraft flying in clear, cold, humid air.

        a. Two distinct types are observed -- exhaust trails and aerodynamic trails.

    2. **Exhaust Contrails.** The exhaust contrail is formed by the addition to the atmosphere of sufficient water vapor from aircraft exhaust gases to cause saturation or supersaturation of the air.

        a. Since heat is also added to the atmosphere in the wake of an aircraft, the addition of water vapor must be of such magnitude that it saturates or supersaturates the atmosphere despite the added heat.

    3. **Aerodynamic Contrails.** In air that is almost saturated, aerodynamic pressure reduction around airfoils, engine nacelles, and propellers cools the air to saturation leaving condensation trails from these components.

        a. This type of trail usually is neither as dense nor as persistent as exhaust trails.

            1) However, under critical atmospheric conditions, an aerodynamic contrail may trigger the formation and spreading of a deck of cirrus clouds.

            2) The induced layer may make necessary the strict use of instruments by a subsequent flight at that altitude.

4. **Dissipation Trails (Distrails).** The term dissipation trail applies to a rift in clouds caused by the heat of exhaust gases from an aircraft flying in a thin cloud layer.

    a. The exhaust gases sometimes warm the air to the extent that it is no longer saturated, and the affected part of the cloud evaporates.

    b. The cloud must be both thin and relatively warm for a distrail to exist; therefore, they are not common.

F. Haze Layers

1. Haze layers not visible from the ground are, at times, of concern at high altitude.

    a. These layers are really cirrus clouds with a very low density of ice crystals.
    b. The tops of these layers generally are very definite and are at the tropopause.
    c. High-level haze occurs in stagnant air; it is rare in fresh outbreaks of cold polar air.

        1) Cirrus haze is common in Arctic winter.
        2) Sometimes ice crystals restrict visibility from the surface to the tropopause.

2. Visibility in the haze sometimes may be near zero, especially when one is facing the sun.

    a. To avoid the poor visibility, climb into the lower stratosphere or descend below the haze.

    b. This change may need to be several thousand feet.

G. Canopy Static

1. Canopy static, similar to the precipitation static sometimes encountered at lower levels, is produced by particles brushing against plastic-covered aircraft surfaces.

    a. The discharge of static electricity results in a noisy disturbance that interferes with radio reception.

    b. Discharges can occur in such rapid succession that interference seems to be continuous.

2. Since dust and ice crystals in cirrus clouds are the primary producers of canopy static, usually you may eliminate it by changing altitude.

H. Icing

1. Although icing at high altitudes is not as common or extreme as at low altitudes, it can occur.

    a. It can form quickly on airfoils and exposed parts of jet engines.
    b. Structural icing at high altitudes usually is rime, although clear ice is possible.

2. High-altitude icing generally forms in tops of tall cumulus buildups, anvils, and even in detached cirrus.

    a. Clouds over mountains are more likely to contain liquid water than those over more gently sloping terrain because of the added lift of the mountains.

        1) Thus, icing is more likely to occur and to be more hazardous over mountainous areas.

3. Because ice generally accumulates slowly at high altitudes, anti-icing equipment usually eliminates any serious problems.

    a. However, anti-icing systems currently in use are not always adequate.

    b. If such is the case, avoid the icing problem by changing altitude or by varying course to remain clear of the clouds.

I.   Thunderstorms

1.   A well-developed thunderstorm may extend upward through the troposphere and penetrate the lower stratosphere.

    a.   Sometimes the main updraft in a thunderstorm may toss hail out the top or the upper portions of the storm.

        1)   An aircraft may thus encounter hail in clear air at a considerable distance from the thunderstorm, especially under the anvil cloud.

    b.   Turbulence may be encountered in clear air for a considerable distance both above and around a growing thunderstorm.

2.   Thunderstorm avoidance rules apply equally at high altitudes.

    a.   When flying in the clear, visually avoid all thunderstorm tops.

        1)   In a severe thunderstorm situation, avoid tops by at least 20 mi.

    b.   When you are on instruments, weather avoidance radar helps you in avoiding thunderstorm hazards.

        1)   If in an area of severe thunderstorms, avoid the most intense echoes by at least 20 mi.

J.   In Closing

1.   High-altitude weather phenomena include

    a.   The tropopause -- the thin layer forming the boundary between the troposphere and the stratosphere

        1)   Characterized by an abrupt change in temperature lapse rate.

    b.   The jet stream -- a river of maximum winds (50 kt. or greater) extending around the globe in a wavelike pattern.

    c.   Cirrus clouds -- when dense, require you to refer constantly to your instruments.

        1)   The jet stream cirrus shield is associated with turbulence.

    d.   Clear air turbulence (CAT) -- a term commonly reserved for high-level wind shear turbulence.

    e.   Condensation trails -- cloud-like streamers that are frequently generated in the wake of aircraft flying in clear, cold, humid air.

    f.   Haze layers -- really cirrus clouds with a low density of ice crystals.

    g.   Canopy static -- discharge of static electricity causing interference with radio reception.

    h.   Icing -- can occur at high altitudes, although it is not as common or extreme as at low altitudes.

    i.   Thunderstorms -- may extend upward through the troposphere and penetrate the lower stratosphere. Thunderstorm avoidance rules apply equally at high altitudes.

2.   Use upper air charts and forecasts (see Part III of this book) to locate areas to avoid.

END OF CHAPTER

CHAPTER FOURTEEN
ARCTIC WEATHER

Please take a few minutes to study each of the concepts listed above and anticipate/imagine what they are and how they relate to the other listed concepts.

A.   Definition

1.   The Arctic is the region shown in Figure 131 below which lies north of the Arctic Circle (66½° latitude).

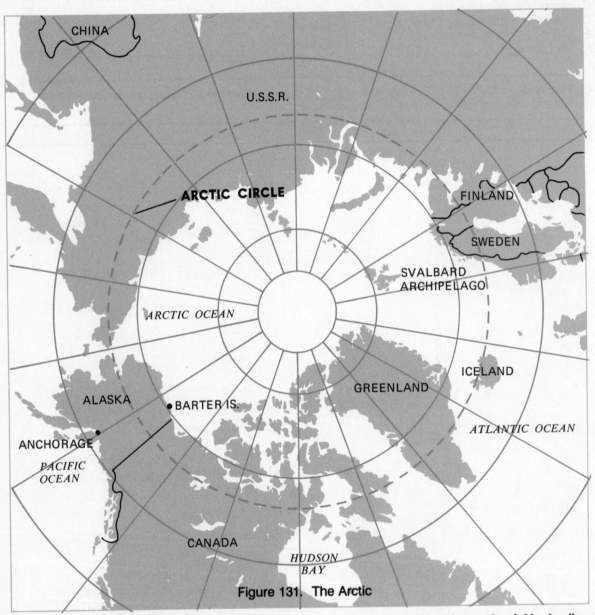

Figure 131. The Arctic

2.   However, this discussion includes Alaskan weather even though much of Alaska lies south of the Arctic Circle.

B.   **Climate, Air Masses, and Fronts**

1.   The climate of any region is largely determined by the amount of energy received from the sun.

a.   Local characteristics of the area also influence climate.

2.   **Long Days and Nights**

a.   A profound seasonal change in length of day and night occurs in the Arctic because of the Earth's tilt and its revolution around the sun.

1)   Figure 132 below shows that any point north of the Arctic Circle has autumn and winter days when the sun stays below the horizon all day and days in spring and summer with 24 hr. of sunshine.

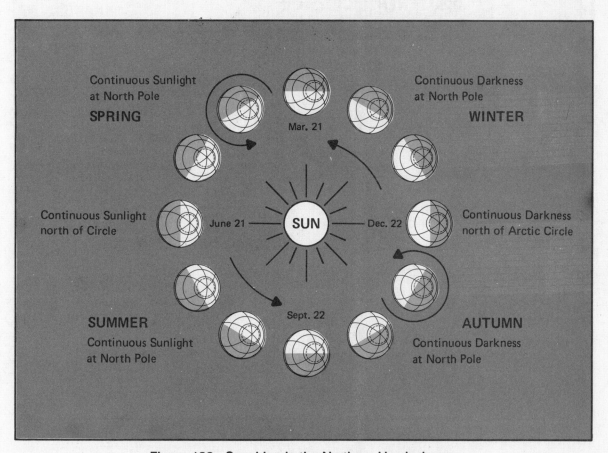

Figure 132. Sunshine in the Northern Hemisphere

2)   The number of these days increases toward the North Pole; there the sun stays below the horizon for 6 months and shines continuously during the other 6 months.

b.   Twilight in the Arctic is prolonged because of the shallow angle of the sun below the horizon.

1)   In more northern latitudes, it persists for days when the sun remains just below the horizon.

2)   This abundance of twilight often makes visual reference possible at night.

3.   Land and Water

   a.   Figure 133 below shows the water and land distribution in the Arctic.

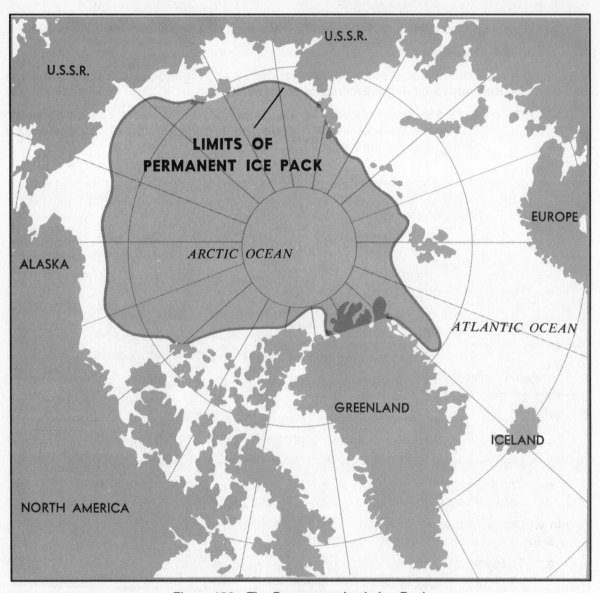

Figure 133. The Permanent Arctic Ice Pack

   1)   Arctic mountain ranges are effective barriers to air movement.
   2)   Large masses of air stagnate over the inland continental areas.

      a)   Thus, the Arctic continental areas are air mass source regions.

   b.   A large portion of the Arctic Ocean is covered throughout the year by a deep layer of ice -- the permanent ice pack as shown in Figure 133 above.

   1)   Even though the ocean is ice-covered through much of the year, the ice and water below contain more heat than the surrounding cold land, thus moderating the climate to some extent.

   2)   Oceanic and coastal areas have a milder climate during winter than would be expected and a cool climate in summer.

      a)   As opposed to large water bodies, large land areas show a more significant seasonal temperature variation.

4. **Temperature**

    a. The Arctic is very cold in winter, but due to local terrain and the movement of pressure systems, occasionally some areas are surprisingly warm.

        1) During winter, coastal areas average about 20° warmer than the interior.

        2) During summer, interior areas are pleasantly warm with many hours of sunshine.

    b. Coastal areas have relatively cool, short summers due to their proximity to water.

5. **Clouds and Precipitation**

    a. Cloudiness over the Arctic is at a minimum during winter, reaching a maximum in summer and fall.

        1) Spring also brings many cloudy days.

        2) During summer afternoons, scattered cumulus clouds forming over the interior occasionally grow into thundershowers.

            a) These thundershowers, usually circumnavigable, move generally from northeast to southwest in the polar easterlies, which is opposite the general movement in midlatitudes.

    b. Precipitation in the Arctic is generally light.

        1) Annual amounts over the ice pack and along the coastal areas are only 3 to 7 in.

            a) The interior is somewhat wetter, with annual amounts of 5 to 15 in.

        2) Precipitation falls mostly in the form of snow over ice caps and oceanic areas and mostly as summer rain over interior areas.

6. **Wind**

    a. Strong winds occur more often along the coasts than elsewhere.

    b. The frequency of high winds in coastal areas is greatest in fall and winter.

    c. Wind speeds are generally light in the continental interior during the entire year, but are normally at their strongest during summer and fall.

7. **In winter, air masses** form over the expanded ice pack and adjoining snow-covered land areas.

    a. These air masses are characterized by very cold surface air, very low humidity, and strong low-level temperature inversions.

    b. Occasionally, air from unfrozen ocean areas flows northward over the Arctic.

        1) These intrusions of moist, cold air account for most of the infrequent wintertime cloudiness and precipitation in the Arctic.

8. **During the summer**, the top layer of the Arctic permafrost melts leaving very moist ground, and the open water areas of the Polar Basin expand markedly.

    a. Thus, the entire area becomes more humid, relatively mild, and semimaritime in character.

    b. The largest amount of cloudiness and precipitation occurs inland during the summer months.

9. **Fronts.** Occluded fronts are the rule.

    a. Weather conditions with occluded fronts are much the same in the Arctic as elsewhere -- low clouds, precipitation, poor visibility, and sudden fog formation.

    b. Fronts are much more frequent over coastal areas than over the interior.

C. Arctic Peculiarities

1. Effects of Temperature Inversion

a. The intense low-level inversion over the Arctic during much of the winter causes sound (including people's voices) to carry over extremely long distances.

b. Light rays are bent as they pass at low angles through the inversion.

    1) This bending creates an effect known as looming -- a form of mirage that causes objects beyond the horizon to appear above the horizon.

    2) Mirages distorting the shape of the sun, moon, and other objects are common with the low-level inversions.

2. Aurora Borealis

a. Certain energy particles from the sun strike the Earth's magnetic field and are carried along the lines of force where they tend to lower and converge near the geomagnetic poles.

    1) The energy particles then pass through rarefied gases of the outer atmosphere, illuminating them in much the same way as an electrical charge illuminates neon gas in neon signs.

b. The Aurora Borealis takes place at high altitudes above the Earth's surface.

    1) The highest frequency of observations is over the northern United States and northward.

    2) Displays of aurora vary from a faint glow to an illumination of the Earth's surface equal to a full moon.

    3) They frequently change shape and form and are also called dancing lights or northern lights.

3. Light Reflection by Snow-Covered Surfaces

a. Much more light is reflected by snow-covered surfaces than by darker surfaces.

b. Snow often reflects Arctic sunlight sufficiently to blot out shadows, thus markedly decreasing the contrast between objects.

    1) Dark distant mountains may be easily recognized, but a crevasse normally directly in view may be undetected due to lack of contrasts.

4. Light from Celestial Bodies

a. Illumination from the moon and stars is much more intense in the Arctic than in lower latitudes.

    1) Pilots have found that light from a half-moon over a snow-covered field may be sufficient for landing.

    2) Even illumination from the stars creates visibility far beyond that found elsewhere.

b. Only under heavy overcast skies does the night darkness in the Arctic begin to approach the degree of darkness in lower latitudes.

D. Weather Hazards

1. **Fog** limits landing and takeoff in the Arctic more than any other visibility restriction.

a. Water-droplet fog is the main hazard to aircraft operations in coastal areas during the summer.

b. Ice fog is the major restriction in winter.

2. **Ice fog** is common in the Arctic.

   a. It forms in moist air during extremely cold, calm conditions in winter, occurring often and tending to persist.

   b. Effective visibility is reduced much more in ice fog when one is looking toward the sun.

   c. Ice fog may be produced both naturally and artificially.

   d. Ice fog affecting aviation operations most frequently is produced by the combustion of aircraft fuel in cold air.

      1) When the wind is very light and the temperature is about −30°F or colder, ice fog often forms instantaneously in the exhaust gases of automobiles and aircraft.

      2) It lasts from as little as a few minutes to days.

3. **Steam fog,** often called "sea smoke," forms in winter when cold, dry air passes from land areas over comparatively warm ocean waters.

   a. Moisture evaporates rapidly from the water surface; but since the cold air can hold only a small amount of water vapor, condensation takes place just above the surface of the water and appears as "steam" rising from the ocean.

   b. This fog is composed entirely of water droplets that often freeze quickly and fall back into the water as ice particles.

   c. Low-level turbulence can occur and icing can become hazardous.

4. **Advection fog,** which may be composed either of water droplets or of ice crystals, is most common in winter and is often persistent.

   a. Advection fog forms along coastal areas when comparatively warm, moist, oceanic air moves over cold land.

      1) If the land areas are hilly or mountainous, lifting of the air results in a combination of low stratus and fog.

         a) The stratus and fog quickly diminish inland.

   b. Lee sides of islands and mountains usually are free of advection fog because of drying due to compressional heating as the air descends downslope.

   c. Icing in advection fog is in the form of rime and may become quite severe.

5. **Blowing Snow**

   a. Over the frozen Arctic Ocean and along the coastal areas, blowing snow and strong winds are common hazards during autumn and winter.

   b. Blowing snow is a greater hazard to flying operations in the Arctic than in midlatitudes because the snow is "dry" and fine and can be picked up easily by light winds.

      1) Winds in excess of 8 kt. may raise the snow several feet off the ground, obliterating objects such as runway markers.

      2) A sudden increase in surface wind may cause an unlimited visibility to drop to near zero in a few minutes.

         a) This sudden loss of visibility occurs frequently without warning in the Arctic.

      3) Stronger winds sometimes lift blowing snow to heights above 1,000 ft. and produce drifts over 30 ft. deep.

6. **Icing** is most likely in spring and fall, but is also encountered in winter.

    a. During spring and fall, icing may extend to upper levels along frontal zones.

    b. While icing is mostly a problem over water and coastal areas, it does exist inland.

        1) It occurs typically as rime, but a combination of clear and rime icing is not unusual in coastal mountains.

7. **Frost**

    a. In coastal areas during spring, fall, and winter, heavy frost and rime may form on aircraft parked outside, especially when fog or ice fog is present.

    b. This frost should be removed; it reduces lift and is especially hazardous if surrounding terrain requires a rapid rate of climb.

8. **Whiteout** is a visibility restriction phenomenon that occurs in the Arctic when a layer of cloudiness of uniform thickness overlies a snow- or ice-covered surface.

    a. Parallel rays of the sun are broken up and diffused when passing through the cloud layer so that they strike the snow surface from many angles.

        1) The diffused light then reflects back and forth countless times between the snow and the cloud eliminating all shadows.

        2) The result is a loss of depth perception.

    b. Buildings, people, and dark-colored objects appear to float in the air, and the horizon disappears.

        1) Low-level flight over icecap terrain or landing on snow surfaces becomes dangerous.

E. **Arctic Flying Weather**

    1. A great number of pilots who fly Alaska and the Arctic are well-seasoned. They are eager to be of help and are your best sources of information.

        a. Before flying the Arctic, be sure to learn all you can about your proposed route.

    2. Generally, flying conditions in the Arctic are good when averaged over the entire year.

        a. However, areas of Greenland compete with the Aleutians for the world's worst weather.

            1) These areas are exceptions.

    3. Whiteouts, in conjunction with overcast skies, often present a serious hazard, especially for visual flight.

        a. Many mountain peaks are treeless and rounded rather than ragged, making them unusually difficult to distinguish under poor visibility conditions.

    4. **Oceanic and Coastal Areas**

        a. In oceanic and coastal areas, predominant hazards change with the seasons.

            1) In summer, the main hazard is fog in coastal areas.

        b. In winter, ice fog is the major restriction to aircraft operation.

            1) Blowing and drifting snow often restrict visibility as well.

            2) Storms and well-defined frontal passages frequent the coastal areas accompanied by turbulence, especially in the coastal mountains.

        c. Icing is most frequent in spring and fall and may extend to high levels in active, turbulent frontal zones.

            1) Fog is also a source of icing when the temperature is colder than freezing.

5. **Continental Areas**

    a.   Over the continental interior, good flying weather prevails much of the year, although during winter, ice fog often restricts aircraft operations.

        1)   In terms of ceiling and visibility, the summer months provide the best flying weather.

        2)   Thunderstorms develop on occasion during the summer, but they usually can be circumnavigated without much interference with flight plans.

F. **In Closing**

1. The Arctic climate is influenced by local characteristics:

    a.   Long days and nights
    b.   Land and water distribution
    c.   Temperature
    d.   Clouds and precipitation
    e.   Wind
    f.   Air masses forming over the expanded ice pack in winter
    g.   Melting of the top layer of the Arctic permafrost
    h.   Fronts

2. Other phenomena peculiar to the Arctic are

    a.   Mirages caused by the effects of low-level temperature inversions.

        1)   Also, sound carries over extremely long distances.

    b.   Aurora Borealis.
    c.   Light reflection by snow-covered surfaces.
    d.   Light from celestial bodies.

3. Weather hazards include

    a.   Fog,
    b.   Blowing snow,
    c.   Icing,
    d.   Frost,
    e.   Whiteout.

4. Interior areas generally have good flying weather, but coastal areas and Arctic slopes often are plagued by low ceiling, poor visibility, and icing.

5. Whiteout conditions over ice and snow covered areas often cause pilot disorientation.

6. Flying conditions are usually worse in mountain passes than at reporting stations along the route.

7. Routes through the mountains are subject to strong turbulence, especially in and near passes.

8. Beware of a false mountain pass that may lead to a dead end.

9. Thundershowers sometimes occur in the interior during May through August. They are usually circumnavigable and generally move from northeast to southwest.

10. Always file a flight plan. Stay on regularly traversed routes, and if downed, stay with your plane.

11. If lost during summer, fly down-drainage, that is, downstream. Most airports are located near rivers, and chances are you can reach a landing strip by flying downstream.

    a.   If forced down, you will be close to water on which a rescue plane can land. In summer, the tundra is usually too soggy for landing.

12. Weather stations are few and far between.

    a.   Adverse weather between stations may go undetected unless reported by a pilot in flight.

    b.   A report confirming good weather between stations is also just as important.

    c.   Help yourself and your fellow pilots by reporting weather en route.

END OF CHAPTER

CHAPTER FIFTEEN
TROPICAL WEATHER

A. Definition

1. Technically, the Tropics lie between latitudes 23½°N and 23½°S.

   a. However, weather typical of this region sometimes extends as much as 45° from the Equator.

2. One generally thinks of the Tropics as uniformly rainy, warm, and humid.

   a. However, the Tropics contain both the wettest and the driest regions of the world.

B. Circulation

1. In Part I, Chapter 4, Wind, beginning on page 29, we learned that wind blowing out of the subtropical high pressure belts toward the Equator form the northeast and southeast trade winds of the two hemispheres.

   a. These trade winds converge in the vicinity of the Equator where air rises.

      1) This convergence zone is the "intertropical convergence zone" (ITCZ).

   b. In some areas of the world, seasonal temperature differences between land and water areas generate rather large circulation patterns that overpower the trade wind circulation.

      1) These areas are monsoon regions.

2. **Subtropical High Pressure Belts**

   a. If the surface under the subtropical high pressure belts were all water of uniform temperature, the high pressure belts would be continuous highs around the globe.

      1) The belts would be areas of descending or subsiding air and would be characterized by strong temperature inversions and very little precipitation.

      2) However, land surfaces at the latitudes of the high pressure belts are generally warmer throughout the year than are water surfaces.

         a) Thus, the high pressure belts are broken into semipermanent high pressure anticyclones over oceans with troughs or lows over continents.

   b. The subtropical highs shift southward during the Northern Hemisphere winter and northward during summer.

      1) The seasonal shift, the height and strength of the inversion, and terrain features determine weather in the subtropical high pressure belts.

    c.   **Continental Weather**

       1)   Along the west coasts of continents under a subtropical high, the air is stable.

          a)   The inversion is strongest and lowest where the east side of an anticyclone overlies the west side of a continent.

          b)   Moisture is trapped under the inversion; fog and low stratus occur frequently.

              i)   However, precipitation is rare since the moist layer is shallow and the air is stable.

          c)   Heavily populated areas also add contaminants to the air which, when trapped under the inversion, create an air pollution problem.

       2)   The situation on eastern continental coasts is just the opposite.

          a)   The inversion is weakest and highest where the west side of an anticyclone overlies the eastern coast of a continent.

          b)   Convection can penetrate the inversion, and showers and thunderstorms often develop.

       3)   Low ceiling and fog often prevent landing at a west coast destination, but a suitable alternate generally is available a few miles inland.

          a)   An alternate selection may be more critical for an eastern coast destination because of widespread instability and associated hazards.

    d.   **Weather over Open Sea**

       1)   Under a subtropical high over the open sea, cloudiness is scant.

       2)   The few clouds that do develop have tops from 3,000 to 6,000 ft., depending on the height of the inversion.

       3)   Ceiling and visibility are generally quite ample for VFR flight.

    e.   **Island Weather**

       1)   An island under a subtropical high receives very little rainfall because of the persistent temperature inversion.

       2)   Surface heating over some larger islands causes light convective showers.

       3)   Cloud tops are only slightly higher than over open water.

       4)   Temperatures are mild, showing small seasonal and diurnal changes.

3.   **Trade Wind Belts**

    a.   Figures 138 and 139 on page 165 show prevailing winds throughout the Tropics for July and January.

       1)   Note that trade winds blowing out of the subtropical highs over ocean areas are predominantly northeasterly in the Northern Hemisphere and southeasterly in the Southern Hemisphere.

    b.   The inversion from the subtropical highs is carried into the trade winds and is known as the "trade wind inversion."

       1)   As in a subtropical high, the inversion is strongest where the trades blow away from the west coast of a continent and weakest where they blow onto an eastern continental shore.

       2)   Daily variations from these prevailing directions are small except during tropical storms.

          a)   As a result, weather at any specific location in a trade wind belt varies little from day to day.

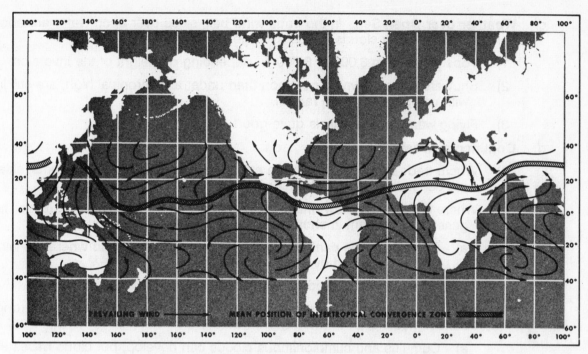

Figure 138. Prevailing Winds throughout the Tropics in July

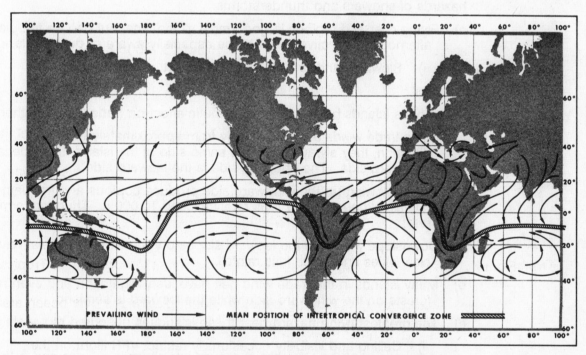

Figure 139. Prevailing Winds in the Tropics in January

c. **Weather over Open Sea.** In the trade wind belt, skies over open water are about one-half covered by clouds on the average.

   1) Tops range from 3,000 to 8,000 ft., depending on height of the inversion.

   2) Showers, although more common than under a subtropical high, are still light with comparatively little rainfall.

   3) Flying weather generally is quite good.

d. **Continental Weather**

   1) Where trade winds blow offshore along the west coasts of continents, skies are generally clear and the area is quite arid.

   2) Where trade winds blow onshore on the east sides of continents, rainfall is generally abundant in showers and occasional thunderstorms.

     a) Rainfall may be carried a considerable distance inland where the winds are not blocked by a mountain barrier.

     b) Inland areas blocked by a mountain barrier are deserts.

   3) Afternoon convective currents are common over arid regions due to strong surface heating.

     a) Cumulus and cumulonimbus clouds can develop, but cloud bases are high and rainfall is scant because of the low moisture content.

   4) Flying weather along eastern coasts and mountains is subject to the usual hazards of showers and thunderstorms.

     a) Flying over arid regions is good most of the time, but can be turbulent in afternoon convective currents; be especially aware of dust devils.

       i) Blowing sand or dust sometimes restricts visibility.

e. **Island Weather**

   1) Mountainous islands have the most dramatic effect on trade wind weather.

     a) Since trade winds are consistently from approximately the same direction, they always strike the same side of the island; this side is the windward side. The opposite side is the leeward side.

     b) Winds blowing up the windward side produce copious and frequent rainfall, although cloud tops rarely exceed 10,000 ft. Thunderstorms are rare.

     c) Downslope winds on the leeward slopes dry the air leaving relatively clear skies and much less rainfall.

     d) Many islands in the trade wind belt have lush vegetation and even rain forests on the windward side while the leeward is semiarid.

     e) The greatest flying hazard near these islands is obscured mountain tops.

       i) Ceiling and visibility occasionally restrict VFR flight on the windward side in showers.

       ii) IFR weather is virtually nonexistent on leeward slopes.

   2) Islands without mountains have little effect on cloudiness and rainfall.

     a) Afternoon surface heating increases convective cloudiness slightly, but shower activity is light.

b) Any island in either the subtropical high pressure belt or trade wind belt enhances cumulus development even though tops do not reach great heights.

    i) Thus, a cumulus top higher than the average tops of surrounding cumulus usually marks the approximate location of an island.

    ii) If it becomes necessary to ditch in the ocean, look for a tall cumulus.

4. **The Intertropical Convergence Zone (ITCZ).** Converging winds in the intertropical convergence zone (ITCZ) force air upward.

    a. The inversion typical of the subtropical high and trade wind belts disappears.

        1) Figures 138 and 139 show the ITCZ and its seasonal shift.

        2) The ITCZ is well marked over tropical oceans but is weak and ill-defined over large continental areas.

    b. **Weather over Islands and Open Water.** Convection in the ITCZ carries huge quantities of moisture to great heights.

        1) Showers and thunderstorms frequent the ITCZ and tops to 30,000 ft. or higher are common, as shown in Figure 137 below.

            a) Precipitation is copious.

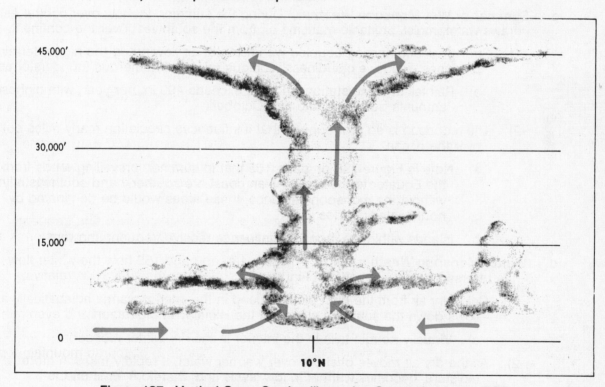

Figure 137. Vertical Cross Section Illustrating Convection in the ITCZ

        2) Since convection dominates the ITCZ, there is little difference in weather over islands and open sea under the ITCZ.

        3) Flying through the ITCZ usually presents no great problem if one follows the usual practice of avoiding thunderstorms.

4)   Since the ITCZ is ill-defined over continents, we will not attempt to describe ITCZ continental weather as such.

   a)   Continental weather ranges from arid to rain forests and is more closely related to the monsoon than to the ITCZ.

5.   **Monsoon**

   a.   Refer to Figures 23 and 24 in Part I, Chapter 4, Wind, on page 35. You can see that over the large land mass of Asia, the subtropical high pressure breaks down completely.

   1)   Asia is covered by an intense high during the winter and a well-developed low during the summer.

   2)   You can also see the same over Australia and central Africa, although the seasons are reversed in the Southern Hemisphere.

   b.   The cold, high pressures in winter cause wind to blow from the deep interior outward and offshore.

   1)   In summer, wind direction reverses, and warm, moist air is carried far inland into the low pressure area.

   2)   This large-scale seasonal wind shift is the monsoon.

   a)   The most notable monsoon is that of southern and southeastern Asia.

   c.   **Summer or Wet Monsoon Weather.** During the summer, the low over central Asia draws warm, moist, unstable maritime air from the southwest over the continent.

   1)   Strong surface heating coupled with rising of air flowing up the higher terrain produces extensive cloudiness, copious rain, and numerous thunderstorms.

   a)   Rainfall at some stations in India exceeds 400 in. per year, with highest amounts between June and October.

   2)   The monsoon is so pronounced that it influences circulation many miles out over the ocean.

   a)   Note in Figure 138 on page 165 that in summer, prevailing winds from the Equator to the south Asian coast are southerly and southeasterly; without the monsoon influence, these areas would be dominated by northeasterly trades.

   b)   Islands within the monsoon influence receive frequent showers.

   d.   **Winter Monsoon Weather.** Note in Figure 139 on page 165 how the winter flow has reversed from that shown in Figure 138.

   1)   Cold, dry air from the high plateau deep in the interior warms adiabatically as it flows down the southern slopes of the Himalayan Mountains.

   a)   Virtually no rain falls in the interior in the dry winter monsoon.

   2)   As the dry air moves offshore over warmer water, it rapidly takes in more moisture, becomes warmer in low levels and, therefore, unstable.

   a)   Rain is frequent over offshore islands and even along coastal areas after the air has had a significant over-water trajectory.

   3)   The Philippine Islands are in an area of special interest.

   a)   During the summer, they are definitely in southerly monsoon flow and are subjected to abundant rainfall.

   b)   In winter, wind over the Philippines is northeasterly -- in the transition zone between the northeasterly trades and the monsoon flow.

        c)   It is academic whether we call the phenomenon the trade winds or monsoon; in either case, it produces abundant rainfall.

           i)   The Philippines have a year-round humid, tropical climate.

  e.   **Other Monsoon Areas.** Australia in July (Southern Hemisphere winter) is an area of high pressure with predominantly offshore winds as shown in Figure 138 on page 165.

    1)   Most of the continent is dry during the winter.

    2)   In January (Figure 139 on page 165), winds are onshore into the continental low pressure.

        a)   However, most of Australia is rimmed by mountains, and coastal regions are wet where the onshore winds blow up the mountain slopes.

        b)   The interior is arid where downslope winds are warmed and dried.

    3)   Central Africa is known for its humid climate and jungles.

        a)   Note in Figures 138 and 139 on page 165 that prevailing wind is onshore much of the year over these regions.

        b)   Some regions are wet the year round; others have the seasonal monsoon shift and have a summer wet season and a winter dry season.

    4)   In the Amazon Valley of South America during the Southern Hemisphere winter (July), southeast trades, as shown in Figure 138 on page 165, penetrate deep into the valley bringing abundant rainfall which contributes to the jungle climate.

        a)   In January, the ITCZ moves south of the valley as shown in Figure 139 on page 165.

        b)   The northeast trades are caught up in the monsoon, cross the Equator, and also penetrate the Amazon Valley.

        c)   The jungles of the Amazon result largely from monsoon winds.

  f.   **Flying Weather in Monsoons**

    1)   During the winter monsoon, excellent flying weather prevails over dry interior regions.

        a)   Over water, you must pick your way around showers and thunderstorms.

    2)   In the summer monsoon, VFR flight over land is often restricted by low ceilings and heavy rain.

        a)   IFR flight must cope with the hazards of thunderstorms.

    3)   Freezing level in the Tropics is quite high -- 14,000 ft. or higher -- so icing is restricted to high levels.

C. Transitory Systems

  1.   A **wind shear line** found in the Tropics mainly results from midlatitude influences.

    a.   In Part I, Chapter 8, Air Masses and Fronts, beginning on page 73, we stated that an air mass becomes modified when it flows from its source region.

      1)   By the time a cold air mass originating in high latitudes reaches the Tropics, temperature and moisture are virtually the same on both sides of the front.

      2)   A shear line, or wind shift, is all that remains.

b.   A shear line also results when a semi-permanent high splits into two cells, inducing a trough, as shown in Figure 140 below.

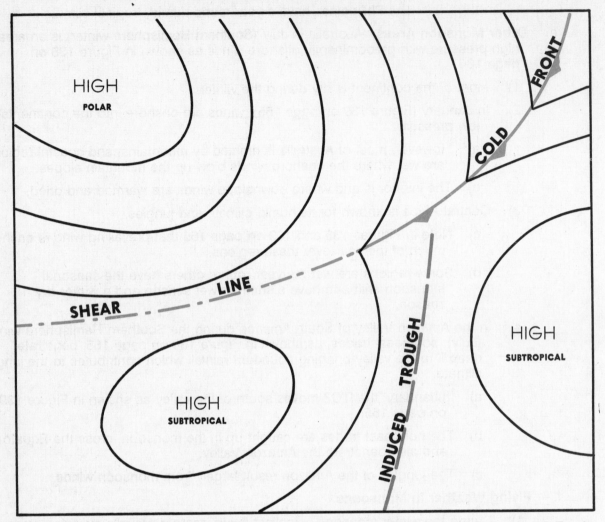

Figure 140. A Shear Line and an Induced Trough Caused by a Polar High Pushing into the Tropics

c.   These shear lines are zones of convergence creating forced upward motion.

1)   Consequently, considerable thunderstorm and rain shower activity occurs along a shear line.

2.   **Troughs** in the atmosphere, generally at or above 10,000 ft. move through the Tropics, especially along the poleward fringes.

    a.   As a trough moves to the southeast or east, it spreads middle and high cloudiness over extensive areas to the east of the trough line.

    b.   Occasionally, a well-developed trough will extend deep into the Tropics, and a closed low forms at the equatorial end of the trough.

        1)   The low then may separate from the trough and move westward, producing a large amount of cloudiness and precipitation.

        2)   If this occurs in the vicinity of a strong subtropical jet stream, extensive and sometimes dense cirrus and some convective and clear air turbulence often develop.

    c.   Troughs and lows aloft produce considerable amounts of rainfall in the Tropics, especially over land areas where mountains and surface heating lift air to saturation.

        1)   Low pressure systems aloft contribute significantly to the record 460 in. average annual rainfall on Mt. Waialeale on Kauai, Hawaii.

        2)   Other mountainous areas of the Tropics are also among the wettest spots on earth.

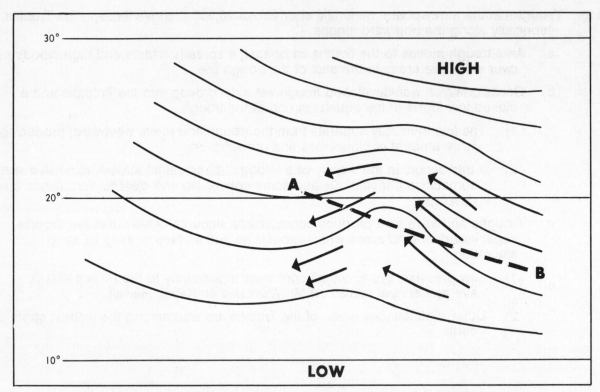

Figure 142. A Northern Hemisphere Easterly Wave Moving from (B) to (A)

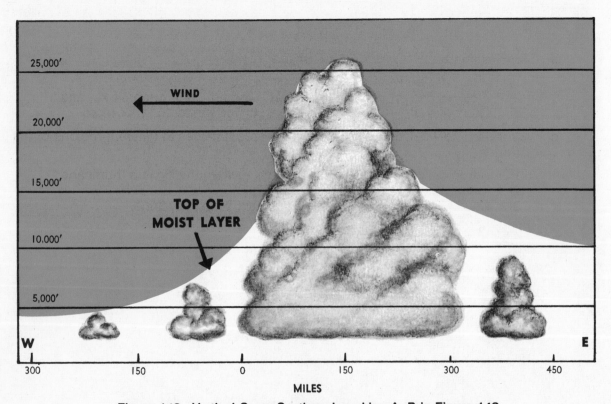

Figure 143. Vertical Cross Section along Line A–B in Figure 142

3. **Tropical waves** (also called easterly waves) are common tropical weather disturbances, normally occurring in the trade wind belt.

    a. In the Northern Hemisphere, they usually develop in the southeastern perimeter of the subtropical high pressure systems.

    b. They travel from east to west around the southern fringes of these highs in the prevailing easterly circulation of the Tropics.

    c. Surface winds in advance of a wave are somewhat more northerly than the usual trade wind direction.

    d. As the wave approaches, as shown in Figure 142 on page 172, pressure falls; as it passes, surface wind shifts to the east-southeast or southeast.

        1) The typical wave is preceded by very good weather but followed by extensive cloudiness, as shown in Figure 143 on page 172, and often by rain and thunderstorms.

        2) The weather activity is roughly in a north-south line.

    e. Tropical waves occur in all seasons, but are more frequent and stronger during summer and early fall.

        1) Pacific waves frequently affect Hawaii.

        2) Atlantic waves occasionally move into the Gulf of Mexico, reaching the U.S. coast.

4. **Tropical cyclone** is a general term for any low that originates over tropical oceans.

    a. Tropical cyclones are classified according to their intensity based on average 1-min. wind speeds.

        1) Wind gusts in these storms may be as much as 50% higher than the average 1-min. wind speeds.

    b. Tropical cyclone international classifications are:

        1) Tropical Depression -- highest sustained winds up to 34 kt.,
        2) Tropical Storm -- highest sustained winds of 35 through 64 kt., and
        3) Hurricane or Typhoon -- highest sustained winds 65 kt. or more.

    c. Strong tropical cyclones are known by different names in different regions of the world.

        1) A tropical cyclone in the Atlantic and eastern Pacific is a "hurricane."
        2) In the western Pacific, "typhoon."
        3) Near Australia and in the Indian Ocean, simply "cyclone."

d.   **Development**. Prerequisite to tropical cyclone development are optimal sea surface temperature under weather systems that produce low-level convergence and cyclonic wind shear.

1)   Favored breeding grounds are tropical (easterly) waves, troughs aloft, and areas of converging northeast and southeast trade winds along the intertropical convergence zone.

2)   The low-level convergence associated with these systems, by itself, will not support development of a tropical cyclone.

   a)   The system must also have horizontal outflow -- divergence -- at high tropospheric levels.

   b)   This combination creates a "chimney," in which air is forced upward causing clouds and precipitation.

   c)   Condensation releases large quantities of latent heat which raises the temperature of the system and accelerates the upward motion.

      i)   The rise in temperature lowers the surface pressure which increases low-level convergence. This draws more moisture-laden air into the system.

   d)   When these chain-reaction events continue, a huge vortex is generated which may culminate in hurricane force winds.

3)   Figure 144 below shows regions of the world where tropical cyclones frequently develop. Notice that they usually originate between latitudes 5° and 20°.

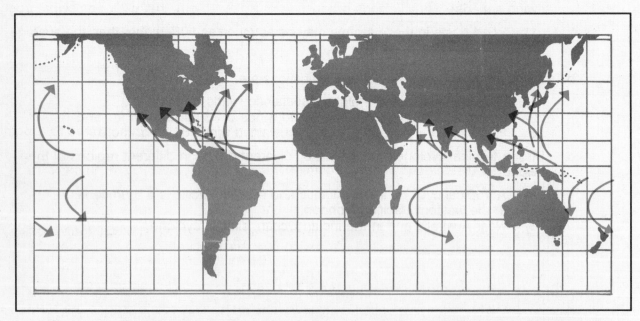

Figure 144.   Principal Regions Where Tropical Cyclones Form and Their Favored Directions of Movement

   a)   Tropical cyclones are unlikely within 5° of the Equator because the Coriolis force is so small near the Equator that it will not turn the winds enough for them to flow around a low-pressure area.

   b)   Winds flow directly into an equatorial low and rapidly fill it.

e. **Movement.** Tropical cyclones in the Northern Hemisphere usually move in a direction between west and northwest while in low latitudes.

1) As these storms move toward the midlatitudes, they come under the influence of the prevailing westerlies.

    a) At this time, the storms are under the influence of two wind systems, i.e., the trade winds at low levels and prevailing westerlies aloft.

    b) Thus, a storm may move very erratically and may even reverse course, or circle.

2) Finally, the prevailing westerlies gain control and the storm recurves toward the north, then to the northeast, and finally to the east-northeast.

    a) By this time, the storm is well into midlatitudes.

f. **Decay.** As the storm curves toward the north or east, it usually begins to lose its tropical characteristics and acquires characteristics of lows in middle latitudes.

1) Cooler air flowing into the storm gradually weakens it.

    a) If the storm tracks along a coastline or over the open sea, it gives up slowly, carrying its fury to areas far removed from the Tropics.

    b) If the storm moves well inland, it loses its moisture source and weakens from starvation and increased surface friction, usually after leaving a trail of destruction and flooding.

2) When a storm takes on middle latitude characteristics, it is said to be "extratropical," meaning "outside the Tropics."

    a) Tropical cyclones produce weather conditions that differ somewhat from those produced by their higher latitude cousins.

g. **Weather in a Tropical Depression**

1) While in its initial developing stage, the cyclone is characterized by a circular area of broken to overcast clouds in multiple layers.

    a) Embedded in these clouds are numerous showers and thunderstorms.

    b) Rain shower and thunderstorm coverage varies from scattered to almost solid.

2) The diameter of the cloud pattern varies from less than 100 mi. in small systems to well over 200 mi. in large ones.

h. **Weather in Tropical Storms and Hurricanes**

1) As cyclonic flow increases, the thunderstorms and rain showers form broken or solid lines paralleling the wind flow that is spiraling into the center of the storm.

    a) These lines are the spiral rain bands frequently seen on radar.

    b) These rain bands continually change as they rotate around the storm.

    c) Rainfall in the rain bands is very heavy, reducing ceiling and visibility to near zero.

2) Winds are usually very strong and gusty and, consequently, generate violent turbulence.

3) Between the rain bands, ceilings and visibilities are somewhat better, and turbulence generally is less intense.

4) The "eye" usually forms in the tropical storm stage and continues through the hurricane stage.

    a) In the eye, skies are free of turbulent cloudiness, and wind is comparatively light.

    b) The average diameter of the eye is between 15 and 20 mi., but sometimes it is as small as 7 mi. and rarely more than 30 mi. in diameter.

    c) Surrounding the eye is a wall of cloud that may extend above 50,000 ft.

        i) This "wall cloud" contains deluging rain and the strongest winds of the storm.

        ii) Maximum wind speeds of 175 kt. have been recorded in some storms.

5) Figure 145 below is a radar display of a mature hurricane.

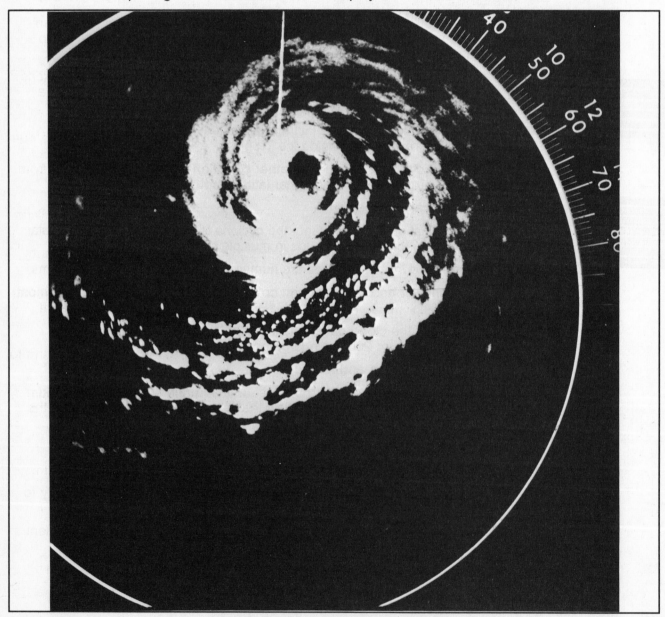

Figure 145. Radar Display of a Hurricane

i.   **Detection and Warning.** The National Weather Service has a specialized hurricane forecast and warning service center at Miami, Florida, which maintains a constant watch for the formation and development of tropical cyclones.

1)   Weather information from land stations, ships at sea, reconnaissance aircraft, long-range radars, and weather satellites is fed into the center.

a)   The center forecasts the development, movement, and intensity of tropical cyclones.

b)   Forecasts and warnings are issued to the public and aviation interests by field offices of the National Weather Service.

j.   **Flying.** All pilots except those especially trained to explore tropical storms and hurricanes should AVOID THESE DANGEROUS STORMS.

1)   Occasionally, jet aircraft have been able to fly over small and less intense storms, but the experience of weather research aircraft shows hazards at all levels within them.

2)   Tops of thunderstorms associated with tropical cyclones frequently exceed 50,000 ft.

a)   Winds in a typical hurricane are strongest at low levels, decreasing with altitude.

i)   However, research aircraft have frequently encountered winds in excess of 100 kt. at 18,000 ft.

b)   Aircraft at low levels are exposed to sustained, pounding turbulence due to the surface friction of the fast-moving air.

c)   Turbulence increases in intensity in spiral rain bands and becomes the most violent in the wall cloud surrounding the eye.

3)   An additional hazard encountered in hurricanes is erroneous altitude readings from pressure altimeters.

a)   These errors are caused by the large pressure difference between the periphery of the storm and its center.

b)   One research aircraft lost almost 2,000 ft. true altitude traversing a storm while the pressure altimeter indicated a constant altitude of 5,000 ft.

4)   In short, tropical cyclones are very hazardous, so avoid them!

a)   To bypass the storm in a minimum of time, fly to the right of the storm to take advantage of the tailwind.

b)   If you fly to the left of the storm, you will encounter strong headwinds which may exhaust your fuel supply before you reach a safe landing area.

D. **In Closing**

1. The circulation basic to weather in the Tropics includes

   a. Subtropical high-pressure belts which shift southward during the Northern Hemisphere winter and northward during summer.

   b. Trade wind belts -- Trade winds blowing out of the subtropical highs over ocean areas are predominantly northeasterly in the Northern Hemisphere.

   c. Intertropical Convergence Zone (ITCZ) -- In the ITCZ, converging winds force air upward.

   1) Showers and thunderstorms are common.

   d. Monsoon regions -- Cold, high pressures over the land in winter cause wind to blow outward and offshore.

   1) In summer, wind direction reverses, and warm, moist air is carried far inland into the low-pressure area.

2. Transitory (or migrating) tropical weather producers include

   a. Wind shear lines, which are zones of convergence creating forced upward motion.

   1) Considerable thunderstorm and rain shower activity occurs along a shear line.

   b. Troughs aloft, which produce large amounts of rainfall in the Tropics, especially in mountainous areas.

   c. Tropical waves, which normally occur in the trade wind belt.

   1) Typically, they are preceded by very good weather but followed by extensive cloudiness, and often by rain and thunderstorms.

   d. Tropical cyclone -- any low that originates over tropical oceans. International classifications based on average wind speeds are

   1) Tropical depression,
   2) Tropical storm,
   3) Hurricane or typhoon.

   a) Tropical storms and hurricanes are extremely hazardous, and you should avoid them.

END OF CHAPTER

CHAPTER SIXTEEN
SOARING WEATHER

Please take a few minutes to study each of the concepts listed above and anticipate/imagine what they are and how they relate to the other listed concepts.

A. **Introduction** -- Soaring bears the relationship to flying that sailing bears to power boating.

1. Soaring has made notable contributions to meteorology.

    a. For example, soaring pilots have probed thunderstorms and mountain waves with findings that have made flying safer for all pilots.

2. A sailplane must have auxiliary power to become airborne, such as a winch, a ground tow, or a tow by a powered aircraft.

    a. Once the sailcraft is airborne and the tow cable released, performance depends on the weather and the skill of the pilot.

    b. Forward thrust comes from gliding downward relative to the air just as thrust is developed in a power-off glide by a conventional aircraft.

        1) Therefore, to gain or maintain altitude, the soaring pilot must rely on upward motion of the air.

3. To a sailplane pilot, "lift" means the rate of climb (s)he can achieve in an up-current, while "sink" denotes his/her rate of descent in a downdraft or in neutral air.

    a. "Zero sink" means that upward currents are just strong enough to enable the pilot to hold altitude but not to climb.

    b. Sailplanes are highly efficient machines; a sink rate of a mere 2 ft. per second (fps) provides an airspeed of about 40 kt., and a sink rate of 6 fps gives an airspeed of about 70 kt.

4. In lift, a sailplane usually flies 35 to 40 kt. with a sink rate of about 2 fps.

    a. Therefore, to remain airborne, the pilot must have an upward air current of at least 2 fps.

    b. There is no point in trying to soar until weather conditions favor vertical speeds greater than the minimum sink rate of the aircraft.

    c. These vertical currents develop from several sources, which categorize soaring into five classes:

        1) Thermal soaring
        2) Frontal soaring
        3) Sea breeze soaring
        4) Ridge or hill soaring
        5) Mountain wave soaring

B. Thermal Soaring

1. A thermal is simply the updraft in a small-scale convective current.

2. A soaring aircraft is always sinking relative to the air.

    a. To maintain or gain altitude, therefore, the soaring pilot must spend sufficient time in thermals to overcome the normal sink of the aircraft as well as to regain altitude lost in downdrafts.

        1) The pilot usually circles at a slow airspeed in a thermal and then darts on a beeline to the next thermal, as shown in Figure 147 below.

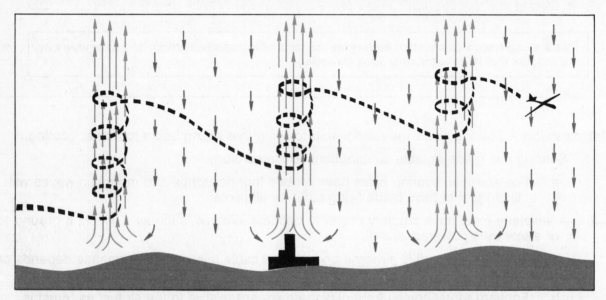

Figure 147. Thermals Generally Occur over a Small Portion of an Area While Downdrafts Predominate

3. Low-level heating is prerequisite to thermals; and this heating is mostly from the sun, although it may be augmented by man-made heat sources such as chimneys, factories, and cities.

    a. Cool air must sink to force warm air upward in thermals.

        1) Therefore, in small-scale convection, thermals and downdrafts coexist side by side.

    b. The net upward displacement of air must equal the net downward displacement.

        1) Fast rising thermals generally cover a small percentage of a convective area, while slower downdrafts predominate over the remaining greater portion, as diagramed in Figure 147 above.

    c. Since thermals depend on solar heating, thermal soaring is virtually restricted to daylight hours with considerable sunshine.

        1) Air tends to become stable at night due to low-level cooling by terrestrial radiation, often resulting in an inversion at or near the surface.

        2) Stable air suppresses convection, and thermals do not form until the inversion "burns off" or lifts sufficiently to allow soaring beneath the inversion.

            a) The earliest that soaring may begin varies from early forenoon to early afternoon, depending on the strength of the inversion and the amount of solar heating.

4. **Locating Thermals.** Since convective thermals develop from uneven heating at the surface, the most likely place for a thermal is above a surface that heats readily.

    a. When the sky is cloudless, the soaring pilot must look for those surfaces that heat most rapidly and seek thermals above those areas.

        1) Barren sandy or rocky surfaces, plowed fields, stubble fields surrounded by green vegetation, cities, factories, chimneys, etc., are good thermal sources.

    b. The angle of the sun profoundly affects location of thermals over hilly landscapes.

        1) During early forenoon, the sun strikes eastern slopes more directly than others; thus, the most favorable areas for thermals are eastern slopes.

        2) The favorable areas move to southern slopes during midday.

        3) In the afternoon, they move to western slopes before they begin to weaken as the evening sun sinks toward the western horizon.

    c. Surface winds must converge to feed a rising thermal; so when you sight a likely spot for a thermal, look for dust or smoke movement near the surface.

        1) If you can see dust or smoke "streamers" from two or more sources converging on the spot as shown in Figure 148 (A), you have chosen wisely.

            a) If, however, the streamers diverge as shown in Figure 148 (B), a downdraft most likely hovers over the spot and it's time to move on.

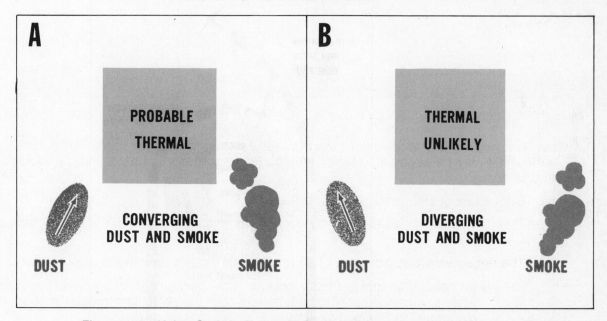

Figure 148. Using Surface Dust and Smoke Movement as Indications of a Thermal

        2) Rising columns of smoke from chimneys and factories mark thermals augmented by man-made sources.

            a) They are good sources of lift if upward speed is great enough to support the aircraft and if they are broad enough to permit circling.

        3) Towns or cities may provide thermals; but to use a thermal over a populated area, the pilot must have sufficient altitude to glide clear of the area in the event the thermal subsides.

d. Dust devils occur under sunny skies over sandy or dusty, dry surfaces and are sure signs of strong thermals with lots of lift.

   1) To tackle this excellent source of lift, you must use caution.

      a) The thermals are strong and turbulent and are surrounded by areas of little lift or possibly areas of sink.

   2) If approaching the dust devil at too low an altitude, an aircraft may sink to an altitude too low for recovery.

      a) A recommended procedure is to always approach the whirling vortex at an altitude 500 ft. or more above the ground.

      b) At this altitude, you have enough airspace for maneuvering in the event you get into a downdraft or turbulence too great for comfort.

   3) A dust devil may rotate either clockwise or counterclockwise.

      a) Before approaching the dusty column, determine its direction of rotation by observing dust and debris near the surface.

   4) You should enter against the direction of rotation.

      a) Figure 149 below diagrams a horizontal cross section of a clockwise rotating dust devil and ways of entering it.

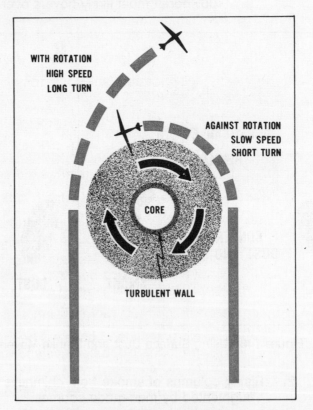

Figure 149. Horizontal Cross Section of a Dust Devil Rotating Clockwise

      b) If you enter with the direction of rotation as on the left, the wind speed is added to your airspeed, giving you a fast circling speed -- probably too great to remain in the thermal.

         i) Against the rotation as on the right, wind speed is subtracted from airspeed, giving you a slow circling speed.

5) Stay out of the eye of the vortex.

   a) Centrifugal force in the center throws air outward, greatly reducing pressure within the hollow center.

   b) The rarified air in the center provides very little lift, and the wall of the hollow center is very turbulent.

6) If you are 500 ft. or more above the ground but having trouble finding lift, the dust devil is well worth a try.

   a) If the thermal is sufficiently broad to permit circling within it, you have it made.

   b) The dust column may be quite narrow, but this fact does not necessarily mean the thermal is narrow; the thermal may extend beyond the outer limits of visible dust.

   c) Approach the dusty column against the direction of rotation at minimum airspeed.

   d) Enter the column near the outer edge of the dust and stay away from the hollow vortex core.

   e) Remain alert; you are circling little more than a wing span away from violent turbulence.

e. Wind causes a thermal to lean with altitude. When seeking the thermal supporting soaring birds or aircraft, you must make allowance for the wind.

1) The thermal at lower levels usually is upwind from your high-level visual cue.

2) A thermal may not be continuous from the surface upward to the soaring birds or sailplane; rather it may be in segments or bubbles.

   a) If you are unsuccessful in finding the thermal where you expect it, seek elsewhere.

f. When convective clouds develop, thermal soaring is usually at its best and the problem of locating thermals is greatly simplified.

1) Cumulus clouds are positive signs of thermals, but thermals grow and die.

   a) A cloud grows with a rising thermal, but when the thermal dies, the cloud slowly evaporates.

   b) Because the cloud dissipates *after* the thermal ceases, the pilot who can spot the difference between a growing and dying cumulus has enhanced his/her soaring skill.

2) The warmest and most rapidly rising air is at the center of the thermal.

   a) Therefore, the cloud base will be highest in the center, giving a concave shape to the cloud base as shown in the left and center of Figure 150 on page 184.

   b) When the thermal ceases, the base assumes a convex shape as shown on the right.

   c) Another cue to look for is the outline of the cloud sides and top.

      i) The outline of the growing cumulus is firm and sharp.

      ii) The dying cumulus has fragmentary sides and lacks the definite outline as shown on the right of Figure 150.

Figure 150. Cumulus Clouds and Thermals

d) You can expect to find a thermal beneath either of the growing cumuli in Figure 150.

    i) On the average, the infant cumulus on the left would be the better choice because of its longer life expectancy.

    ii) This, of course, is a gamble, since all cumuli do not grow to the same size.

3) As a cumulus cloud grows, it may shade the surface that generated it.

    a) The surface cools, temporarily arresting the thermal.

    b) As the cloud dissipates or drifts away with the wind, the surface again warms and regenerates the thermal.

    c) This intermittent heating is one way in which thermals occur as segments or bubbles.

4) Although abundant convective cloud cover reduces thermal activity, we cannot quote a definite amount that renders thermals too weak for soaring.

    a) About 5/10 cover seems to be a good average approximation.

    b) Restriction of thermals by cumulus cloudiness first becomes noticeable at low levels.

    c) A sailplane may be unable to climb more than a few hundred feet at a low altitude while pilots at higher levels are maintaining height in or just beneath 6/10 to 8/10 convective cloud cover.

g. When air is highly unstable, the cumulus cloud can grow into a more ambitious towering cumulus or cumulonimbus.

   1) The energy released by copious condensation can increase buoyancy until the thermals become violent.

     a) Towering cumulus can produce showers.

   2) The cumulonimbus is the thunderstorm cloud producing heavy rain, hail, and icing.

     a) Well-developed *towering cumulus and cumulonimbus are for the experienced pilot only.*

     b) Some pilots find strong lift in or near convective precipitation, but they avoid hail, which can seriously batter the aircraft.

   3) Violent thermals just beneath and within these highly developed clouds often are so strong that they will continue to carry a sailplane upward even with nose down and airspeed at the redline.

h. Dense, broken, or overcast middle and high cloudiness shade the surface, cutting off surface heating and convective thermals.

   1) On a generally warm, bright day but with thin or patchy middle or high cloudiness, cumulus may develop, but the thermals are few and weak.

   2) The high-level cloudiness may drift by in patches.

     a) Thermals may surge and wane as the cloudiness decreases and increases.

   3) Never anticipate optimal thermal soaring when plagued by these mid- and high-level clouds.

   4) Altocumulus castellanus clouds, which are middle-level convective clouds, develop in updrafts at and just below the cloud levels.

     a) They do not extend upward from the surface.

     b) If a sailplane can reach levels near the cloud bases, the updrafts with altocumulus castellanus can be used in the same fashion as thermals formed by surface convection.

     c) The problem is reaching the convective level.

i. Wet ground favors thermals less than dry ground since wet ground heats more slowly.

   1) Some flat areas with wet soil such as swamps and tidewater areas have reputations for being poor thermal soaring areas.

     a) Convective clouds may be abundant but thermals generally are weak.

   2) Showery precipitation from scattered cumulus or cumulonimbus is a sure sign of unstable air favorable for thermals.

     a) When showers have soaked the ground in localized areas, however, downdrafts are almost certain over these wet surfaces. Avoid shower-soaked areas when looking for lift.

5.   **Thermal Structure.** Thermals are as varied as trees in a forest. No two are exactly alike.

   a.   When surface heating is intense and continuous, a thermal, once begun, continues for a prolonged period in a steady column as in Figure 153 below.

   1)   Sometimes called the "chimney thermal," this type seems from experience to be most prevalent.

   2)   In the chimney thermal, lift is available at any altitude below a climbing sailplane or soaring birds.

   b.   When heating is slow or intermittent, a "bubble" may be pinched off and forced upward; after an interval ranging from a few minutes to an hour or more, another bubble forms and rises as in Figure 154 below.

   1)   As explained earlier, intermittent shading by cumulus clouds forming atop a thermal is one reason for the bubble thermal.

   2)   A sailplane or birds may be climbing in a bubble, but an aircraft attempting to enter the thermal at a lower altitude may find no lift.

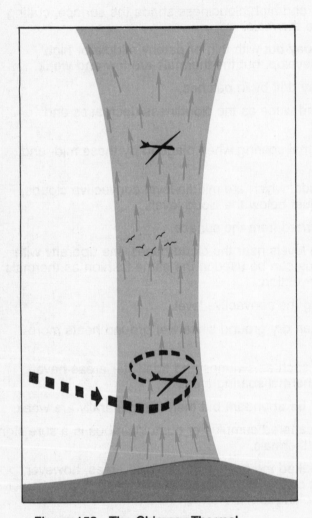

Figure 153. The Chimney Thermal

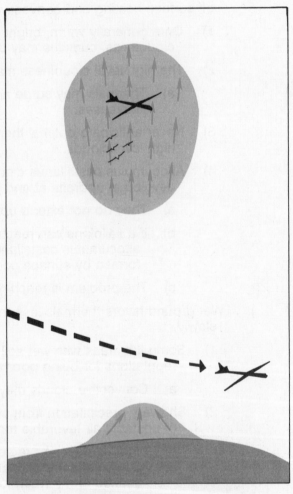

Figure 154. Thermal Bubble

c.   A favored theoretical structure of some bubble thermals is the vortex shell which is much like a smoke ring blown upward as diagrammed in Figure 155 below.

1)   Lift is strongest in the center of the ring; downdrafts may occur in the edges of the ring or shell; and outside the shell, one would expect weak downdrafts.

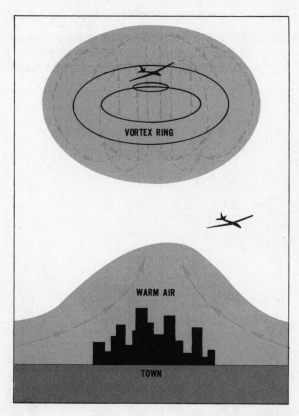

Figure 155. Vortex Ring of a Bubble Thermal

d.   Wind and wind shear. Thermals develop with a calm condition or with light, variable wind. However, it seems that a surface wind of 5 to 10 kt. favors more organized thermals.

1)   A surface wind in excess of 10 kt. usually means stronger winds aloft, resulting in vertical wind shear.

a)   This shear causes thermals to lean noticeably.

b)   When seeking a thermal under a climbing sailplane and you know or suspect that thermals are leaning in shear, look for lift upwind from the higher aircraft as shown in Figure 156 below.

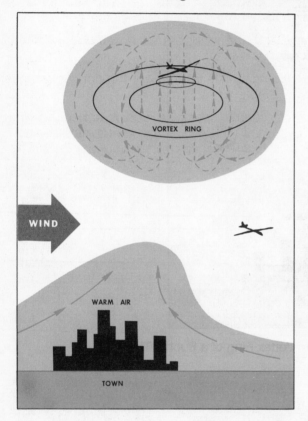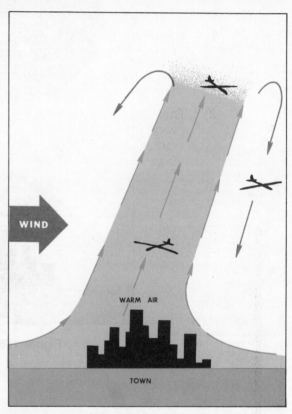

Figure 156. Wind Causes Thermals to Lean

2)   Effect of shear on thermals depends on the relative strength of the two.

a)   Strong thermals can remain fairly well organized with strong vertical wind shear; surface wind may even be at the maximum that will allow a safe launch.

b)   Weak thermals are disorganized and ripped to shreds by strong vertical wind shear; individual thermal elements become hard to find and often are too small to use for lift.

c)   A shear in excess of 3 kt. per 1,000 ft. distorts thermals to the extent that they are difficult to use.

3)   No critical surface wind speed can tell us when to expect such a shear. However, shearing action often is visible in cumulus clouds.

a)   A cloud sometimes leans but shows a continuous chimney.

b)   At other times, the clouds are completely segmented by the shear.

c)   Remember, however, that this shearing action is at cloud level; thermals below the clouds may be well organized.

4) We must not overlook one other vital effect of the low-level wind shear.

   a) On final approach for landing, the aircraft is descending into decreasing headwind.

   b) Inertia of the aircraft into the decreasing wind causes a drop in airspeed. The decrease in airspeed may result in loss of control and perhaps a stall.

   c) A good rule is to add one knot airspeed to normal approach speed for each knot of surface wind.

e. Thermal streets are bands of thermals which become organized into straight lines parallel to each other, often providing lift over a considerable distance.

   1) Generally, these streets are parallel to the wind, but on occasion they have been observed at right angles to the wind.

   2) They form when wind direction changes little throughout the convective layer and the layer is capped by very stable air.

   3) The formation of a broad system of evenly spaced streets is enhanced when wind speed reaches a maximum within the convective layer; that is, wind increases with height from the surface upward to a maximum and then decreases with height to the top of the convective layer.

   4) Figure 158 below diagrams conditions favorable for thermal streeting.

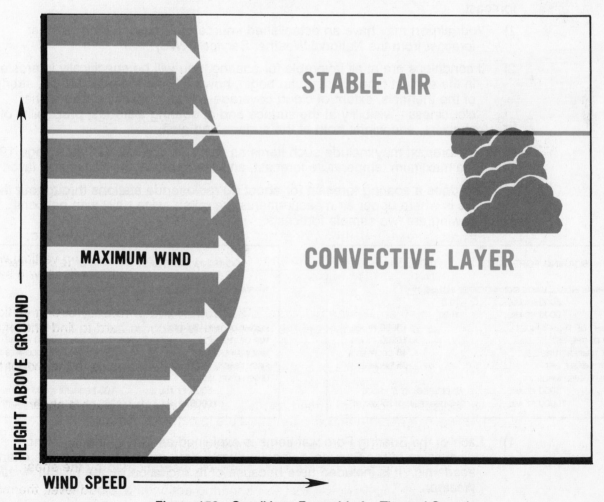

Figure 158. Conditions Favorable for Thermal Streeting

5) Thermal streets may occur in clear air, or they may be indicated by cumulus clouds, which appear as long, narrow, parallel bands.

   a) If cumulus clouds mark thermal streets, the top of the convective layer is approximately the height of the cloud tops.

6) The distance between streets in such a system is two to three times the general depth of the convective layer.

   a) If convective clouds are present, this distance is two to three times the height of the cloud tops.

7) Downdrafts between these thermal streets are usually at least moderate and sometimes strong.

8) Cumulus cloud streets frequently form in the United States behind cold fronts in the cold air of polar outbreaks in which relatively flat cumuli develop.

9) Cloud streets are advantageous for sailplane pilots because, rather than circling in isolated thermals and losing height between them, the pilot soaring under a thermal street can maintain almost continuous, straight flight.

6. **Height and Strength of Thermals**

   a. Since thermals are a product of instability, height of thermals depends on the depth of the unstable layer, and their strength depends on the degree of instability.

   b. Most likely you will be soaring from an airport with considerable soaring activity -- possibly the home base of a soaring club -- and you are interested in a soaring forecast.

      1) Your airport may have an established source of a daily soaring weather forecast from the National Weather Service (NWS).

      2) If conditions are at all favorable for soaring, you will be specifically interested in the earliest time soaring can begin, how high the thermals will be, strength of the thermals, extent of cloud coverage -- both convective and higher cloudiness -- visibility at the surface and at soaring altitudes, probability of showers, and winds both at the surface and aloft.

      3) The forecast may include such items as the thermal index (TI) (see page 194), the maximum temperature forecast, and the depth of the convective layer.

   c. The NWS does a soaring forecast for about 60 radiosonde stations throughout the U.S. (this is where upper air measurements are taken twice daily with balloons). The following are two sample forecasts:

SOARING FORECAST	DATE...3/14/1986...122
THERMAL INDEX...MINUS SIGN INDICATES INSTABILITY	
5000 FT ASL .... −10.5	
10000 FT ASL .... −6.0	
HEIGHT OF THE −3 INDEX .... 13600 FT ASL	
TOP OF THE LIFT .... 16900 FT ASL	
MAX TEMPERATURE .... 46 DEGREES F	
FIRST USABLE LIFT .... 35 DEGREES F	
UPPER LEVEL WINDS	
5000 FT ASL .... /// DEGREES AT // KNOTS	
10000 FT ASL .... 015 DEGREES AT 15 KNOTS	

SOARING FORECAST	DATE...3/14/1986...122
THERMAL INDEX...MINUS SIGN INDICATES INSTABILITY	
5000 FT ASL .... −6.0	
10000 FT ASL 0.0	
HEIGHT OF THE −3 INDEX .... 7400 FT ASL	
TOP OF THE LIFT .... 9900 FT ASL	
MAX TEMPERATURE .... 46 DEGREES F	
FIRST USABLE LIFT .... 42 DEGREES F	
UPPER LEVEL WINDS	
5000 FT ASL .... 105 DEGREES AT 03 KNOTS	
10000 FT ASL .... 260 DEGREES AT 05 KNOTS	

1) Each of the Soaring Forecast items is explained as part of the following discussion of the Pseudo-Adiabatic chart. Your author feels that this chart is academic. It is included here because of its inclusion by the FAA in *Aviation Weather*.

2) The FAA is going to delete or modify "Top of the Lift" and "First Usable Lift."

7.   **The Pseudo-Adiabatic Chart**

a.   The pseudo-adiabatic chart is used to graphically compute adiabatic changes in vertically moving air and to determine stability. It can be used to explain how the information in the Soaring Forecast is computed.

1)   It has five sets of lines shown in Figure 160 below. These lines are:

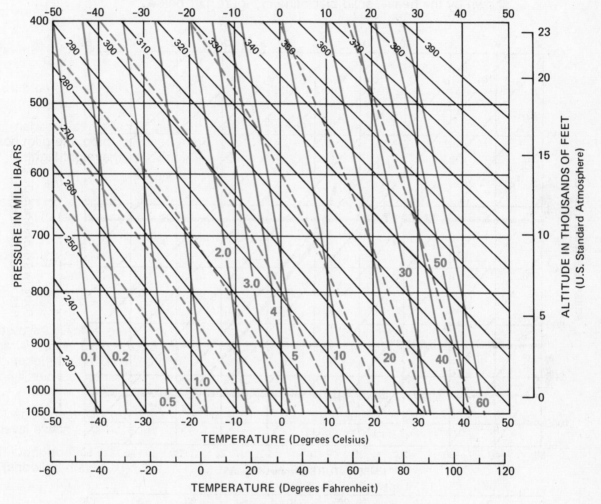

Figure 160. The Pseudo-Adiabatic Chart

a)   Pressure in millibars (horizontal lines),

b)   Temperature in degrees Celsius (vertical lines),

c)   Dry adiabats (sloping black lines),

d)   Lines of constant water vapor or mixing ratio of water vapor to dry air (solid red lines), and

e)   Moist adiabats (dashed red lines).

2)   The chart also has an altitude scale in thousands of feet above sea level (ASL) in a standard atmosphere along the right margin and a Fahrenheit temperature scale across the bottom.

3)   The chart used in actual practice has a much finer grid than the one shown in Figure 160 above.

4)   The following examples deal with dry thermals; and since the red lines in Figure 160 on page 191 concern moist adiabatic changes, they are omitted from the examples.

a)   If you care to delve deeper into use of the chart, you will find moist adiabatic processes even more complicated than dry processes.

b.   An upper air observation, or sounding, is plotted on the pseudo-adiabatic chart as shown by the heavy, solid black line in Figure 161 below.

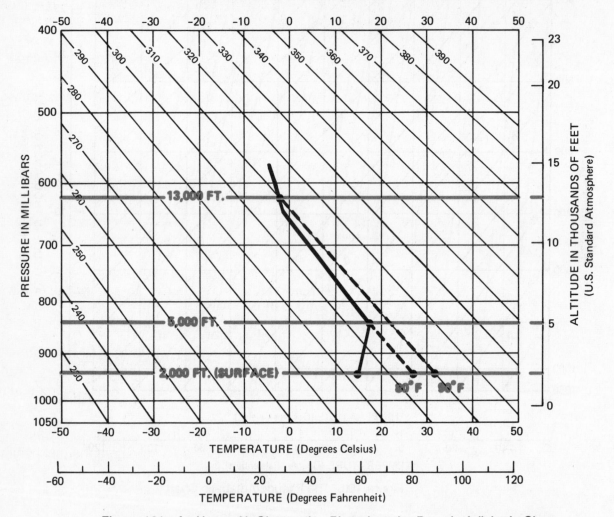

Figure 161. An Upper Air Observation Plotted on the Pseudo-Adiabatic Chart

1)   This plotting is the vertical temperature profile at the time the radiosonde observation was taken.

2)   It is the actual or existing lapse rate.

3)   Blue lines are added to the illustration showing appropriate altitudes to help you interpret the chart.

c. Depth of Convective Layer (Height of Thermals)

1) We know that for air to be unstable, the existing lapse rate must be equal to or greater than the dry adiabatic rate of cooling.

    a) In other words, in Figure 161 on page 192, the solid black line representing the plotted existing lapse rate would slope parallel to or slope more than the dry adiabats. Obviously it does not from the surface to 5,000 ft. ASL.

    b) Therefore, at the time the sounding was taken, the air was stable; there was no convective or unstable layer, and thermals were nonexistent. Thermal soaring was impossible.

2) As the surface temperature rises during the day, air is warmed and forced upward, (i.e., convection), cooling at the dry adiabatic rate.

    a) This movement continues until the temperature of the air moving upward is the same as the surrounding air.

3) Assume that the sounding (as shown by the solid black line) in Figure 161 on page 192 was made at sunrise with a surface temperature of 59°F (15°C). The forecast temperature at noon is 80°F and the maximum temperature of the day is forecast at 90°F. By using the chart in Figure 161 on page 192, you can determine the height of the thermals at these times.

    a) Plot 80°F (about 27°C) at the surface elevation and draw a dashed line parallel to the dry adiabats (sloping lines) to the point at which it intersects the sounding (solid line), which is 5,000 ft. ASL.

        i) Convection lifts the warmer air to a level at which it cools adiabatically (represented by the dashed line) to the temperature of the surrounding air.

        ii) The thermal height is 5,000 ft. ASL (3,000 ft. AGL).

    b) Repeat the process using the maximum temperature of 90°F (about 30°C). The thermal height is 13,000 ft. ASL (11,000 ft. AGL).

        i) The Soaring Forecast provides the maximum forecast temperature for a specific station.

4) Remember that we are talking about dry thermals.

    a) If convective clouds form below the indicated maximum thermal height, they will greatly distort the picture.

    b) However, if cumulus clouds do develop, thermals below the cloud base should be strengthened.

    c) If more higher clouds develop than were forecast, they will curtail surface heating, and the maximum temperature will most likely be cooler than forecast.

        i) Thermals will be weaker and will not reach as high an altitude.

5) The Soaring Forecast provides you with the thermal height. It is listed as "top of the lift" (see the Soaring Forecast on page 190).

6) The Soaring Forecast also provides you with the surface temperature that will provide usable lift, and it is listed as "first usable lift."

d.   Thermal Index (TI)

1)   Since thermals depend on sinking cold air forcing warm air upward, strength of thermals depends on the temperature difference between the sinking air and the rising air -- the greater the temperature difference the stronger the thermals.

   a)   To arrive at an approximation of this difference, the forecaster computes a thermal index (TI).

2)   A thermal index may be computed for any level; but for the Soaring Forecast they are computed for 5,000 ft. ASL (850-mb) and 10,000 ft. ASL (700-mb) as shown on the Soaring Forecast samples on page 190.

   a)   These levels are selected because they are in the altitude domain of routine soaring and because temperature data are routinely available for these two levels (i.e., Constant Pressure Analysis Chart).

3)   Three temperature values are needed -- the observed 850-mb and 700-mb temperatures and the forecast maximum temperature.

4)   Assume a sounding as in Figure 162 on page 195 with an 850-mb temperature of 15°C, a 700-mb temperature of 10°C, and forecast maximum of 86°F (30°C).

   a)   Plot the three temperatures, using care to place the maximum temperature plot at field elevation (2,000 ft. in Figure 162 on page 195).

   b)   Now draw a line (the black dashed line) through the maximum temperature parallel to the dry adiabats.

   c)   Note that the dashed line intersects the 850-mb level at 20°C and the 700-mb level at 4°C.

   d)   Subtract these temperatures from actual sounding temperatures at corresponding levels.

   e)   Note the difference is −5°C at 850 mb (15 − 20 = −5) and +6°C at 700 mb (10 − 4 = +6).

   f)   These values are the TI's at the two levels.

5)   Strength of thermals is proportional to the magnitude of the negative value of the TI.

   a)   A TI of −8 or −10 predicts very good lift and a long soaring day.

   b)   Thermals with this high a negative value will be strong enough to hold together even on a windy day.

   c)   A TI of −3 indicates a very good chance of sailplanes reaching the altitude of this temperature difference.

      i)   The height of the −3 index is given on the Soaring Forecast, as shown on page 190.

   d)   A TI of −2 to zero leaves much doubt; and a positive TI offers even less hope of thermals reaching the altitude.

6) Remember that the TI is a forecast value.

    a) A miss in the forecast maximum or a change in temperature aloft can alter the picture considerably.

    b) The example in Figure 162 below should promise fairly strong thermals to above 5,000 ft. (TI of −5).

        i) The "top of the lift" is indicated at the intersection of the dashed line and solid line. This value is given in the Soaring Forecast.

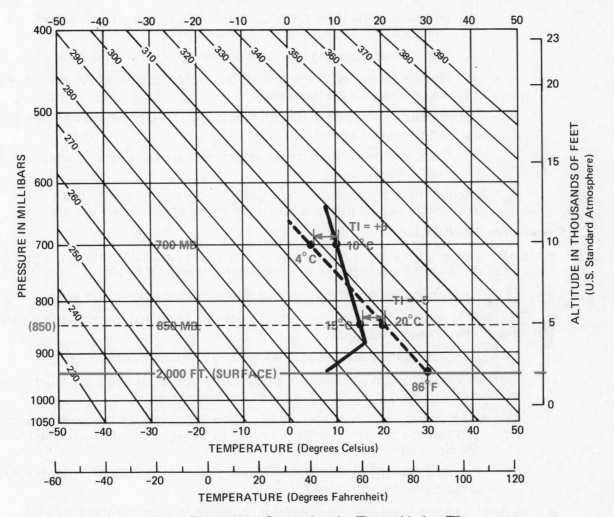

Figure 162. Computing the Thermal Index (TI)

7) The Soaring Forecast also provides the winds from the upper air sounding at the 5,000-ft. and 10,000-ft. levels. See the Soaring Forecast on page 190.

8) Often the National Weather Service will have no upper air sounding taken near a soaring base. Forecasts must be based on a simulated sounding derived from distant observations.

    a) At other times, for some reason a forecast may not be available.

    b) Furthermore, you can often augment the forecast with local observations.

e. **Do It Yourself.** The first step in determining height and strength of thermals is to obtain a local sounding.

  1) Send your tow aircraft aloft about sunrise and simply read outside air temperatures from the aircraft thermometer and altitudes from the altimeter.

   a) Read temperatures at 500-ft. intervals for about the first 2,000 ft. and at 1,000-ft. intervals at higher altitudes.

   b) The information may be radioed back to the ground, or may be recorded in flight and analyzed after landing.

   c) When using the latter method, read temperatures on both ascent and descent and average the temperatures at each level.

   d) This type of sounding is an airplane observation or APOB.

   e) Plot the sounding on the pseudo-adiabatic chart using the altitude scale rather than the pressure scale.

  2) Next you need a forecast maximum temperature. Perhaps you can pick this up from the local forecast.

   a) If not, you can use your best judgment comparing today's weather with yesterday's.

  3) Although these procedures are primarily for dry thermals, they work reasonably well for thermals below the bases of convective clouds.

8. **Convective Cloud Bases**

  a. Soaring experience suggests a shallow, stable layer immediately below the general level of convective cloud bases through which it is difficult to soar.

   1) This layer is 200 to 600 ft. thick and is known as the *sub-cloud layer*.

  b. The layer appears to act as a filter allowing only the strongest thermals to penetrate it and form convective clouds.

   1) Strongest thermals are beneath developing cumulus clouds.

  c. Thermals intensify within a convective cloud, but evaporation cools the outer edges of the cloud causing a downdraft immediately surrounding it.

   1) Add to this the fact that downdrafts predominate between cumulus clouds, and you can see the slim chance of finding lift between clouds above the level of the cloud base.

   2) In general, thermal soaring during convective cloud activity is practical only at levels below the cloud base.

  d. In Part I, Chapter 6, we learned to estimate height in thousands of feet of a convective cloud base by dividing the surface temperature-dew point spread by 4.

   1) If the rising column were self-contained -- that is, if no air were drawn into the sides of the thermal -- the method would give a fairly accurate height of the base. However, this is not the case.

   2) Air is entrained or drawn into the sides of the thermal, which lowers the water vapor content of the thermal, allowing it to reach a somewhat higher level before condensation occurs.

   a) Bases of the clouds are generally 10% to 15% higher than the computed height.

3) Entrainment is a problem; observers and forecasters can only estimate its effect.

a) Until a positive technique is developed, heights of cumulus bases will tend to be reported and forecast too low.

9. **Cross-Country Thermal Soaring**

a. A pilot can soar cross-country using either isolated thermals or thermal streets.

1) When using isolated thermals, (s)he gains altitude circling in thermals and then proceeds toward the next thermal in the general direction of his/her cross-country.

2) Under a thermal street, (s)he may be able to proceed with little if any circling if his/her chosen course parallels the thermal streets.

3) (S)he can obtain the greatest distance by flying in the direction of the wind.

b. In the central and eastern United States, the most favorable weather for cross-country soaring occurs behind a cold front.

1) Four factors contribute to making this pattern ideal.

a) The cold polar air is usually dry, and thermals can build to relatively high altitudes.

b) The polar air is colder than the ground, and thus, the warm ground aids solar radiation in heating the air.

i) Thermals begin earlier in the morning and last later in the evening.
ii) On occasions, soarable lift has been found at night.

c) Quite often, colder air at high altitudes moves over the cold, low-level outbreak, intensifying the instability and strengthening the thermals.

d) The wind profile frequently favors thermal streeting -- a real boon to speed and distance.

c. The same four factors may occur with cold frontal passages over mountainous regions in the western United States.

1) However, rugged mountains break up the circulation; and homogeneous conditions extend over smaller areas than over the eastern parts of the country.

2) The western mountain regions and particularly the desert southwest have one decided advantage.

a) Air is predominantly dry with more abundant daytime thermal activity favoring cross-country soaring, although it may be for shorter distances.

d. Among the world's most favorable tracks for long distance soaring is a high plains corridor along the east slope of the Rocky Mountains stretching from southwest Texas to Canada.

1) Terrain in the corridor is relatively flat and high with few trees; terrain surface ranges from barren to short grass.

a) These surface features favor strong thermal activity.

2) Prevailing wind is southerly and moderately strong, giving an added boost to northbound cross-country flights.

C. **Frontal Soaring**

1. Warm air forced upward over cold air above a frontal surface can provide lift for soaring.

    a. However, good frontal lift is transitory and accounts for a very small portion of powerless flight.

    b. Seldom will you find a front parallel to your desired cross-country route, and seldom will it stay in position long enough to complete a flight.

        1) A slowly moving front provides only weak lift.

        2) A fast moving front often plagues the soaring pilot with cloudiness and turbulence.

2. A front can occasionally provide excellent lift for a short period.

    a. On a cross-country, you may be riding wave or ridge lift and need to move over a flat area to take advantage of thermals.

        1) A front may offer lift during your transition.

3. Fronts are often marked by a change in cloud type or amount.

    a. However, the very presence of clouds may deter you from flying into the front.

        1) Spotting a dry front is difficult.

    b. Knowing that a front is in the vicinity and studying your aircraft reaction can tell you when you are in the frontal lift.

        1) Staying in the lift is another problem.
        2) Observing ground indicators of surface wind helps.

4. An approaching front may enhance thermal or hill soaring.

    a. An approaching front or a frontal passage most likely will disrupt a sea breeze or mountain wave.

D. **Sea Breeze Soaring**

1. In many coastal areas during the warm seasons, a pleasant breeze from the sea occurs almost daily.

    a. Caused by the heating of land on warm, sunny days, the sea breeze usually begins during early forenoon, reaches a maximum during the afternoon, and subsides around dusk after the land has cooled.

    b. The leading edge of the cool sea breeze forces warmer air inland to rise as shown in Figure 165 on page 199.

        1) Rising air from the land returns seaward at higher altitude to complete the convective cell.

    c. A sailplane pilot operating in or near coastal areas often can find lift generated by this convective circulation.

        1) The transition zone between the cool, moist air from the sea and the warm, drier air inland is often narrow and is a shallow, ephemeral kind of pseudo-cold front.

2. Sometimes the wedge of cool air is called a **sea breeze front**.

    a. If sufficient moisture is present, a line of cumuliform clouds just inland may mark the front.

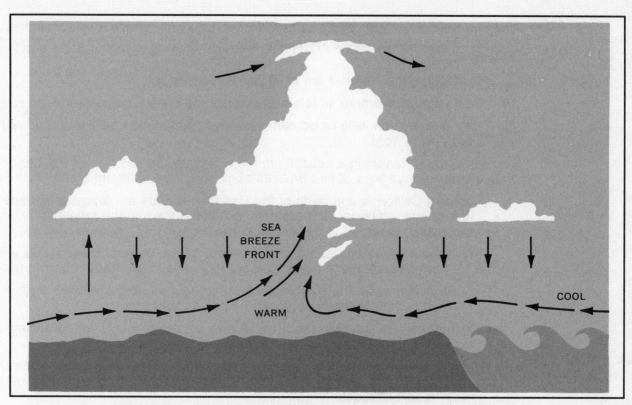

Figure 165. Schematic Cross Section through a Sea Breeze Front

1) Whether marked by clouds or not, the upward moving air at the sea breeze front occasionally is strong enough to support soaring flight.

b. Within the sea breeze (i.e., between the sea breeze front and the ocean) the air is usually stable, and normally no lift may be expected at lower levels.

1) However, once airborne, pilots occasionally have found lift at higher levels in the return flow aloft.

2) A visual indication of this lift is cumulus extending seaward from the sea breeze front.

c. A large difference in land and sea water temperature intensifies the convective cell generating a sea breeze.

1) Where coastal waters are quite cool, such as along the California coast, and land temperatures warm rapidly in the daytime, the sea breeze becomes pronounced, penetrating perhaps 50 to 75 mi. inland at times.

2) Copious sunshine and cool sea waters favor a well-developed sea breeze front.

d. The sea breeze is a local effect.

1) Strong pressure gradients with a well-developed pressure system can overpower the sea breeze effect.

2) Winds will follow the direction and speed dictated by the strong pressure gradient.

a) Therefore, a sea breeze front is most likely when pressure gradient is weak and wind is light.

e. When convection is very deep, the frontal effect of a sea breeze may sometimes trigger cumulonimbus clouds, provided the lifted air over land contains sufficient moisture.

   1) More often, the cumulus are of limited vertical extent.

   2) Over vegetation where air is usually moist, sea breeze cumulus are the rule.

   3) Over arid regions, little or no cumulus development may be anticipated with a sea breeze front.

f. Irregular or rough terrain in a coastal area may amplify the sea breeze front and cause convergence lines of sea breezes originating from different areas.

   1) Southern California and parts of the Hawaiian Islands are favorable for sea breeze soaring because orographic lift (lift induced by the presence of mountains) is added to the frontal convection.

      a) Sea breezes occasionally may extend to the leeward sides of hills and mountains unless the ranges are high and long without abrupt breaks.

      b) In either case, the sea breeze front converges on the windward slopes, and upslope winds augment the convection.

   2) Where terrain is fairly flat, sea breezes may penetrate inland for surprising distances but with weaker lift along the sea breeze front.

      a) In the Tropics, sea breezes sometimes penetrate as much as 150 mi. inland, whereas an average of closer to 50 mi. inland is more usual in middle latitudes.

   3) Sea breezes reaching speeds of 15 to 25 kt. are not uncommon.

g. When a sea breeze front develops, visual observations may provide clues to the extent of lift that you may anticipate.

   1) Expect little or no lift on the seaward side of the front when the sea air is markedly devoid of convective clouds or when the sea breeze spreads low stratus inland.

      a) However, some lift may be present along the leading edge of the sea breeze or just ahead of it.

   2) Expect little or no lift on the seaward side of the front when visibility decreases markedly in the sea breeze air.

      a) This is an indicator of stable air within the sea breeze.

   3) A favorable visual indication of lift along the sea breeze front is a line of cumulus clouds marking the front; cumuli between the sea breeze front and the ocean also indicate possible lift within the sea breeze air, especially at higher levels.

      a) Cumulus bases in the moist sea air are often lower than along the front.

   4) When a sea breeze front is devoid of cumulus but converging streamers of dust or smoke are observed, expect convection and lift along the sea breeze front.

   5) Probably the best combination to be sighted is cumuli and converging dust or smoke plumes along the sea breeze front as it moves upslope over hills or mountains.

      a) The upward motion is amplified by the upslope winds.

   6) A difference in visibility between the sea air and the inland air often is a visual clue to the leading edge of the sea breeze.

      a) Visibility in the sea air may be restricted by haze while visibility inland is unrestricted.

b) On the other hand, the sea air may be quite clear while visibility inland is restricted by dust or smoke.

3. Local Sea Breeze Explorations

a. Unfortunately, a sea breeze front is not always easy to find, and it is likely that many an opportunity for sea breeze soaring goes unnoticed.

   1) As yet, little experience has been accrued in locating a belt of sea breeze lift without visual clues such as clouds, haze, or converging smoke or dust plumes.

b. The sea breeze front moving from the Los Angeles coastal plain into the Mojave Desert has been dubbed the "Smoke Front."

   1) It has intense thermal activity and offers excellent lift along the leading edge of the front.

   2) Associated with the sea breeze that moves inland over the Los Angeles coastal plain are two important zones of convergence, shown in Figure 166 below.

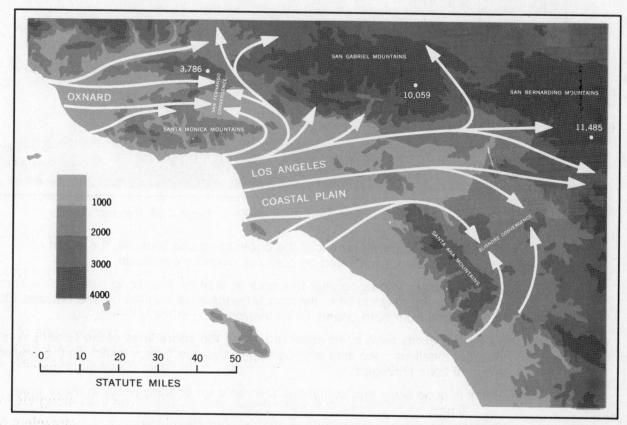

Figure 166. Sea Breeze Flow into the San Fernando Valley

a) One convergence line is the "San Fernando Convergence Zone."

b) A larger scale zone is in the Elsinore area, also shown in Figure 166.

   3) This convergence zone apparently generates strong vertical currents because soaring pilots fly back and forth across the valley along the line separating smoky air to the north from relatively clear air to the south.

   4) Altitudes reached depend upon the stability, but usually fall within the 6,000 ft. to 12,000 ft. MSL range for the usual dry thermal type lift.

a) Seaward, little or no lift is experienced in the sea breeze air marked by poor visibility.

c.   Figure 167 below shows converging air between sea breezes flowing inland from opposite coasts of the Cape Cod Peninsula.

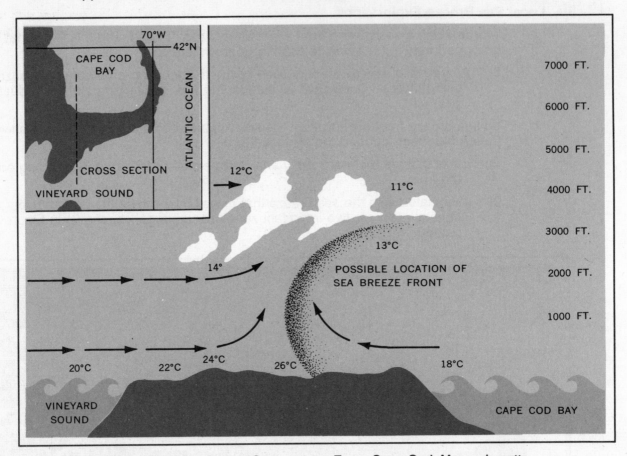

Figure 167. Sea Breeze Convergence Zone, Cape Cod, Massachusetts

1)   Later in the development of the converging sea breezes, the onset of convection is indicated by cumulus over the peninsula.

2)   Sailplane pilots flying over this area as well as over Long Island, New York, have found good lift in the convergence lines caused by sea breezes blowing inland from both coasts of the narrow land strips.

d.   Sea breeze fronts have been observed along the shore lines of the Great Lakes. Weather satellites have also photographed this sea breeze effect on the western shore of Lake Michigan.

1)   It is quite likely that conditions favorable for soaring occur in these areas at times.

E.   **Ridge or Hill Soaring**

1.   Wind blowing toward hills or ridges flows upward, over, and around the abrupt rises in terrain. The upward-moving air creates lift which is sometimes excellent for soaring.

a.   Figure 168 on page 203 is a schematic showing area of best lift.

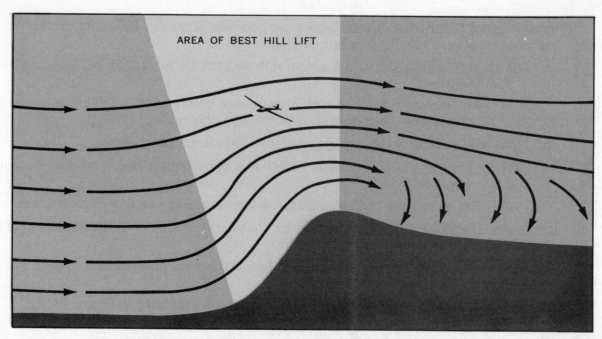

AREA OF BEST HILL LIFT

Figure 168. Schematic Cross Section of Airflow over a Ridge

2. **Wind.** To create lift over hills or ridges, wind direction should be within about 30° to 40° of being perpendicular to the ridge line.

    a. A sustained speed of 15 kt. or more usually generates enough lift to support a sailplane.

    b. Height of the lift usually is two or three times the height of the rise from the valley floor to the ridge crest.

    c. Strong winds tend to increase turbulence and low-level eddies without an appreciable increase in the height of the lift.

3. **Stability** affects the continuity and extent of lift over hills or ridges.

    a. Stable air allows relatively streamlined upslope flow.

        1) A pilot experiences little or no turbulence in the steady, uniform area of best lift shown in Figure 168 above.

        2) Since stable air tends to return to its original level, air spilling over the crest and downslope is churned into a snarl of leeside eddies, also shown in Figure 168 above.

        3) Thus, stable air favors smooth lift but troublesome leeside low-altitude turbulence.

    b. When the airstream is moist and unstable, upslope lift may release the instability, generating strong convective currents and cumulus clouds over windward slopes and hill crests.

        1) The initially laminar flow is broken up into convective cells.

        2) While the updrafts produce good lift, strong downdrafts may compromise low-altitude flight over rough terrain.

        3) As with thermals, the lift will be transitory rather than smooth and uniform.

4. **Steepness of Slope.** Very gentle slopes provide little or no lift. A smooth, moderate slope is most favorable for soaring.

    a. An ideal slope is about 1 to 4 which, with an upslope wind of 15 kt., creates lift of about 6 fps.

       1) With the same slope, a high-performance sailcraft with a sinking speed of 2 fps presumably could remain airborne with only a 5-kt. wind!

    b. Very steep escarpments or rugged slopes induce turbulent eddies.

       1) Strong winds extend these eddies to a considerable height, usually disrupting any potential lift.

       2) The turbulent eddies also increase the possibility of a low-altitude upset.

5. **Continuity of Ridges.** Ridges extending for several miles without abrupt breaks tend to provide uniform lift throughout their length.

    a. In contrast, a single peak diverts wind flow around the peak as well as over it and thus is less favorable for soaring.

    b. Some wind flow patterns over ridges and hills are illustrated in Figure 170 on page 205.

       1) Deviations from these patterns depend on wind direction and speed, on stability, on slope profile, and on general terrain roughness.

6. **Soaring in Upslope Lift**

    a. The soaring pilot, always alert, must remain especially so in seeking or riding hill lift.

    b. When air is unstable, do not venture too near the slope.

       1) You can identify unstable air either by the updrafts and downdrafts in dry thermals or by cumulus building over hills or ridges.

       2) Approaching at too low an altitude may suddenly put you in a downdraft, forcing an inadvertent landing.

    c. When winds are strong, surface friction may create low-level eddies even over relatively smooth slopes.

       1) Also, friction may drastically reduce the effective wind speed near the surface.

       2) When climbing at low altitude toward a slope under these conditions, be prepared to turn quickly toward the valley if you lose lift.

          a) Renew your attempt to climb, farther from the hill.

    d. If winds are weak, you may find lift only very near the sloping surface. Then you must hug the slope to find needed lift.

       1) However, avoid this procedure if there are indications of up- and downdrafts.

       2) In general, for any given slope, keep your distance from the slope proportional to wind speed.

    e. Leeward of hills and ridges is an area where wind is blocked by the obstruction. In soaring circles this area is called the "wind shadow."

       1) In the wind shadow, downdrafts predominate as shown in Figure 168 on page 203.

       2) If you stray into the wind shadow at an altitude near or below the altitude of the ridge crest, you may be forced into an unscheduled and possibly rough landing.

       3) Try to stay within the area of best lift shown in Figure 168 on page 203.

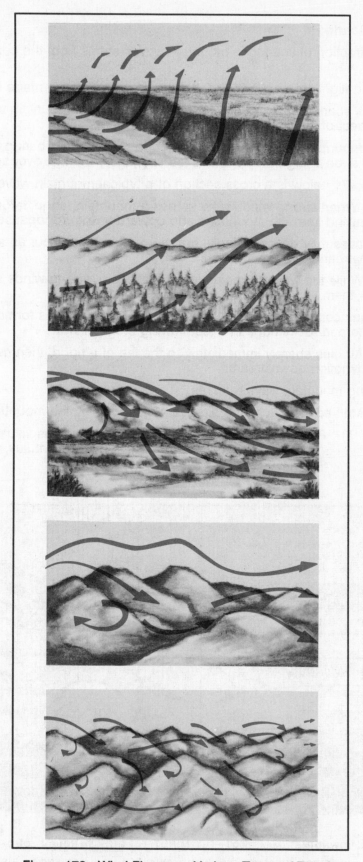

Figure 170. Wind Flow over Various Types of Terrain

F. Mountain Wave Soaring

1. The main attraction of soaring in mountain waves stems from the continuous lift to great heights.

    a. Soaring flights to above 35,000 ft. have frequently been made in mountain waves.

    b. Once a soaring pilot has reached the rising air of a mountain wave, (s)he has every prospect of maintaining flight for several hours.

    c. While mountain wave soaring is related to ridge or hill soaring, the lift in a mountain wave is on a larger scale and is less transitory than lift over smaller rises in terrain.

    d. Figure 171 below is a cross section of a typical mountain wave.

2. **Formation.** When strong winds blow across a mountain range, large standing waves occur downwind from the mountains and upward to the tropopause.

    a. The waves may develop singly, but more often they occur as a series of waves downstream from the mountains.

        1) While the waves remain almost stationary, strong winds are blowing through them.

    b. You may compare a mountain wave to a series of waves formed downstream from a submerged rocky ridge in a fast flowing creek or river.

        1) Air dips sharply immediately to the lee of a ridge, then rises and falls in a wave motion downstream.

    c. A strong mountain wave requires:

        1) Marked stability in the airstream disturbed by the mountains.

            a) Rapidly building cumulus over the mountains visually marks the air unstable; convection, evidenced by the cumulus, tends to deter wave formation.

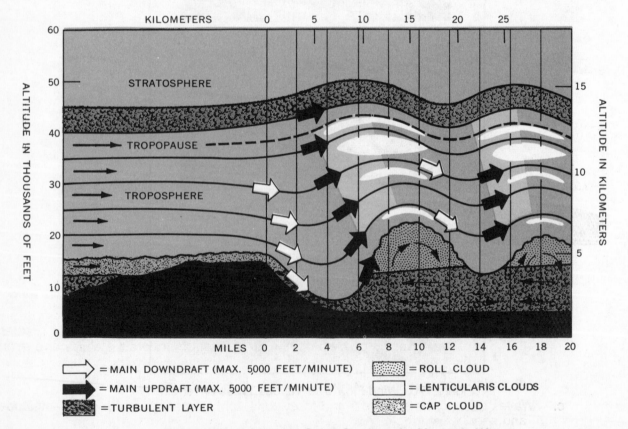

Figure 171. Schematic Cross Section of a Mountain Wave

      2)   Wind speed at the level of the summit should exceed a minimum which varies from 15 to 25 kt. depending on the height of the range.

          a)   Upper winds should increase or at least remain constant with height up to the tropopause.

      3)   Wind direction should be within 30° normal to the range. Lift diminishes as winds more nearly parallel the range.

3.   Wave Length and Amplitude

    a.   Wave length is the horizontal distance between crests of successive waves and is usually between 2 and 25 mi.

      1)   In general, wave length is controlled by wind component perpendicular to the ridge and by stability of the upstream flow.

          a)   Wave length is directly proportional to wind speed and inversely proportional to stability.

      2)   Figure 172 below illustrates wave length and also amplitude.

    b.   Amplitude of a wave is the vertical dimension and is half the altitude difference between the wave trough and crest.

      1)   In a typical wave, amplitude varies with height above the ground.

          a)   It is least near the surface and near the tropopause.

      2)   Greatest amplitude is roughly 3,000 to 6,000 ft. above the ridge crest.

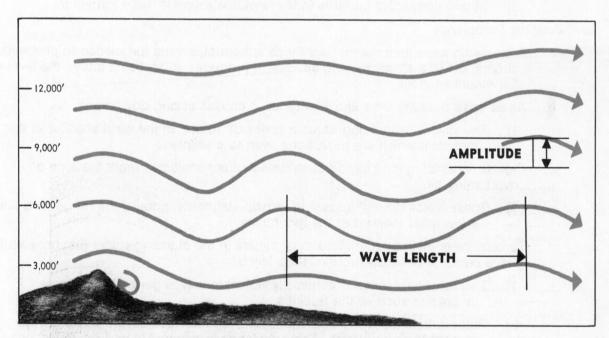

Figure 172. Wave Length and Amplitude

      3)   Wave amplitude is controlled by size and shape of the ridge as well as wind and stability.

          a)   A shallow layer of great stability and moderate wind produces a greater wave amplitude than does a deep layer of moderate stability and strong winds.

      4)   Also, the greater the amplitude, the shorter is the wave length.

    c.   Waves offering the strongest and most consistent lift are those with great amplitude and short wave length.

4. **Visual Indicators**

   a. If the air has sufficient moisture, lenticular (lens-shaped) clouds mark wave crests.

      1) Cooling of air ascending toward the wave crest saturates the air, forming clouds.

      2) Warming of air descending beyond the wave crest evaporates the cloud.

      3) Thus, by continuous condensation windward of the wave crest and evaporation leeward, the cloud appears stationary although wind may be blowing through the wave at 50 kt. or more.

      4) Lenticular clouds in successive bands downstream from the mountain mark a series of wave crests.

   b. Spacing of lenticulars marks the wave length.

      1) Clearly identifiable lenticulars also suggest larger wave amplitude than clouds which barely exhibit lenticular form.

      2) These cloud types along with stratiform clouds on the windward slopes and along the mountain crest indicate the stability favorable to mountain wave soaring.

   c. Thunderstorms or rapidly building cumulus over mountains mark the air unstable.

      1) As they reach maturity, the thunderstorms often drift downwind across leeward valleys and plains.

      2) Strong convective currents in the unstable air deter wave formation.

5. **Soaring Turbulence**

   a. A mountain wave, in a manner similar to a thermal, means turbulence to powered aircraft, but to a slowly moving sailcraft, it produces lift and sink above the level of the mountain crest.

   b. As air spills over the crest like a waterfall, it causes strong downdrafts.

      1) The violent overturning forms a series of "rotors" in the wind shadow of the mountain which are hazardous even to a sailplane.

   c. Clouds resembling long bands of stratocumulus sometimes mark the area of overturning air.

      1) These "rotor clouds" appear to remain stationary, parallel the range, and stand a few miles leeward of the mountains.

   d. Turbulence is most frequent and most severe in the standing rotors just beneath the wave crests at or below mountain-top levels.

      1) This rotor turbulence is especially violent in waves generated by large mountains such as the Rockies.

      2) Rotor turbulence with lesser mountains is much less severe but is always present to some extent. The turbulence is greatest in well-developed waves.

6. **Favored Areas**

   a. Mountain waves occur most frequently along the central and northern Rockies and the northern Appalachians.

      1) Occasionally, waves form to the lee of mountains in Arkansas, Oklahoma, and southwestern Texas.

   b. Weather satellites have observed waves extending great distances downwind from the Rocky Mountains; one series extended for nearly 700 mi.

      1) The more usual distance is 150 to 300 mi.

    c.   While Appalachian waves are not as strong as those over the Rockies, they occur frequently, and satellites have observed them at an average of 115 mi. downwind.

        1)   Wave length of these waves averages about 10 NM.

7.   **Riding the Waves.** You often can detect a wave by the uncanny smoothness of your climb.

    a.   On first locating a wave, turn into the wind and attempt to climb directly over the spot where you first detected lift *provided* you can remain at an altitude above the level of the mountain crest.

        1)   After cautiously climbing well up into the wave, attempt to determine dimensions of the zone of lift.

            a)   If the wave is over rugged terrain, it may be impossible and unnecessary to determine the wave length.

                i)   Lift over such terrain is likely to be in patchy bands.

            b)   Over more even terrain, the wave length may be easy to determine and use in planning the next stage of flight.

    b.   Wave clouds are a visual clue in your search for lift.

        1)   The wave-like shape of lenticulars is usually more obvious from above than from below.

        2)   Lift should prevail from the crest of the lenticulars upwind about one-third the wave length.

        3)   When your course takes you across the waves, climb on the windward side of the wave and fly as quickly as possible to the windward side of the next wave.

        4)   Wave lift of 300 to 1,200 fpm is not uncommon.

            a)   Soaring pilots have encountered vertical currents exceeding 3,000 fpm, the strongest ever reported being 8,000 fpm.

G. In Closing

1.   Thermal soaring -- The pilot circles in a thermal (updraft) to gain height and heads to the next thermal to regain altitude lost in the downdrafts between them.

    a.   Thermals depend on solar heating, so thermal soaring is virtually restricted to daylight hours with considerable sunshine.

        1)   The most likely place for a thermal is above a surface that heats rapidly, e.g., barren sandy or rocky surfaces, plowed fields, stubble fields surrounded by green vegetation, cities, factories, and chimneys.

        2)   Dust devils are excellent sources of lift, but you must use caution because the thermals are strong and turbulent and are surrounded by areas of little lift or possibly of sink.

        3)   Thermal soaring is usually at its best when convective clouds develop.

            a)   Cumulus clouds are positive signs of thermals.

        4)   Wet ground is less favorable for thermals than dry ground because wet ground heats more slowly.

    b.   No two thermals are exactly alike.

        1)   Some form as chimneys whereas others form bubbles of warm air that rise.

        2)   Wind shear can cause the thermal to lean or, at times, can completely break it up.

        3)   Thermals may become organized into thermal streets.

            a)   A pilot can maintain generally continuous flight under a thermal street and seldom have to circle.

    c.   Daily soaring weather forecasts are available from the National Weather Service.

        1)   You can determine height and strength of thermals by obtaining a local sounding and plotting it on a pseudo-adiabatic chart.

    d.   You can soar cross-country using isolated thermals or thermal streets.

2.   Frontal soaring accounts for a very small portion of powerless flight.

    a.   A front can provide excellent lift for a short period.

    b.   An approaching front may enhance thermal or hill soaring.

3.   Southern California and parts of the Hawaiian Islands are favorable for sea breeze soaring because orographic lift augments convection from the sea breeze front.

    a.   Sea breeze fronts are not always easy to find, and visual clues may not be present.

        1)   A sea breeze front moves from the Los Angeles coastal plain into the Mojave Desert.

        2)   Sailplane pilots have found good lift in the convergence lines caused by sea breezes over Cape Cod, MA and Long Island, NY.

        3)   Sea breeze fronts have also been observed along the shorelines of the Great Lakes.

4.   Ridge or hill soaring -- Wind blowing toward hills or ridges flows upward, over, and around the abrupt rises in terrain. Upward-moving air with a wind speed of 15 kt. usually generates enough lift to support a sailplane.

    a.   A smoother, moderate slope is most favorable for soaring.

    b.   When soaring in upslope lift, you must remain alert for unstable air.

5.   Mountain wave soaring is related to ridge or hill soaring, but the lift in a mountain wave is on a larger scale and is less transitory than lift over smaller rises in terrain.

    a.   Mountain waves occur most frequently along the central and northern Rockies and the northern Appalachians.

6.   Records are made to be broken. Altitude and distance records are a prime target of many sailplane enthusiasts.

    a.   Distance records may be possible by flying a combination of lift sources such as thermal, frontal, ridge, or wave.

    b.   Altitude records are set in mountain waves.

        1)   Altitudes above 46,000 ft. have been attained over the Rocky Mountains.

        2)   Soaring flights to more than 24,000 ft. have been made in Appalachian waves.

        3)   Flights as high as 20,000 ft. have been recorded from New England to North Carolina.

7.   We hope that this chapter has given you an insight into the minute variations in weather that profoundly affect a soaring aircraft.

    a.   When you have remained airborne for hours without power, you have met a unique challenge and experienced a singular thrill of flying.

END OF CHAPTER
END OF PART II

PART III
AVIATION WEATHER SERVICES

Aviation Weather Services is published periodically by the FAA/NWS to keep pilots abreast of weather maps and other services available from their FSS. AC 00-45D replaced its predecessor, AC 00-45C, in 1993.

<table>
<tr><td colspan="2" align="center">**Table of Contents**
AC 00-45D*</td><td colspan="2" align="center">**Table of Contents**
This Part III</td></tr>
<tr><td>1.</td><td>The Aviation Weather Service Program</td><td>1.</td><td>The Aviation Weather Service Program</td></tr>
<tr><td>2.</td><td>Surface Aviation Weather Reports</td><td>2.</td><td>Surface Aviation Weather Reports</td></tr>
<tr><td>3.</td><td>Pilot and Radar Reports and Satellite Pictures</td><td>3.</td><td>Pilot Weather Reports (PIREPs)</td></tr>
<tr><td></td><td></td><td>4.</td><td>Radar Weather Report (SD)</td></tr>
<tr><td></td><td></td><td>5.</td><td>Satellite Weather Pictures</td></tr>
<tr><td>4.</td><td>Aviation Weather Forecasts</td><td>6.</td><td>Terminal Forecasts (FT and TAF)</td></tr>
<tr><td></td><td></td><td>7.</td><td>Area Forecast (FA)</td></tr>
<tr><td></td><td></td><td>8.</td><td>TWEB Route Forecasts and Synopsis</td></tr>
<tr><td></td><td></td><td>9.</td><td>In-Flight Aviation Weather Advisories (WST, WS, WA)</td></tr>
<tr><td></td><td></td><td>10.</td><td>Winds and Temperatures Aloft Forecast (FD)</td></tr>
<tr><td></td><td></td><td>11.</td><td>Special Flight Forecast</td></tr>
<tr><td></td><td></td><td>12.</td><td>Center Weather Service Unit (CWSU) Products</td></tr>
<tr><td></td><td></td><td>13.</td><td>Hurricane Advisory (WH)</td></tr>
<tr><td></td><td></td><td>14.</td><td>Convective Outlook (AC)</td></tr>
<tr><td></td><td></td><td>15.</td><td>Severe Weather Watch Bulletin (WW)</td></tr>
<tr><td>5.</td><td>Surface Analysis Chart</td><td>16.</td><td>Surface Analysis Chart</td></tr>
<tr><td>6.</td><td>Weather Depiction Chart</td><td>17.</td><td>Weather Depiction Chart</td></tr>
<tr><td>7.</td><td>Radar Summary Chart</td><td>18.</td><td>Radar Summary Chart</td></tr>
<tr><td>8.</td><td>Significant Weather Prognostics</td><td>19.</td><td>U.S. Low-Level Significant Weather Prog</td></tr>
<tr><td></td><td></td><td>20.</td><td>High-Level Significant Weather Prog</td></tr>
<tr><td>9.</td><td>Winds and Temperatures Aloft</td><td>21.</td><td>Winds and Temperatures Aloft Charts</td></tr>
<tr><td>10.</td><td>Composite Moisture Stability Chart</td><td>22.</td><td>Composite Moisture Stability Chart</td></tr>
<tr><td>11.</td><td>Severe Weather Outlook Chart</td><td>23.</td><td>Severe Weather Outlook Chart</td></tr>
<tr><td>12.</td><td>Constant Pressure Analysis Charts</td><td>24.</td><td>Constant Pressure Analysis Charts</td></tr>
<tr><td>13.</td><td>Tropopause Data Chart</td><td>25.</td><td>Tropopause Data Chart</td></tr>
<tr><td>14.</td><td>Tables and Conversion Graphs</td><td>26.</td><td>Tables and Conversion Graphs</td></tr>
</table>

AC 00-45D has 14 chapters (listed above). We have divided AC 00-45D into 26 chapters in our book to facilitate study of various weather reports and weather forecasts. The chapter titles are aligned above so you can see which AC 00-45D chapters were broken up into additional chapters in this book.

* AC 00-45D was not yet published when the first printing of this book went to press. We have researched and updated the topics in AC 00-45C with the expectation that this book will summarize AC 00-45D.

CHAPTER ONE
THE AVIATION WEATHER SERVICE PROGRAM

> Please take a few minutes to study each of the concepts listed above and anticipate/imagine what they are and how they relate to the other listed concepts.

A. **Weather Service: Aviation Effort** -- Weather service to aviation is a joint effort of the National Weather Service (NWS), the Federal Aviation Administration (FAA), the Department of Defense (DOD) Weather Service, and other aviation-oriented groups and individuals.

 1. Because of international flights and a need for world-wide weather forecasts, foreign weather services also have a vital input into our service.

 2. This section follows the development and flow of observations, reports, and forecasts through the service to the users.

B. **Weather Observations** are measurements and estimates of existing weather, both at the surface and aloft. When recorded and transmitted, a weather observation becomes a report, and these reports are the basis for analyses and forecasts.

 1. **Surface Aviation Observations** include weather elements pertinent to flying.

     a. A network of airport stations provides routine up-to-date aviation weather reports.

     b. Most of the stations in the network are either NWS or FAA; however, the military services and contracted civilians are also included.

     c. A major change in the surface weather observation network is ongoing with the installation of an Automated Surface Observing System (ASOS) across the country.

         1) These automated stations are expected to become a major part of the network in the near future.

 2. **Radar Observation.** Precipitation reflects radar signals and the reflected signals are displayed as echoes on the radarscope.

     a. NWS radar covers nearly all of the U.S. east of the Rocky Mountains.

         1) Radar coverage over the remainder of the U.S. is performed largely by Air Route Traffic Control radars.

         2) Except for some western mountainous terrain, radar coverage is nearly complete over the contiguous 48 states.

         3) Figure 1-1 on page 213 maps the radar observing network.

             a) The WSR-57 and the newer WSR-74S radar sites are used for detecting coverage, intensity, and movement of precipitation.

b)   The WSR-74C are local warning sites that augment the network by operating on an as-needed basis to support warning and forecast programs.

c)   In the western sections of the country, the network is supplemented by FAA Air Route Traffic Center (ARTC) radar sites.

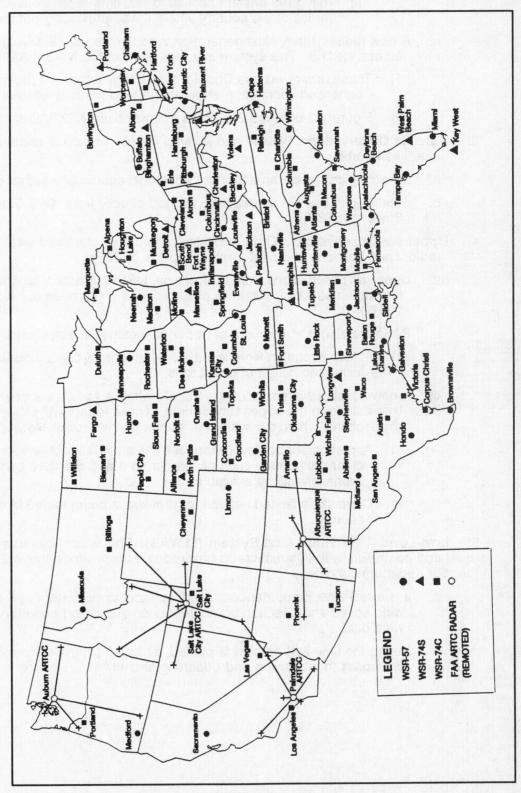

Figure 1-1. The Radar Observing Network

b.   The Radar Remote Weather Display System (RRWDS) is specifically designed to provide real-time radar weather information from many different radars.

1)   It is connected to FAA and Air Force Air Route Surveillance radars as well as NWS weather radars.

a)   This gives briefers access to real-time radar weather information from areas of the country where it was previously not available.

c.   A new radar system, Next-generation Weather Radar (NEXRAD) is being installed across the U.S. This system is a joint effort of the NWS, FAA, and DOD.

1)   These radars will use Doppler technology which will provide greater detail and enhanced information about thunderstorms and weather systems.

2)   Figure 1-2 on page 215 maps the proposed NEXRAD sites.

3.   **Satellite Observations.** Visible and infrared images of clouds are available from weather satellites in orbit.

a.   Satellite pictures are an important additional source of weather observations.

b.   Satellite pictures are available by fax and directly from NWS Satellite Field Service Stations (SFSS).

4.   **Upper Air Observations.** Other important sources of observed weather data are radiosonde balloons and pilot weather reports (PIREPs).

a.   Upper air observations from radiosonde, taken twice daily at specified stations, furnish temperature, humidity, pressure, and wind, often to heights above 100,000 ft.

b.   Pilots are also a vital source of upper air weather observations.

1)   In fact, aircraft in flight are the only means of directly observing turbulence, icing, and height of cloud tops.

c.   A new sensing system utilizing vertically arrayed radars will provide increased real-time data from the upper atmosphere. These radars will provide profiles of the atmosphere. Thus, the system is known as the Profiler Network.

1)   At present, upper level winds are the only data obtained from this network, but other parameters such as temperature and moisture content at various levels eventually will be available.

2)   Currently, a limited network of profilers is being tested in the central part of the country.

5.   **Low-Level Wind Shear Alert System (LLWAS).** This airport-based system provides pilots and controllers with information on hazardous surface winds that create unsafe landing or departure conditions.

a.   It is a real-time, computer-controlled, surface winds sensor system which compares wind speed and direction from sensors on the airport periphery with center field wind data.

b.   During the time that an alert is posted, air traffic controllers provide wind shear advisories to all arriving and departing aircraft.

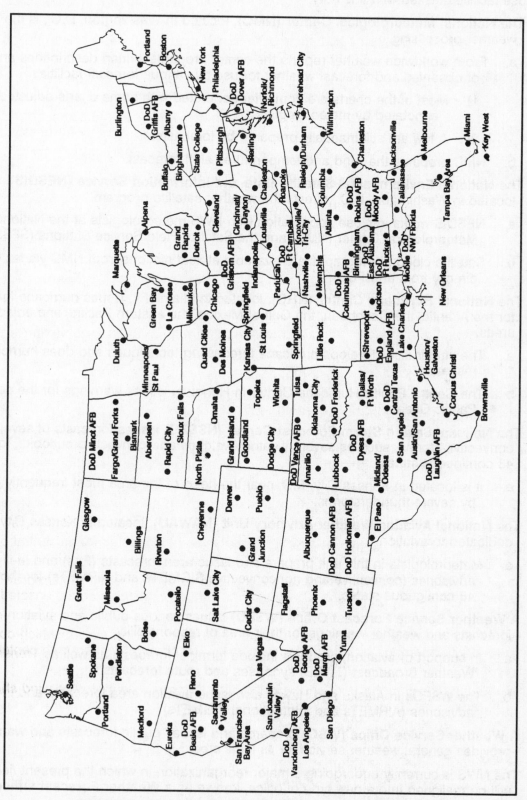

Figure 1-2. The Proposed NEXRAD Radar Observing Network

C. **National Oceanic and Atmospheric Administration (NOAA)** collects and analyzes data and prepares forecasts on a national, hemispheric, and global basis. The following is a description of those facilities tasked with this duty.

   1. The **National Meteorological Center (NMC)**, located in Washington, D.C., is the hub of weather processing.

      a. From worldwide weather reports the center prepares written descriptions and charts of observed and forecast weather for use by various forecast facilities.

         1) Most of the charts are computer-generated with some charts adjusted and annotated by meteorologists.

         2) A few are still manually prepared by forecasters.

      b. NMC prepares the wind and temperatures aloft forecast.

   2. The **National Environmental Satellite Data and Information Service (NESDIS)**, also located in Washington, D.C., directs the weather satellite program.

      a. NESDIS works in close cooperation with NWS meteorologists at the National Meteorological Center (NMC) and the Satellite Field Service Stations (SFSS).

      b. Satellite cloud photographs are available at field stations from NMC via fax or directly from an SFSS.

   3. The **National Hurricane Center (NHC)**, located in Miami, FL, issues hurricane forecasts for the Atlantic, the Caribbean, the Gulf of Mexico, the eastern Pacific, and adjacent land areas.

      a. The center also develops hurricane forecasting techniques and does hurricane research.

      b. The Central Pacific Hurricane Center in Honolulu issues warnings for the central Pacific Ocean.

   4. The **National Severe Storm Forecast Center (NSSFC)** issues forecasts of severe convective storms, such as severe weather watches and convective outlooks, for the 48 contiguous states.

      a. It is located in Kansas City, MO, near the heart of the area most frequently affected by severe thunderstorms.

   5. The **National Aviation Weather Advisory Unit (NAWAU)**, located in Kansas City, MO, is dedicated to aviation.

      a. Meteorologists in this unit prepare and issue area forecasts (FAs) and in-flight advisories (convective and nonconvective SIGMETs and AIRMETs) for the 48 contiguous states.

   6. A **Weather Service Forecast Office (WSFO)** issues various public and aviation-oriented forecasts and weather warnings for their area of responsibility.

      a. In support of aviation, products include terminal forecasts as well as Transcribed Weather Broadcast (TWEB) synopses and route forecasts.

      b. The WSFOs in Alaska and Hawaii also issue aviation area forecasts and in-flight advisories (AIRMETs and international SIGMETs).

   7. A **Weather Service Office (WSO)** prepares and issues public forecasts and warnings and provides general weather service for its local area.

   8. The NWS is currently undergoing a major reorganization in which the present field offices will be realigned into a new type of office, known as a **Weather Forecast Office (WFO)**.

      a. The WFO is designed to take advantage of NEXRAD and other new technology to improve weather services.

      b.   The WFOs will serve a smaller area than current WSFOs now do.

      c.   This reorganization is expected to be completed by the mid-1990s.

D. **Weather Service Outlet** -- defined as any facility, either government or nongovernment, that provides aviation weather service. Only FAA and NWS outlets are discussed here.

    1. **Flight Service Station (FSS).** The FAA operates two types of FSSs.

        a.   Nonautomated FSSs are located at a large number of airports around the country.

        b.   Automated FSSs (AFSSs) are replacing the older FSSs, usually with one per state. As more AFSSs become operational, the older FSSs will be closed.

           1)   You will continue to receive the same service from the AFSSs as from the older FSSs.

    2. The FSS provides more aviation weather briefing service than any other government service outlet.

        a.   It provides preflight and in-flight briefings, makes scheduled and unscheduled weather broadcasts, and furnishes weather advisories to known flights in the FSS area.

        b.   Because of the increasing number of flight operations, selected FSSs also provide transcribed (recorded) weather briefings.

           1)   These transcribed briefings provide weather information in a specified format, along with instructions on how to obtain a more detailed, person-to-person briefing.

        c.   There are three types of transcribed briefings.

           1)   The Transcribed Weather Broadcast (TWEB) is a continuous broadcast on selected NDB and VOR frequencies.

              a)   The TWEB is based on a route-of-flight concept with the order and content of the TWEB transcription as follows:

                  i)   Introduction
                 ii)   Synopsis
                iii)   Adverse Conditions
                iv)   TWEB Route Forecasts
                 v)   Outlook (optional)
                vi)   Winds Aloft
               vii)   Radar Reports
             viii)   Surface Weather Reports
              ix)   Pilot Reports (PIREPs)
               x)   Notice to Airmen (NOTAMs)
              xi)   Military Training Activity
             xii)   Density Altitude
            xiii)   Closing Announcement

              b)   The first five items are forecasts prepared by the NWS.

              c)   The synopsis and route forecast are prepared specifically for the TWEB by the WSFOs.

              d)   Flight precautions, outlook, and winds aloft are adapted, respectively, from in-flight advisories, area forecasts, and the NMC winds aloft forecast.

              e)   At selected locations, telephone access to the TWEB has been provided (TEL-TWEB). Telephone numbers for this service are found in the FSS and National Weather Service Telephone Numbers section of the *Airport/Facility Directory (A/FD)*.

2)   Pilot's Automatic Telephone Weather Answering Service (PATWAS) is provided by a nonautomated FSS and is a recorded telephone briefing service with the forecast for the local area, usually within a 50-NM radius of the station. A few selected stations also include route forecasts similar to the TWEB.

   a)   The order and the content of the PATWAS recording are as follows:

      i)   Introduction (describing PATWAS area)
      ii)   Adverse Conditions
      iii)   Recommendation (VFR flight not recommended, if appropriate)
      iv)   Synopsis
      v)   Current Conditions
      vi)   Surface Winds
      vii)   Forecast
      viii) Winds Aloft
      ix)   NOTAMs
      x)   Military Training Activity
      xi)   Request for PIREPs
      xii)   Closing Announcements
      xiii) Suspension Announcement

   b)   FSSs providing PATWAS place a high operational priority on it to ensure the information is current and accurate.

   c)   Detailed PATWAS information is usually prepared at selected time intervals between 0500 and 2200 local time, with updates issued as needed.

      i)   A general outlook for the PATWAS area is available between 2200 and 0500 local time if service is reduced during the period.

   d)   The *Airport/Facility Directory* lists PATWAS telephone numbers of FSS and NWS briefing offices.

3)   Telephone Information Briefing Service (TIBS) is provided by AFSSs and provides continuous telephone recordings of meteorological and/or aeronautical information.

   a)   Specifically, TIBS provides area and/or route briefings, airspace procedures, and special announcements (if applicable) concerning aviation interests.

   b)   The order and content of the TIBS recording is similar to the PATWAS.

   c)   The *Airport/Facility Directory* lists TIBS locations which can be called nationwide on one standard telephone number: (800) WX-BRIEF.

d.   The Hazardous In-flight Weather Advisory Service (HIWAS) is a continuous broadcast service of in-flight weather advisories, i.e., SIGMETs, Convective SIGMETs, AIRMETs, CWAs, and AWWs, over selected VORs. Also, hazardous weather not yet covered by an advisory is included.

   1)   In areas where HIWAS is already being utilized, controllers and specialists have discontinued their routine broadcast of in-flight advisories, but continue broadcasting a short alerting message.

e.   The En Route Flight Advisory Service (EFAS), also known as Flight Watch, is a weather service on a nationwide frequency of 122.0 MHz from selected FSSs.

   1)   The Flight Watch specialist maintains a continuous weather watch, provides time-critical assistance to en route pilots facing hazardous or unknown weather, and may recommend alternate or diversionary routes.

   2)   Additionally, Flight Watch is a focal point for rapid receipt and dissemination of pilot reports.

3)   Figure 1-7 below indicates the sites where EFAS and associated outlets are located.

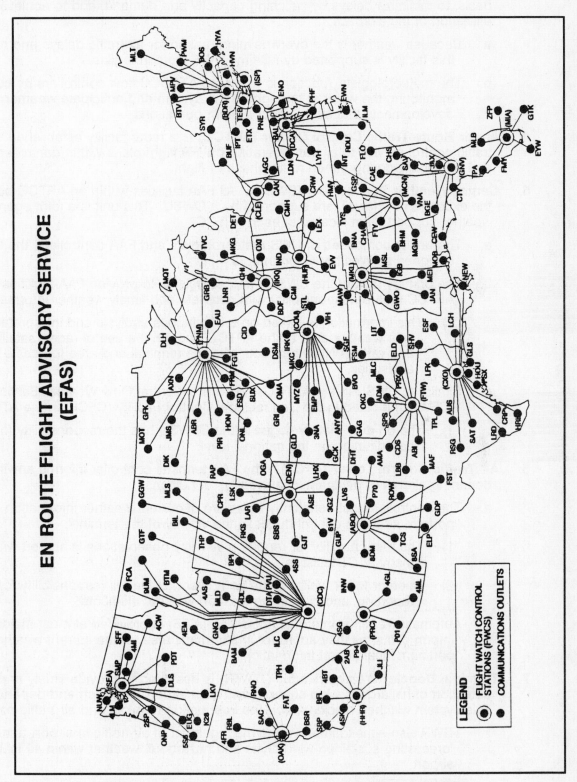

Figure 1-7. En Route Flight Advisory (Flight Watch Facilities)

3. **Air Traffic Control System Command Center (ATCSCC)** is located at FAA headquarters in Washington, D.C. Its objective is to manage the flow of air traffic on a system-wide basis, to minimize delays by watching capacity and demand, and to achieve maximum utilization of the airspace.

    a. Because weather is the overwhelming reason for air traffic delays and reroutings, this facility is supported by full-time NWS meteorologists.

    b. The meteorologists' function is to advise ATCSCC flow controllers by continuously monitoring the weather throughout the system and anticipate weather developments that might affect system operations.

4. An **Air Route Traffic Control Center (ARTCC)** is a radar facility established to provide air traffic control service to aircraft operating on IFR flight plans within controlled airspace and principally during the en route phase of flight.

5. **Center Weather Service Unit (CWSU).** All FAA facilities within an ARTCC boundary, with the exception of Hawaii, are supported by a CWSU. This unit is a joint agency aviation weather support team located at each ARTCC.

    a. The unit is composed of NWS meteorologists and FAA controllers, the latter being assigned as Weather Coordinators.

    b. The primary task of the CWSU meteorologist is to provide FAA facilities within the ARTCC area of responsibility with accurate and timely weather information.

        1) This information is based on a continuous analysis and interpretation of real-time weather data at the ARTCC through the use of radar, satellite, PIREPs, and various NWS products such as terminal and area forecasts and in-flight advisories.

    c. Similar to CWSU in the ARTCCs, there is a Central Flow Weather Service Unit (CFWSU) located in the Central Flow Control Facility (CFCF) in the ATCSCC.

        1) The on-duty meteorologist in the CFWSU has the responsibility for the weather coordination on the national level.

6. **Air Traffic Control Tower (ATCT).** The FAA terminal controller informs arriving and departing aircraft of pertinent local weather conditions.

    a. The controller must constantly be aware of current weather information needed to perform air traffic control duties in the vicinity of the terminal.

        1) The responsibility for reporting visibility observations is shared with the NWS at many ATCT facilities.

        2) At other tower facilities, the controller has the full responsibility for observing, reporting, and classifying aviation weather elements.

    b. Automatic Terminal Information Service (ATIS) is provided at most major airports to inform pilots, as they approach the terminal area, of the current weather and other pertinent local airport information.

7. **Terminal Doppler Weather Radar (TDWR)** is designed to provide timely and accurate detection of hazardous wind shear in and near airport approach and departure corridors. This system will then quickly report the information to pilots and air traffic controllers.

    a. TDWR also aims to improve the management of air traffic in airport terminal areas by forecasting sustained wind shifts and hazardous weather within 40 NM of an airport.

    b. TDWR systems will be installed at 47 airports through the rest of the 1990s. The first, at Memphis, TN, became operational in 1993.

8. **Weather Service Office (WSO).** NWS Offices provide selective aviation weather information, but not certified weather briefings, and provide local warnings to aviation.

9. **Weather Service Forecast Office (WSFO).** NWS Forecast Offices provide some selective pilot weather information, but not certified weather briefings.

    a. When getting a briefing from an FSS, you may, if necessary, request a telephone "patch in" to the WSFO forecaster.

E. AM WEATHER

1. AM WEATHER is a 15-min. weather program broadcast Monday through Friday mornings nationally on approximately 305 Public Broadcast Television Stations. See Appendix B on page 397 for more detailed information.

2. Professional meteorologists from the NWS and the NESDIS provide weather information primarily for pilots to enable them to make a better go or no-go flight decision.

3. National and Regional Weather Maps are provided, along with satellite sequences, radar reports, winds aloft, and weather watches.

    a. Extended forecasts are provided daily and on Fridays to cover the weekend.

    b. AM WEATHER also serves many other interest groups that depend upon weather information.

4. The program draws upon the U.S. weather observation network, geostationary and polar orbiting satellite data, and computer analysis to produce daily forecasts with 85% to 90% accuracy.

5. Alaska has its own aviation weather program. A meteorologist from the WSFO Anchorage conducts a 30-min. program, Monday through Friday, on Alaska's Public Television. This broadcast airs at 6 PM local time throughout the state.

F. Communications Systems

1. High-speed communications and automated data processing have improved the flow of weather data and products through the aviation weather network.

    a. The flow of weather information among the NOAA weather facilities is accomplished through the Automation of Field Operations and Services (AFOS) communications system.

    b. Alphanumeric and graphic products are displayed on a cathode ray tube (CRT), similar to a television screen, which eliminates the need for the slower teletypewriters and facsimile machines.

2. A new computer-based Automated Weather Information Processing System (AWIPS) is being developed for the NWS. This system will replace the current AFOS system and will allow quicker dissemination of weather information between the weather offices and their users.

    a. This system will be linked with the NEXRAD system to provide better detection, observing, and forecasting of weather systems, especially severe weather.

    b. The AWIPS network is scheduled to be in place in NWS offices by the mid-1990s.

3. The flow of alphanumeric weather information to the FAA Service Outlets is accomplished through the Leased Terminal Equipment (LTE), which also displays data on a CRT and so eliminates the need for teletypewriters.

4. Exchange of weather information between the NWS and FAA Service Outlets is generally accomplished in two ways.

    a. Graphic products (weather maps) are received by FAA Service Outlets from NMC in Washington, D.C., through a private sector contractor.

    b. Alphanumeric information is exchanged through the Weather Message Switching Center (WMSC) in Kansas City, MO.

        1) This switching facility serves as the gateway for the flow of alphanumeric information from one communication system to another (i.e., between the various FAA facilities, NWS, and other users).

G. Users

1. The ultimate users of the aviation weather service are pilots and dispatchers.

    a. As a user of the service, you should also contribute to it. Send pilot weather reports (PIREPs) to help briefers, forecasters, and your fellow pilots.

    b. The service can be no better or more complete than the information that goes into it.

2. In the interest of safety, you should get a complete briefing before each flight.

    a. You can get a preliminary briefing on your telephone by listening to the PATWAS, TIBS, or TEL-TWEB at your home or place of business.

    b. Many times the weather situation may be complex, and you may not completely comprehend the recorded message.

    c. If you need additional information after listening to the PATWAS or TIBS, you should contact an FSS for a more complete briefing tailored for your specific flight.

3. **How to Get a Good Weather Briefing**

    a. Call an AFSS (or an FSS) by dialing (800) WX-BRIEF [(800) 992-7433].

    b. When requesting a briefing, identify yourself as a pilot. Give clear and concise facts about your flight:

        1) Type of flight, VFR or IFR
        2) Aircraft number or pilot's name
        3) Aircraft type
        4) Departure point
        5) Route of flight
        6) Destination
        7) Flight altitude(s)
        8) Estimated time of departure
        9) Estimated time en route or estimated time of arrival

    c. With this background, the briefer can proceed directly with the briefing and concentrate on weather relevant to your flight.

    d. The weather briefing you receive depends on the type requested.

        1) A STANDARD briefing should include:

            a) Adverse conditions (you may elect to cancel at this point);

            b) VFR flight not recommended (VNR);

            c) Weather synopsis (positions and movements of lows, highs, fronts, and other significant causes of weather);

            d) Current weather;

            e) Forecast weather (en route and destination);

            f) Forecast winds aloft;

            g) Alternate routes (if any);

            h) Aeronautical information (NOTAMs);

            i) ATC delays;

            j) Request for PIREPs.

    2)   An ABBREVIATED briefing will be provided when the user requests information to

        a)   Supplement mass disseminated data (e.g., AM WEATHER, TIBS, PATWAS).

        b)   Update a previous briefing, or

        c)   Be limited to specific information.

    3)   An OUTLOOK briefing will be provided when the briefing is 6 or more hours in advance of proposed departure.

        a)   Briefing will be limited to applicable forecast data for the proposed flight.

  e.   The FSSs are here to serve you. You should not hesitate to discuss factors that you do not fully understand.

    1)   You have a complete briefing only when you have a clear picture of the weather to expect.

    2)   It is to your advantage to make a final weather check immediately before departure if at all possible.

4.   **Request/Reply Service** is available at all FSSs, WSOs, and WSFOs.

  a.   You may request through the service any reports or forecasts not routinely available at your service outlet.

    1)   These include route forecasts used in TWEB and PATWAS, recorder briefings, and RADAR plots.

  b.   You can request a forecast for any numbered TWEB route or any of the longer cross-country numbered routes.

    1)   See Part III, Chapter 8, TWEB Route Forecasts and Synopsis, beginning on page 275.

5.   **Have an Alternate Plan of Action**

  a.   When weather is questionable, get a picture of expected weather over a broader area.

    1)   Preplan a route to take you rapidly away from the weather if it goes sour.

  b.   When you fly into weather through which you cannot safely continue, you must act quickly.

    1)   Without preplanning, you may not know the best direction in which to turn; a wrong turn could lead to disaster.

  c.   A preplanned diversion beats panic. It is better to be safe than sorry.

END OF CHAPTER

CHAPTER TWO
SURFACE AVIATION WEATHER REPORTS

Please take a few minutes to study each of the concepts listed above and anticipate/imagine what they are and how they relate to the other listed concepts.

On July 1, 1993, the U.S. began issuing the new Aviation Routine Weather Report (METAR) on a supplemental test basis for about 250 international "landing rights" airports in the U.S.A. Complete conversion to the METAR format in the U.S. is expected in 1996. See Appendix E, New International Aerodrome Meteorological Codes, beginning on page 427 for a discussion of the METAR format.

A.   **Elements: 10 of Them** -- When a Surface Aviation Observation (SAO) is reported and transmitted, it is called a weather *report*. An SAO report contains some or all of the following elements:

1.   Station designator
2.   Type and time of report
3.   Sky condition and ceiling
4.   Visibility
5.   Weather and obstructions to vision
6.   Sea level pressure
7.   Temperature and dew point
8.   Wind direction, speed, and character
9.   Altimeter setting
10.   Remarks and coded data

B.   **Omitted Elements.** If one or more elements did not occur at observation time or are not pertinent to the observation, they are omitted from the report.

1.   When an element should be included but is unavailable, the letter "M" is transmitted in lieu of the missing element.

2.   Those elements that are included are transmitted in the above sequence.

C. **Example** of a Surface Aviation Observation Report (SAO).

SLC SA 1251 M50 BKN 3 K 175/75/68/3010/003/VIRGA ALQDS

To aid in the discussion, we have divided the report into the 10 elements.

SLC   SA 1251   M50 BKN   3   K   175 / 75/68 / 3010 / 003 / VIRGA ALQDS
 1   2   3   4   5   6   7   8   9   10

1.   Salt Lake City
2.   Record observation taken at 1251Z
3.   Measured ceiling 5,000 ft. broken
4.   Visibility 3 SM
5.   Smoke
6.   Sea level pressure of 1017.5 millibars
7.   Temperature 75°F, dewpoint 68°F
8.   Wind from 300° at 10 kt.
9.   Altimeter setting 30.03
10. Remarks: Virga exists in all quadrants

D. **Station Designator** (element 1). The station designator is the three-letter location identifier for the reporting station.

E. **Type and Time of Report** (element 2). There are two basic types of reports:

1.   Record observations (SA) are reports taken on the hour.

2.   Special reports (RS or SP) are observations that report significant changes in weather.

a.   A record special (RS) is a record (hourly on the hour) observation that reports a significant change in weather.

b.   A special (SP) is an observation taken other than on the hour to report a significant change in weather.

3.   The time that the observation was taken is given in Universal Coordinated Time (UTC or Z).

F. **Sky Condition and Ceiling** (element 3)

1.   A clear sky or a layer of clouds or obscuring phenomena aloft is reported by one of the first seven *sky cover designators* in Table 2-1 on page 227.

2.   A layer is defined as clouds or obscuring phenomena whose bases are all at approximately the same level.

3.   Height of the base of a layer precedes the sky cover designator. Height is in hundreds of feet *above ground level*.

4.   When more than one layer is reported, layers are given in ascending order according to height. For each layer above a lower layer or layers, the sky cover designator for that layer represents the total sky covered by that layer and all lower layers.

a.   EXAMPLE: 10 SCT M25 BKN 80 OVC, reports three layers:

1)   A scattered layer at 1,000 ft.

2)   A broken layer (ceiling) measured at 2,500 ft.

3)   A top layer at 8,000 ft. In this case it is assumed that the total sky covered by all the layers exceeds 9/10. Therefore, the upper layer is reported as overcast.

4)   See Figure 2-4 on page 227.

Designator	Meaning	Spoken
CLR	Clear (less than 0.1 sky cover)	CLEAR
SCT	Scattered Layer Aloft (0.1 through 0.5 sky cover)	SCATTERED
BKN*	Broken Layer Aloft (0.6 through 0.9 sky cover)	BROKEN
OVC*	Overcast Layer Aloft (more than 0.9 or 1.0 sky cover)	OVERCAST
–SCT	Thin Scattered. At least ½ of the sky cover aloft is transparent at and below the level of the layer aloft.	THIN SCATTERED
–BKN	Thin Broken. At least ½ of the sky cover aloft is transparent at and below the level of the layer aloft.	THIN BROKEN
–OVC	Thin Overcast. At least ½ of the sky cover aloft is transparent at and below the level of the layer aloft.	THIN OVERCAST
X*	Surface-Based Obstruction. (All of sky is hidden by surface-based phenomena.)	SKY OBSCURED
–X	Surface-Based Partial Obscuration. (0.1 or more, but not all, of sky is hidden by surface based phenomena.)	SKY PARTIALLY OBSCURED

*Sky condition represented by this designator will constitute a ceiling layer.

Table 2-1. Summary of Sky Cover Designators

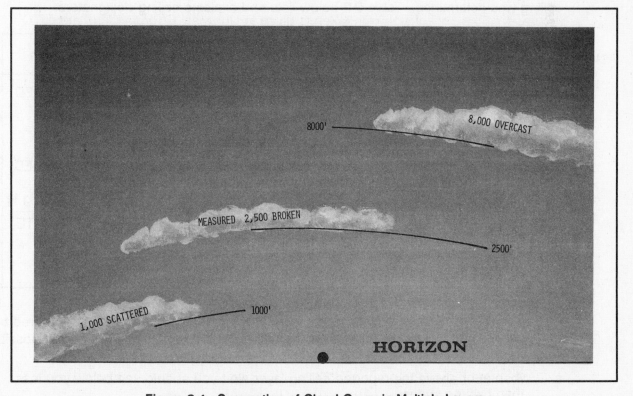

Figure 2-4. Summation of Cloud Cover in Multiple Layers

5. "Transparent" sky cover is clouds or obscuring phenomena aloft through which blue sky or higher sky cover is visible.

    1) A scattered, broken, or overcast layer may be reported as "thin." To be classified as thin, a layer must be half or more transparent (remember that sky cover of a layer includes all sky cover below the layer).

    a. EXAMPLE: If the blue sky is observed at a station as being visible through half or more of the total sky cover reported by the higher layer, the sky report could appear as 8 SCT 350 –SCT which reads "800 ft. scattered, 35,000 ft. thin scattered."

6. Any phenomenon based at the surface and hiding all or part of the sky is reported as SKY OBSCURED (X) or SKY PARTIALLY OBSCURED (–X).

    a. An obscuration or partial obscuration may be precipitation, fog, dust, blowing snow, etc.

    b. No height value precedes the designator for partial obscuration if vertical visibility is not restricted overhead; e.g., –X M40 OVC reads as "sky partially obscured measured ceiling 4,000 ft. overcast."

    c. A height value precedes the designator for an obscuration and denotes the vertical visibility into the phenomena; e.g., W5 X reads as "indefinite ceiling 500 ft. sky obscured."

7. Ceiling is defined as either

    a. Height above the ground of the lowest layer of clouds or obscuring phenomena aloft that is reported as broken or overcast and not classified as thin, or

    b. Vertical visibility into a surface-based obscuring phenomenon that hides all the sky (i.e., cannot be classified as partial).

8. A *ceiling designator* always precedes the height of the ceiling layer and describes how the ceiling was determined. Table 2-2 below lists and explains ceiling designators.

Coded	Meaning	Spoken
M	Measured. Identifies a ceiling height for a layer aloft determined by a ceiling light, ceilometer, or based on the known height of isolated objects in contact with the ceiling layer 1½ mi. or less from any runway.	MEASURED CEILING
E	Estimated. Identifies a ceiling height for a layer aloft determined by any other method not meeting criteria for measured ceiling.	ESTIMATED CEILING
W	Indefinite. Vertical visibility into a surface-based obstruction. Regardless of the method of determination, vertical visibility is classified as an indefinite ceiling.	INDEFINITE CEILING

Table 2-2. Ceiling Designators

9. The sky cover and ceiling, as determined from the ground, represent as nearly as possible what you should experience in flight.

    a. When at or above the reported ceiling layer aloft you should see below you less than half the surface.

      b.   When descending through a surface-based total obscuration you should first see the ground directly below you from the height reported as vertical visibility into the obscuration.

      c.   However, because of the differing viewing points of you and the observer, these surface reported values do not always exactly agree.

      d.   Figure 2-5 below illustrates the effect of an obscured sky on the vision from a descending aircraft.

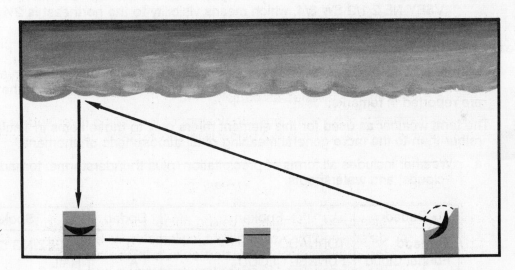

Figure 2-5. Effect of Obscured Sky on Aircraft

      1)   Vertical visibility is the altitude above the ground from which a pilot should first see the ground directly below the airplane (top).

      2)   The real concern is slant range visibility, which most often is less than vertical visibility. Thus, a pilot must descend to a lower altitude (bottom) before seeing a representative surface and being able to fly by visual reference to the ground.

10.   The letter "V" appended to the ceiling height indicates a variable ceiling. The range of the variability is shown in the remarks element of the report.

      a.   Variable ceiling is reported only when the ceiling height is below 3,000 ft.

      b.   EXAMPLE: M15V OVC and in the remarks CIG 15V18 means "measured ceiling 1,500 ft. variable overcast; ceiling variable between 1,500 and 1,800 ft."

G. **Visibility** (element 4)

1. Prevailing visibility at the observation site immediately follows sky and ceiling in the report. Prevailing visibility is the greatest distance at which objects can be seen and identified through at least 180 degrees of the horizon. It is reported in statute miles (SM) and fractions.

    a. EXAMPLE: 1 1/2 means "visibility 1½ SM."

2. When the prevailing visibility is less than 4 SM, the weather observation station and the control tower (if available) both take visibility observations. The lower of the two observations is the prevailing visibility and the other is reported in the remarks.

    a. EXAMPLE: TWR VSBY 1/4 means "tower visibility ¼ SM."

3. The letter "V" suffixed to prevailing visibility denotes a variable visibility. The range of variability is shown in remarks. Variable visibility is reported only when critical to aircraft operations.

    a. EXAMPLE: 3/4V and in the remarks, VSBY 1/2V1 means "visibility three quarters variable; visibility variable between ½ and 1 SM."

4. Visibility in some directions may differ significantly from prevailing visibility. These significant differences are reported in remarks.

    a. EXAMPLE: Prevailing visibility is reported as 1½ SM with a remark, VSBY NE 2 1/2 SW 3/4, which means visibility to the northeast is 2½ SM and to the southwest, it is ¾ SM.

H. **Weather and Obstructions to Vision** (element 5)

1. Weather and obstructions to vision when occurring at the station at observation time are reported immediately following visibility. If observed at a distance from the station, they are reported in remarks.

2. The term *weather* as used for this element refers only to those items in Table 2-3 below rather than to the more general meaning of all atmospheric phenomena.

    a. Weather includes all forms of precipitation, plus thunderstorms, tornadoes, funnel clouds, and waterspouts.

Coded	Spoken	Coded	Spoken
Tornado	TORNADO	ZL	FREEZING DRIZZLE
Funnel Cloud	FUNNEL CLOUD	A	HAIL
Waterspout	WATERSPOUT	IP	ICE PELLETS
T	THUNDERSTORM	IPW	ICE PELLET SHOWER
T+	SEVERE THUNDERSTORM	S	SNOW
R	RAIN	SW	SNOW SHOWER
RW	RAIN SHOWER	SP	SNOW PELLETS
L	DRIZZLE	SG	SNOW GRAINS
ZR	FREEZING RAIN	IC	ICE CRYSTALS

Table 2-3. Weather Symbols and Meanings

3. Precipitation is reported in one of three intensities. The intensity symbol follows the weather symbol.

      LIGHT   –
      MODERATE   (No sign)
      HEAVY   +

4. No intensity is reported for hail (A) or ice crystals (IC).

5. A thunderstorm is reported as "T" and a severe thunderstorm as "T+."

   a. A *severe thunderstorm* is one in which surface wind is 50 kt. or greater and/or hail is ¾ in. or more in diameter.

6. Obstructions to vision include the phenomena listed in Table 2-4 below. No intensities are reported for obstructions to vision.

   a. Obstructions to vision are only reported for visibilities of 6 SM or less.
   b. Weather symbols will be used regardless of visibility.

Coded	Spoken
BD	BLOWING DUST
BN	BLOWING SAND
BS	BLOWING SNOW
BY	BLOWING SPRAY
D	DUST
F	FOG
GF	GROUND FOG
H	HAZE
IF	ICE FOG
K	SMOKE
VOLCANIC ASH	VOLCANIC ASH

Table 2-4. Obstructions to Vision Symbols and Meanings

7. When the obscuring phenomenon is surface-based and partially obscures the sky, a remark reports tenths of sky hidden.

   a. EXAMPLES:

      1) D3 means dust is obscuring 3/10 of the sky.
      2) RF2 means rain and fog are obscuring 2/10 of the sky.

8. A layer of obscuring phenomena aloft is reported in the sky and ceiling portion the same as a layer of cloud cover. A remark identifies the layer as an obscuring phenomenon.

   a. EXAMPLE: 20 −BKN and a remark K20 −BKN means that a thin broken layer of smoke based at 2,000 ft. above the surface.

I. **Sea Level Pressure** (element 6)

   1. Sea level pressure is separated from the preceding elements by a space and is transmitted in record observation (SA) reports only. It is in three digits to the nearest tenth of a millibar (mb) with the decimal point omitted.

      a. Sea level pressure usually is greater than 960.0 mb and less than 1050.0 mb. The first 9 or 10 is omitted.

      b. To decode, prefix a 9 or 10, whichever brings it closer to 1000.0 mb.

   2. EXAMPLE:

As Reported	Decoded
980	998.0 mb
191	1019.1 mb
752	975.2 mb
456	1045.6 mb

J.   **Temperature and Dew Point** (element 7)

1.   Temperature and dew point are in whole degrees Fahrenheit (°F).

   a.   A slash (/) separates the temperature element from the sea level pressure element.

      1)   If the sea level pressure is not transmitted, a space separates the temperature from the preceding elements.

   b.   Temperature and dew point are also separated by a slash.

   c.   A minus sign precedes the temperature or dew point when either of these temperatures is below 0°F.

2.   EXAMPLE: 82/59 means "temperature 82°F, dewpoint 59°F."

K.   **Wind** (element 8)

1.   Wind follows dew point and is separated from it by a slash.

2.   The wind is observed for 1 min., and the average direction and speed are reported in a four-digit group.

   a.   The first two digits are the direction FROM which the wind is blowing. It is in tens of degrees referenced to TRUE north.

      1)   On the other hand, wind direction when verbally broadcast by a control tower or advisory station is referenced to magnetic north. This is done so that you can more closely relate the wind direction to the landing runway.

      2)   To conserve space on the report the last digit (0) is omitted, i.e., 020° appears as 02, and 220° appears as 22.

   b.   The second two digits of the wind element are the wind speed in kt. A calm wind is reported as 0000.

   c.   EXAMPLE: 3010 means the wind is from 300° true at 10 kt.

3.   If the wind speed is 100 kt. or greater, 50 is added to the direction code and the hundreds digit of the speed is omitted.

   a.   EXAMPLE: 8315 can be recognized as a wind speed of 100 kt. or more because the first two digits are greater than 36 (36 represents a direction of 360°, which is the largest number of degrees on the compass).

   b.   To decode, subtract 50 from 83 (83 − 50 = 33) and add 100 to the last two digits (15 + 100 = 115). The wind is from 330° at 115 kt.

4.   A *gust* is a variation in wind speed of at least 10 kt. between peak winds and lulls. A *squall* is a sudden increase of at least 15 kt. in average wind speed to a sustained speed of 20 kt. or more which lasts for at least 1 min.

   a.   Gusts or squalls are reported by the letter "G" or "Q" respectively, following the average 1-min. wind speed. The peak speed of the gust or squall in kt. follows the letter.

   b.   EXAMPLE: 2123G38 means wind from 210° at 23 kt. gusting to peak speed of 38 kt.

5. When any part of the wind report is estimated (direction, speed, peak speed in gusts or squalls), the letter "E" precedes the wind group.

    a. EXAMPLE: E3122Q27 means estimated wind from 310° at 22 kt. with peak speed in squalls at 27 kt.

L. Altimeter Setting (element 9)

1. Altimeter setting follows the wind group and is separated from it by a slash.

    a. Normal range of altimeter settings is from 28.00 in. to 31.00 in. Hg.

    b. The last three digits are transmitted with the decimal point omitted.

    c. To decode, prefix the coded value in the report with either a 2 or 3, whichever brings it closer to 30.00 in.

2. EXAMPLES:

    a. 998 means "altimeter setting 29.98 (in. Hg)."
    b. 025 means "altimeter setting 30.25 (in. Hg)."

M. Remarks (element 10)

1. Remarks, if any, follow altimeter setting and are separated from it by a slash. Certain remarks are reported routinely, and others are included when considered significant to aviation by the observer.

    a. Often, some of the most important information in an observation may be the remarks portion.

2. The first remark, when transmitted, is runway visibility or runway visual range.

    a. *Runway visibility* -- the visibility from a particular location along an identified runway reported in miles and fractions of a mile.

    b. *Runway visual range* -- the maximum horizontal distance a pilot can see down a specified instrument runway where the pilot can identify standard high intensity runway lights. Visual range is reported in hundreds of feet.

    c. The report consists of a runway designator and the contraction "VV" or "VR" followed by the appropriate visibility or visual range. Both the VV and the VR report are for a 10-min. period preceding observation time. The remark usually reports the 10-min. extremes separated by the letter "V."

        1) However, if the visual range or visibility has not changed significantly during the 10 min., a single value is sent indicating that the value has remained constant.

        2) The following examples show several reports and their decoding.

            a) R36VV11/2 means RUNWAY THREE SIX, VISIBILITY VALUE ONE AND ONE-HALF. (Visibility remained constant during the 10-min. period.)

            b) R18VR20V30 means RUNWAY ONE EIGHT, VISUAL RANGE VARIABLE BETWEEN TWO THOUSAND FEET AND THREE THOUSAND FEET.

    d. Runway visual range in excess of 6,000 ft. is written 60+ and the minimum value is encoded as the minimum suffixed by a minus sign.

        1) EXAMPLE: R36LVR10-V25 means RUNWAY THREE SIX LEFT, VISUAL RANGE VARIABLE FROM LESS THAN ONE THOUSAND FEET TO TWO THOUSAND FIVE HUNDRED FEET.

3.   The following, by category, are coded remarks clarifying or expanding on coded elements contained within the report.

a.   Sky and Ceiling

Coded Elements	Coded Remarks
FEW CU	Few cumulus clouds
HIR CLDS VSB	Higher clouds visible
BINOVC	Breaks in overcast
ACCAS ALQDS*	Altocumulus castellanus all quadrants
ACSL SW-NW*	Altocumulus standing lenticular southwest through northwest
ROTOR CLDS NW*	Rotor clouds northwest
VIRGA E-SE*	Virga (precipitation not reaching the ground) east through southeast
TCU W*	Towering cumulus clouds west
CB N MOVG E*	Cumulonimbus north moving east
CLDS TPG MTNS SW	Clouds topping mountains southwest
RDGS OBSCD W-N	Ridges obscured west through north

*These cloud types are highly significant and the observer should always report them. A pilot in flight should also report them when observed.

b.   Obscuring Phenomena

Coded Elements	Coded Remarks
S7	Snow obscuring 7/10 of the sky
BS3	Blowing snow obscuring 3/10 of the sky

c.   Visibility

Coded Elements	Coded Remarks
TWR VSBY 3/4	Tower visibility 3/4 SM
SFC VSBY 1/2	Surface visibility 1/2 SM

d.   Weather and Obstruction to Vision

Coded Elements	Coded Remarks
T W MOVG E FQT LTGCG	Thunderstorm west moving east, frequent lightning cloud to ground
RB30	Rain began 30 min. after the hour
SB15E40	Snow began 15, ended 40 min. after the hour
T OVHD MOVG E	Thunderstorm overhead, moving east
OCNL DSNT LTG NW	Occasional distant lightning northwest
HLSTO 2	Hailstones 2 in. in diameter
INTMT R–	Intermittent light rain
OCNL RW	Occasional moderate rain shower
SNOINCR 1/4/8	Snow increased 1 in. in past hour, 4 in. since last 6 hourly and 8 in. total on ground at time of observation
R– OCNLY R+	Light rain occasionally heavy rain
RWU	Rain shower of unknown intensity
KOCTY	Smoke over city
PTCHY GF S	Patchy ground fog south

e.   Wind

Coded Elements	Coded Remarks
WSHFT 30	Wind shifted at 30 min. past the hour
WND 27V33	Wind variable between 270° and 330°
PK WND 3348/22	Peak wind within the past hour from 330° at 48 kt. occurred 22 min. past the hour

f.   Pressure

Coded Elements	Coded Remarks
PRESRR	Pressure rising rapidly
PRESFR	Pressure falling rapidly

4.   **Freezing Level Data**

    a.   Upper air observation stations (rawinsonde) provide *freezing level data*, which appear after any remarks.

        1)   The coded remark is appended to the first record report transmitted after the information becomes available.

    b.   Code for the remark is as follows: RADAT UU (D) $(h_p h_p h_p)$ (/n)

        1)   RADAT -- a contraction identifying the remark as freezing level data.

            a)   UU -- relative humidity at the freezing level in percent. When more than one level is sent, UU is the highest relative humidity observed at any of the levels transmitted.

            b)   (D) -- a coded letter "L", "M", or "H" to indicate that relative humidity is for the "lowest", "middle", or "highest" level coded. This letter is omitted when only one level is sent.

            c) $(h_p h_p h_p)$ -- a height in hundreds of feet above MSL at which the sounding crossed the 0°C isotherm. No more than three levels are coded. If the sounding crosses the 0°C isotherm more than three times, the levels coded are the lowest, highest, and the intermediate crossing with the highest relative humidity (RH).

            d)   (/n) -- an indicator to show the number of crossings of the 0°C isotherm, other than those coded. The indicator is omitted when all levels are coded.

    c.   EXAMPLES:

        1)   RADAT 87045 means relative humidity 87%, only crossing of 0°C isotherm was 4,500 ft. MSL.

        2)   RADAT 87L024105 means relative humidity 87% at the lowest (L) crossing. Two crossings occurred at 2,400 and 10,500 ft. MSL.

            a)   Temperature was below 0°C below 2,400 ft. MSL; above 0°C between 2,400 ft. MSL and 10,500 ft. MSL; and below 0°C above 10,500 ft. MSL.

        3)   RADAT 84M019045051/1 means relative humidity 84% at the middle (M) crossing of the three coded crossings. Coded crossings were at 1,900, 4,500, and 5,100 ft. "/1" indicates one additional crossing.

        4)   RADAT MISG means the sounding terminated below the first crossing of the 0°C isotherm. All temperatures were above freezing.

        5)   RADAT ZERO means the entire sounding was below 0°C.

N.   **Automated Surface Observations**

    1.   The Automatic Meteorological Observing Station (AMOS) is a solid-state system capable of automatically observing temperature, dew point, wind direction and speed, pressure (altimeter setting), peak wind speed, and precipitation accumulation.

        a.   The field sensors are tied in directly to the FAA observation network.

            1)   At a staffed AMOS, the observer can manually enter additional information to give a more complete observation.

        b.   Figure 2-12 on page 237 shows the breakdown of an unstaffed AMOS.

        c.   Figure 2-13 on page 237 shows the breakdown of a staffed AMOS.

            1)   Note the addition of sky condition, visibility, and remarks made by an observer.

d. The information is coded in the report in the same manner as a Surface Aviation Observation.

MDO SA 1548 AMOS 33/29/3606/975/PK WND 08 001

MDO	STATION IDENTIFICATION: (Middleton Island, AK) Identifies report using FAA identifiers.
SA	TYPE OF REPORT: (Record) SP = Special, SA = Record, RS = Record Special.
1548	TIME OF REPORT: Coordinated Universal Time -- UTC.
AMOS	AUTOMATIC STATION IDENTIFIER.
33	TEMPERATURE: (33 degrees F.) Minus sign indicates sub-zero temperatures.
29	DEW POINT: (29 degrees F.) Minus sign indicates sub-zero temperatures.
3606	WIND: (360 degrees true at 6 knots.) Direction is first two digits and is reported in tens of degrees. To decode, add a zero to the first two digits. The last digits are speed; e.g., 2524 is 250 degrees at 24 knots.
975	ALTIMETER SETTING: (29.75 inches) The tens digits and decimal are omitted from the report. To decode, prefix a 2 or 3 to the coded value, whichever brings it closer to 30.00 inches.
PK WND 08	PEAK WIND SPEED: (8 knots) Reported speed is highest detected wind speed since last hourly observation.
001	PRECIPITATION ACCUMULATION: (0.01 inch) The amount of precipitation since the last synoptic time (00, 06, 12, 1800 UTC).

Figure 2-12. Decoding Unstaffed AMOS

SMP SP 0056 AMOS -X M20 BKN 7/8 L-FK 046/66/65/2723/967 PK WND 36 027/VSBY S 1/4

SMP	STATION IDENTIFICATION: (Stampede Pass, WA) Identifies report using FAA identifiers.
SP	TYPE OF REPORT: (Special) See Figure 2-12 for explanation.
0056	TIME OF REPORT: UTC
AMOS	AUTOMATIC STATION IDENTIFIER
-X M20 BKN	SKY AND CEILING: (partly obscured sky, ceiling measured 2,000 feet broken) Figures are height in 100s of feet above ground level (AGL). The letter preceding height indicates the method used to determine the height. The symbol after the height is the amount of sky cover (SCT, BKN, OVC).
7/8	PREVAILING VISIBILITY: (seven-eights statute miles)
L-FK	WEATHER AND OBSTRUCTIONS TO VISION: (Light drizzle, Fog and Smoke) The algebraic signs indicate intensity (+ heavy, - light, no symbol means moderate).
046	SEA-LEVEL PRESSURE: (1004.6 millibars) Only the tens, units and tenths digits are reported. To decode, prefix the value with a 9 or 10, whichever brings the value closest to 1000 millibars.
66	TEMPERATURE: (66 degrees F.) See Figure 2-12 for explanation.
65	DEW POINT: (65 degrees F.) See Figure 2-12 for explanation.
2723	WIND: (270 degrees true north at 23 knots.) See Figure 2-12 for explanation.
967	ALTIMETER SETTING: (29.67 inches) See Figure 2-12 for explanation.
PK WND 36	PEAK WIND SPEED: (36 knots) Highest speed detected since last hourly observation.
027	PRECIPITATION ACCUMULATION: (0.27 inch) Amount of precipitation received since the last synoptic observation (00, 06, 12, 1800 UTC).
VSBY S 1/4	MISCELLANEOUS REMARKS AND NOTAMs: (Visibility to the south 1/4 mile). Remarks are entered here using standard contractions.

Figure 2-13. Decoding Staffed AMOS

2.   The Automatic Observing Station (AUTOB) is an AMOS with added capability to *automatically* report sky conditions, visibility, and precipitation occurrence. AUTOB is polled at 20 min. intervals.

  a.   The upper limit of cloud amount and height measurements is 6,000 ft. AGL. Visibility in statute miles (SM) is determined by a backscatter sensor with reportable categories of 0 to 8 (see Table 2-5 below).

  1)   If a visibility report consisting of three values is encountered, it is decoded as shown in the following example:

"BV786," 7 = present visibility, 8 = maximum visibility during past 10 min., and 6 = minimum visibility during past 10 min.

Index of visibility	When visibility is (SM):	Index of visibility	When visibility is (SM):
0	less than 15/16	5	4 1/2 – 5 1/2
1	1 – 1 7/8	6	5 1/2 – 6 1/2
2	2 – 2 7/8	7	6 1/2 – 7 1/2
3	3 – 3 1/2	8	above 7 1/2
4	3 1/2 – 4 1/2		

Table 2-5. Reportable Visibility Categories

  b.   AUTOB may indicate no cloud layers in either a clear situation or when the sensor is unable to penetrate a surface-based obscuration. To distinguish the two, the following rules apply.

  1)   If the visibility is less than 2 SM, either a partial obscuration "–X" or indefinite obscuration "WX" is reported.

  a)   A "–X" implies some cloud returns and a "WX" implies no cloud returns.
  b)   A vertical visibility value for "WX" is not measured.

  2)   When visibility is 2 SM or greater and no cloud returns are detected a "CLR BLO 60" is used which indicates a clear sky below 6,000 ft. "E" is the ceiling designator. A maximum of 3 (lowest) cloud layers will be reported.

  c.   A remark "HIR CLDS DETECTED" is included if clouds are detected above an overcast, with the higher clouds "HIR CLDS" being less than 6,000 ft. AGL.

  1)   Note that an AUTOB makes no distinction between a thin and opaque cloud layer. "E30 OVC" may be a thin overcast, but is reported as a ceiling.

  d.   Figure 2-14 on page 239 shows the breakdown of an AUTOB message.

3.   The Remote Automatic Meteorological Observing Station (RAMOS) is similar to the unstaffed AMOS observation except that a 3-hour pressure change, maximum/minimum temperature, and 24-hour precipitation accumulation are also included at designated times.

  a.   Figure 2-15 on page 239 shows the breakdown of a RAMOS message.

ENV SA 1648 AUTOB E25 BKN BV7 P 33/29/3606/975/ PK WND 08 001

ENV	STATION IDENTIFICATION: (Wendover, UT) Identifies report using FAA identifiers.
SA	TYPE OF REPORT: (Record) See Figure 2-12 for explanation.
1648	TIME OF REPORT: UTC.
AUTOB	AUTOMATIC STATION IDENTIFIER.
E25 BKN	SKY AND CEILING: (Estimated 2,500 feet) Figures are height in 100s of feet above ground level (AGL). Letter preceding height indicates ceiling.
BV7	PRESENT VISIBILITY: (Visibility 7 miles) Reported in whole statute miles between 0 and 8 inclusive. (Table 2-5)
P	PRECIPITATION OCCURRENCE: (P means precipitation has occured in the past 10 minutes).
33	TEMPERATURE: (33 degrees F.) See Figure 2-12 for explanation.
29	DEW POINT: (29 degrees F.) See Figure 2-12 for explanation.
3606	WIND: (360 degrees true at 6 knots.) See Figure 2-12 for explanation.
975	ALTIMETER SETTING: (29.75 inches) See Figure 2-12 for explanation.
PK WND 08	PEAK WIND SPEED: (8 knots) See Figure 2-12 for explanation.
001	PRECIPITATION ACCUMULATION: (0.01 inches) See Figure 2-12 for explanation.

NOTE: If no clouds are detected below 6,000 feet and the visibility is greater than 2 miles, the reported sky condition will be CLR BLO 60.

Figure 2-14. Decoding AUTOB

P67 SA 2356 RAMOS 046/66/65/2723/967 PK WND 36 0002 027 83 20043

P67	STATION IDENTIFICATION: (Lidgerwood, ND) Identifies report using FAA identifiers.
SA	TYPE OF REPORT: (Record) See Figure 2-12 for explanation.
2356	TIME OF REPORT: UTC.
RAMOS	AUTOMATIC STATION IDENTIFIER.
046	SEA-LEVEL PRESSURE: (1004.6 millibars) Only the tens, units and tenths digits are reported. Prefix a 9 or 10 to the code, whichever brings the value closer to 1000 millibars.
66	TEMPERATURE: (66 degrees F.) See Figure 2-12 for explanation.
65	DEW POINT: (65 degrees F.) See Figure 2-12 for explanation.
2723	WIND: (270 degrees true at 23 knots) See Figure 2-12 for explanation.
967	ALTIMETER SETTING: (29.67 inches) See Figure 2-12 for explanation.
PK WND 36	PEAK WIND SPEED: (36 knots) Reported speed is the highest detected since last hourly observation.
0002	THREE-HOUR PRESSURE CHANGE: (rising then falling, 0.02 millibars higher now than three hours ago.) See note below.
027	PRECIPITATION ACCUMULATION: (0.27 inches) See Figure 2-12 for explanation.
83	TEMPERATURE: (Maximum temperature for day of 83 degrees F.) Maximum tempertures are reported at 00 and 06z, Minimum temperatures are reported at 12 and 18z.
20043	PRECIPITATION ACCUMULATION PAST 24-HOURS 2RRRR: (0.43 inches)

NOTE: In THREE-HOUR PRESSURE CHANGE the first digit is the barometer tendency. The tendency is higher than three hours ago if the digit is 0, 1, 2 or 3. The pressure is lower than three hours ago if the digit is 5, 6, 7 or 8 and if the digit is a 4, no change has occurred in the last three hours.

Figure 2-15. Decoding RAMOS

4.   The Automated Weather Observing System (AWOS III) is being installed by the FAA at selected airports around the country.

   a.   AWOS can operate in one of four modes:

      1)   Mode 1 -- Full-time automated operation

      2)   Mode 2 -- Full-time automated operation with local NOTAMs.

      3)   Mode 3 -- Full-time automated operation with manual weather augmentation and local NOTAM option.

      4)   Mode 4 -- Part-time manual operation

   b.   The AWOS consist of automated reports of

      1)   Ceiling/sky conditions,
      2)   Visibility,
      3)   Temperature and dew point,
      4)   Wind direction/speed/gusts,
      5)   Altimeter setting, and
      6)   If certain conditions are met, automated remarks containing density altitude, variable visibility, and variable wind direction.

   c.   Automated observations are broadcast on a ground-to-air radio and made available on a telephone answering device.

      1)   Selected sites will have the capability to transmit observations to the Service A teletype network.

   d.   Figure 2-16 shows is the breakdown of an AWOS III message.

FOD SA 0055 AWOS M5 OVC 1/2 70/68/3325G30/992/ P110/WND 30V36/OBSERVER WEA: TRW+ T OVHD

FOD	STATION IDENTIFICATION: (Fort Dodge, IA)
SA	TYPE OF REPORT: (Record) See Figure 2-12 for explanation.
0055	TIME OF REPORT: UTC.
AWOS	AUTOMATIC STATION IDENTIFIER.
M5 OVC	SKY AND CEILING: (Measured 500 feet overcast) Determined from sensor outputs every 30 seconds and integrated over a 30-minute sampling period.
1/2	PREVAILING VISIBILITY: (One-half statute mile) Determined from sensor outputs every 10-second intervals that are used to compute a one-minute average. The one-minute visibility values are averaged over a 10-minute period to determine the reported visibility.
70	TEMPERATURE: (70 degrees F.) A 5-minute average reading.
68	DEW POINT: (68 degrees F.) A 5-minute average reading.
3325G30	WIND: (330 degrees true at 25 knots, gust to 30 knots) Direction is the first two digits and is reported in tens of degrees and the second two digits is the speed. The "G" stands for wind gust and the last two digits is the highest 5-second average wind speed for the past 10 minutes.
992	ALTIMETER SETTING: (29.92 inches) See Figure 2-12 for an explanation.
P110	PRECIPITATION ACCUMULATION: (1.10 inches) The cumulative amount or liquid equivalent precipitation. Reported on each observation it is an accumulation reported to the nearest 0.01 inch in inches/tenths/hundredths and prefixed with the letter "P".
WND 30V36	AUTOMATED REMARK: (Wind direction variable between 300 degrees and 360 degrees true) Variable wind direction is reported when the wind direction varies around the reported direction by 60 degrees or more with the wind speed is 7 knots or greater.
OBSERVER WEA: TRW+ T OVHD	MANUAL WEATHER AUGMENTATION: (Thunderstorm with heavy rain shower directly over the station) This example indicates that the weather observer at the station reports a thunderstorm with a heavy rain shower directly over the station. The data that can be added to the report is limited to thunderstorms (their intensity and direction), obstructions to visibility (dependent on the visibility being 3 miles or less) and precipitation (type and intensity).

Figure 2-16. Decoding AWOS

5. The Automated Surface Aviation Observing System (ASOS) is designed to support aviation operations, weather forecasting activities, and at the same time, support the general needs of the hydrometeorological, climatological, and meteorological research communities.

a. ASOS provides continuous minute-by-minute observations and performs the basic observing functions necessary to generate an SAO.

1) When fully implemented, ASOS will more than double the number of full-time surface aviation weather locations (between 900 and 1,700 locations).

b. While the automated and nonautomated systems may differ in their methods of data collection and interpretation, both produce an observation that is similar in form and content.

1) For the "objective" elements (i.e., pressure, temperature, dew point, wind, and precipitation accumulation), both the ASOS and the observer use a fixed location and time-averaging technique.

2) For the "subjective" elements, observers use a fixed time, spatial averaging technique to describe the visual elements (i.e., sky condition, visibility, and present weather), while the ASOS uses a fixed location, time-averaging technique.

a) Although this is a fundamental change, the manual and automated techniques yield remarkably similar results within the limits of the respective capabilities.

c. All ASOS locations will prepare and disseminate SAOs.

1) Some locations operate completely unattended with no human intervention in the generation of the observation. This type of station is designated as "AO2" in the body of the SAO message.

2) Other locations operate with human on-station oversight and possible intervention in the form of augmentation and/or backup of the SAO message.

a) This type of station is designated as "AO2A" in the body of the SAO message and means that an observer is present at the ASOS location to provide general oversight of the observation.

d. Among the basic strengths of the ASOS observation is the fact that critical weather parameters (e.g., sky condition and visibility) are measured at specific locations where they are needed most.

1) ASOS data are updated once each minute and can be accessed through a variety of media.

a) Computer-generated voice messages can be made available through the telephone and directly to the pilot through the ground-to-air radio.

e. The ASOS will automatically report the following surface weather elements in the SAO.

1) Sky condition: Cloud height and amount (CLR, SCT, BKN, OVC) up to 12,000 ft. AGL.

2) Visibility up to 10 SM.

3) Basic present weather information (type and intensity): Rain, snow, and freezing rain.

4) Obstructions to vision: Fog and haze.

5) Pressure: Sea-level pressure in millibars and altimeter setting in inches of mercury.

6) Ambient temperature and dew point temperature (°F).

7) Wind: Direction (tens of degrees referenced TRUE north), speed (kt.), and character (gusts, squalls).

8) Selected significant remarks including: Variable cloud height, variable visibility, precipitation beginning and ending times, rapid pressure changes, pressure change tendency, wind shift, peak wind, etc.

9) Precipitation accumulation (hundredths of an inch).

f.   Figure 2-17 below provides a breakdown of an ASOS observation.

IAD SA 1155 AO2 CLR BLO 120 6F 101/42/41/2804/991/ 52102 70125 10060 20041

IAD	STATION IDENTIFICATION: (Dulles International Airport) Identifies report using FAA identifiers.
SA	TYPE OF REPORT: (Record) See Figure 2-12 for explanation.
1155	TIME OF REPORT: UTC
AO2	TYPE OF STATION: Automated station with precipitation discriminator and no augmentation.
CLR BLO 120	SKY AND CEILING: (No clouds were detected below 12,000 feet AGL) Determined from a laser ceilometer every 30 seconds over a 30 minute period.
6	PREVAILING VISIBILITY: (Measured visibilities between 5 and less than 7 statute miles) Determined from a visibility sensor once every 30 seconds and is used to compute a one-minute average. The one-minute visibility values are then averaged over a 10-minute period to determine the reported (prevailing) visibility.
F	WEATHER AND OBSTRUCTIONS TO VISION: (fog) When visibility drops below 7 statute miles, the current Dew Point Depression (DD) will be used to distinguish between Fog and Haze. If the DD is less than or equal to 4 degrees F, then F will be reported as the obstruction to vision.
101	SEA-LEVEL PRESSURE: (1010.1 millibars) Only the tens, units, and tenths digits are reported. To decode, prefix the value with a 9 or 10, whichever brings the value closest to 1000 millibars.
42	TEMPERATURE: (42 degrees F.) See Figure 2-12 for explanation.
41	DEW POINT: (41 degrees F.) See Figure 2-12 for explanation.
2804	WIND: (280 degrees true north at 4 knots) See Figure 2-12 for explanation. The current 2-minute average wind is computed once every 5 seconds.
991	ALTIMETER SETTING: (29.91 inches) See Figure 2-12 for explanation.
52102	PRESSURE TENDENCY, 5appp: (steadily rising, 10.2 millibars higher than three hours ago.) See note on bottom of Figure 2-15.
70125	24-HOUR PRECIPITATION 7RRRR: (1.25 inches of precipitation, liquid equivalent, in the past 24 hours.)
10060	MAXIMUM TEMPERATURE 1sTTT: (60 degrees F.) The maximum temperature during the past 6 hours.
20041	MINIMUM TEMPERATURE 2sTTT: (41 degrees F.) The minimum temperature during the past 6 hours.

Figure 2-17. Decoding ASOS

END OF CHAPTER

CHAPTER THREE
PILOT WEATHER REPORTS (PIREPs)

A. **Purpose** -- No observation is more timely than the one made from the cockpit. In fact, aircraft in flight are the **only** means of directly observing cloud tops, icing, and turbulence.

1. PIREPs are also valuable in that they help fill the gaps between reporting stations.

2. Pilots should help themselves, the aviation public, and the aviation weather service by sending pilot reports.

B.   **Format** -- A PIREP is usually transmitted in a prescribed format shown in Figure 3-1 below.

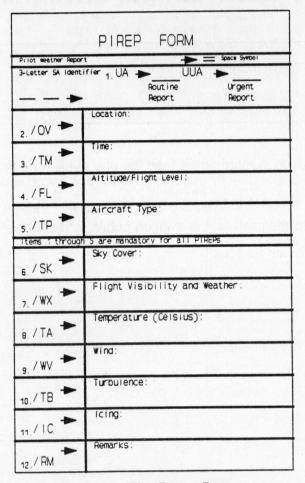

Figure 3-1. Pilot Report Format

1.   The letters "UA" identify the message as a pilot report. The letters "UUA" identify an urgent PIREP.

2.   Required elements for all PIREPs are message type, location, time, flight level, type of aircraft, and at least one weather element encountered.

3.   When not required, elements without reported data are omitted.

4.   All altitude references are MSL unless otherwise noted.

5.   Distances are in NM, and time is in UTC.

C. **Examples**

1. A PIREP is usually transmitted as part of a group of PIREPs collected by state or as a remark appended to a surface aviation weather report. The weather being reported is coded in contractions and symbols.

    a. EXAMPLE (refer to Figure 3-1 as a guide):

        UA /OV MRB-PIT/TM 1600/FL 100/TP BE55/SK 024 BKN 032/042
        BKN-OVC/WX FV02 R H/TA -12/WV 270020/TB LGT/IC LGT-MDT RIME
        055-080/RM MDT RAIN

        1) The PIREP decodes as follows:

            Pilot report, Martinsburg to Pittsburgh at 1600 UTC at 10,000 ft. MSL.
            Type of aircraft is a Beechcraft Baron (BE55). First cloud layer has a
            base at 2,400 ft. MSL broken with tops at 3,200 ft. MSL. The second
            cloud layer has a base at 4,200 ft. MSL broken occasionally overcast
            with no tops reported. Flight visibility is 2 NM with rain and haze.
            Outside air temperature is −12°C. Wind is from 270° magnetic at 20 kt.
            Light turbulence. Light to moderate rime icing is reported between 5,500
            and 8,000 ft. MSL. Currently in moderate rain shower.

    b. The following is an example of how a PIREP would be appended to a Surface
        Aviation Observation:

        DSM SA 1755 M8 OVC 3R-F 132/45/44/3213/922/UA /OV DSM 320012/TM
        1735/FL UNKN/TP UNKN /SK OVC 065/080 OVC 140.

        1) The PIREP decodes as follows:

            Pilot report, 12 NM on the 320° radial from the Des Moines VOR, at
            1735 UTC. The flight level and type of aircraft are unknown. The top of
            the lower overcast layer of clouds is 6,500 ft. MSL. The base of the
            second overcast layer is at 8,000 ft. MSL with tops at 14,000 ft. MSL.

D. **Contractions Used** -- Most contractions in PIREP messages are self-explanatory.

1. Icing and turbulence reports state intensities using standard terminology when possible.

2. Intensity tables for turbulence and icing are in Part III, Chapter 26, Tables and Conversion
    Graphs, beginning on page 373.

    a. If a pilot's description of an icing or turbulence encounter cannot readily be
        translated into standard terminology, the pilot's description is transmitted verbatim.

3. Pilot reports of individual cloud layers, bases, and tops should be reported in standard
    contractions. The height of a cloud base will precede the sky cover symbol and the
    height of the cloud top will follow the symbol.

    a. EXAMPLE: 038 BKN 070 means the base of a broken layer is at 3,800 ft. MSL and
        the top is at 7,000 ft. MSL.

4. Outside air temperature, in degrees Celsius, is given in two digits.

    a. If the temperature is below zero, the value will be preceded by a hyphen (-).

5. Wind is given in six digits with the first three digits being direction and the last three digits
    being speed in knots.

END OF CHAPTER

CHAPTER FOUR
RADAR WEATHER REPORT (SD)

A. **Introduction** -- Thunderstorms and general areas of precipitation can be observed by radar. Most radar stations report each hour at H+35 (i.e., 35 min. past the hour) with intervening special reports as required.

   1. The report includes the type, intensity, intensity trend, and location of the precipitation.

   2. Also included is the echo top (i.e., height) of the precipitation and, if significant, the echo base.

   3. All heights are reported above Mean Sea Level (MSL).

B. **Interpretation** of a radar weather report (SD)

   1. Precipitation Intensity and Intensity Trend

      a. See Table 3-1 below for corresponding rainfall rates defining intensities.
      b. Note that intensity and intensity trend are not applicable to frozen precipitation.

Intensity		Intensity Trend	
Symbol	Intensity	Symbol	Trend
–	Light	+	Increasing
(none)	Moderate		
+	Heavy	–	Decreasing
++	Very Heavy		
X	Intense	NC	No change
XX	Extreme		
U	Unknown	NEW	New echo

Table 3-1. Precipitation Intensity and Intensity Trend

2.   EXAMPLE:

```
OKC 1934 LN 8TRW++/+ 86/40 164/60 199/115 15W L2425
MT 570 AT 159/65 2 INCH HAIL RPRTD THIS CELL
^MO1 NO1 ON3 PM34 RL2 SL9=
```

```
OKC 1934  LN    8    TRW ++/+    86/40 164/60 199/115
    a.        b.  c.      d.                        e.


    15W    L2425     MT 570 AT 159/65
    f.        g.             h.


         2 INCH HAIL RPRTD THIS CELL
                    i.

      ^M01 N01 ON3 PM34 QM3 RL2 SL9=
                    j.
```

Table 3-2. Ordered Content of a Radar Weather Report

a.   Location identifier and time of radar observation (Oklahoma City SD at 1934 UTC).

b.   Echo pattern (LN). The radar echo pattern or configuration may be a

1)   Line (LN) -- a line of precipitation echoes at least 30 NM long, at least five times as long as it is wide and at least 30% coverage within the line.

2)   Fine Line (FINE LN) -- a unique clear air echo (usually precipitation-free and cloud-free) in the form of a thin or fine line on the radarscope. It represents a strong temperature/moisture boundary such as an advancing dry cold front.

3)   Area (AREA) -- a group of echoes of similar type and not classified as a line.

4)   Spiral Band Area (SPRL BAND AREA) -- an area of precipitation associated with a hurricane that takes on a spiral band configuration around the center.

5)   Single Cell (CELL) -- a single isolated convective echo such as a rain shower.

6)   Layer (LYR) -- an elevated layer of stratiform precipitation not reaching the ground.

c.   Coverage in tenths (8/10 in the example).

d.   Type, intensity, and intensity trend of weather. In the example, the radar depicted thunderstorms (T) and very heavy rainshowers (RW++) that are increasing in intensity (+). Note that the intensity is separated from intensity trend by a slash.

e.   Azimuth (reference true north) and range in NM from the radar station of points defining the echo pattern (86°, 40 NM to 164°, 60 NM to 199°, 115 NM in the example).

f.   Dimension of echo pattern (15W in the example). The dimension of an echo pattern is given when azimuth and range define only the center line of the pattern. In this example, 15W means the line has a total width of 15 NM, 7½ mi. either side of a center line drawn from the points given. D15 would mean a convective echo is 15 NM in diameter around a given center point.

g.   Pattern movement (the LINE is moving *from* 240° at 25 kt. in the example). This element may also show movement of individual storms or cells with a "C" or movement of an area with an "A."

h.   Maximum top (MT) and location (57,000 ft. MSL on radial 159° at 65 NM in the example).

i.   Remarks are self-explanatory using plain-language contractions.

j.   The digital section is used for preparing the radar summary chart.

C.   **Remarks** -- A radar weather report may contain remarks in addition to the coded observation.

1.   Certain types of severe storms may produce distinctive patterns on the radarscope.

a.   A "hook-shaped echo" may be associated with a tornado.

b.   A "line echo wave pattern" (LEWP) in which one portion of a squall line bulges out ahead of the rest of the line may produce strong gusty winds at the bulge.

c.   A "vault" on the Range-Height Indicator scope may be associated with a severe thunderstorm that is producing large hail and strong gusty winds at the surface.

2.   If hail, strong winds, tornado activity, or other adverse weather is known to be associated with identified echoes on the radarscope, the location and type of these phenomena are included as remarks.

3.   Examples of remarks are "HAIL REPORTED THIS CELL," "TORNADO ON GROUND AT 338/15," and "HOOK ECHO 243/18."

4.   As far as indicating precipitation **not** reaching the ground, two commonly used contractions are "MALF" and "PALF."

a.   These two contractions mean that the precipitation is "mostly aloft" and "partly aloft," respectively.

1)   That is, most or some of the precipitation is **not** reaching the ground.

b.   Bases of the precipitation will be reported in hundreds of feet MSL.

c.   EXAMPLE: "PALF BASE 040" means part of the precipitation detected is evaporating at 4,000 ft. MSL.

D. Precipitation Amounts

1.   Radar weather reports also contain groups of digits. For example: MO1 NO1 ON3 PM34.

    a.   These groups of digits are the final entry on the SD. This digitized radar information is used primarily by meteorologists and hydrologists for estimating the amount of rainfall and in preparing the radar summary chart.

    b.   Using a grid overlay chart for the radar site, each digit represents the intensity level of precipitation within a box on the grid.

        1)   A code of PM34 indicates precipitation in two consecutive boxes in the same row, i.e., PM = 3, PN = 4.

        2)   See Figure 3-3 below for an example of a plotted digital radar report.

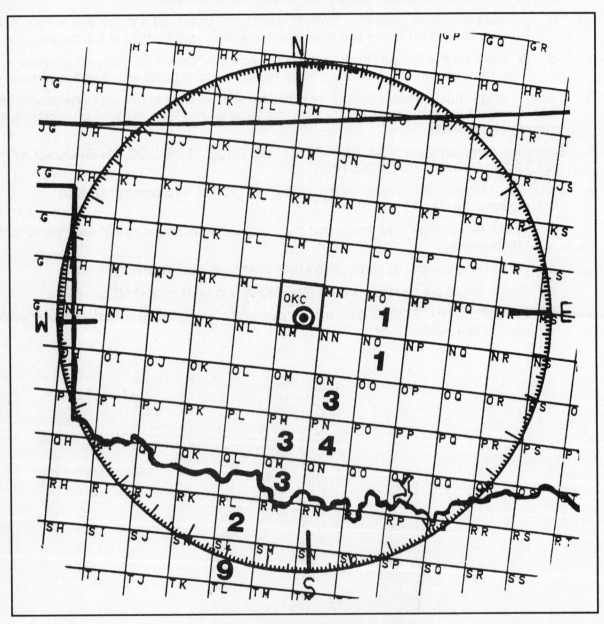

Figure 3-3. Plotted Digital Radar Report

E. **Blank Reports** -- When a radar weather report is transmitted, but does not contain any encoded weather observation, a contraction which indicates the operational status of the radar is sent.

Contraction	Operational Status
PPINE	Equipment normal and operating in PPI (Plan Position Indicator) mode; no echoes observed.
PPIOM	Radar inoperative or out-of-service for preventative maintenance.
PPINA	Observations not available for reasons other than PPINE or PPIOM.
ROBEPS	Radar operating below performance standards.
ARNO	"A" scope or azimuth/range indicator inoperative.
RHINO	Radar cannot be operated in RHI (Range-Height Indicator) mode. Height data not available.

Table 3-3. Contractions of Radar Operational Status

1. EXAMPLE: OKC 1135 PPINE means that the Oklahoma City, OK radar at 1135 UTC detects no echoes.

F. **Using SDs** -- When using hourly and special radar weather reports in preflight planning, note the location and coverage of echoes, the type of weather reported, the intensity trend, and especially the direction of movement.

1. A WORD OF CAUTION -- Remember that, when the National Weather Service radar detects objects in the atmosphere, it only detects those of precipitation size or greater. It is **not** designed to detect ceilings and restrictions to visibility.

    a. An area may be blanketed with fog or low stratus but unless precipitation is also present, the radarscope will be clear of echoes.

2. Pilots should use SDs along with SAO reports and forecasts when planning a flight.

3. SDs help pilots to plan ahead to avoid thunderstorm areas. Once airborne, however, you must depend on visual sighting or airborne radar to avoid individual storms.

END OF CHAPTER

CHAPTER FIVE
SATELLITE WEATHER PICTURES

> Please take a few minutes to study each of the concepts listed above and anticipate/imagine what they are and how they relate to the other listed concepts.

A. **Types of Satellites** -- There are two types of weather satellites in use today: GOES (a geostationary satellite) and NOAA (a polar orbiting satellite).

   1. Two U.S. GOES (Geostationary Operational Environmental Satellite) are used to take pictures. One is stationed over the Equator at 75°W longitude and the other at 135°W longitude.

      a. Together they cover North and South America and surrounding waters.

         1) These satellites normally transmit a picture of this area from pole to pole every half hour.

      b. When disastrous weather threatens the U.S., these satellites can scan small areas rapidly so a picture can be received as often as every 3 min.

         1) Data from these rapid scans are used at national warning centers.

   2. Since the GOES are positioned over the Equator, the pictures poleward of about 50° latitude become greatly distorted. For pictures above 50°, the NOAA satellite is employed.

      a. The NOAA satellite is a polar orbiter and orbits the earth on a track which nearly crosses the north and south poles.

      b. A high resolution picture is produced covering about 800 mi. either side of its track on the journey from pole to pole.

      c. The NOAA pictures are vital to weather personnel in Alaska and Canada.

B. **Types of Imagery** -- Basically, two types of imagery are available, and when combined, give a great deal of information about clouds.

    1.   Through interpretation, the analyst can determine the type of cloud, the temperature of cloud tops (from this the approximate height of the cloud can be determined), and the thickness of cloud layers.

        a.   From this information, the analyst gets a good idea of the associated weather.

    2.   One type of imagery is visible imagery. With a visible picture, we are looking at clouds and the Earth reflecting sunlight to the satellite sensor.

        a.   The greater the reflected sunlight reaching the sensor, the whiter the object is on the picture.

            1)   The amount of reflectivity reaching the sensor depends upon the height, thickness, and ability of the object to reflect sunlight.

            2)   Since clouds are much more reflective than the Earth, clouds will usually show up white on the picture, especially thick clouds.

            3)   Thus, the visible picture is primarily used to determine the presence of clouds and the type of cloud from shape and texture.

    3.   The second type of imagery is infrared (IR) imagery. With an IR picture, we are looking at heat radiation being emitted by the Earth and clouds.

        a.   The images show temperature differences between cloud tops and the ground, as well as temperature gradations of cloud tops and along the Earth's surface.

            1)   Ordinarily, cold temperatures are displayed light gray or white.

            2)   High clouds appear whitest.

            3)   However, various computer-generated enhancements are sometimes used to sharply illustrate important temperature changes.

        b.   IR pictures are used to determine cloud top temperatures that can approximate the height of the cloud.

            1)   From this, you can see the importance of using visible and IR imagery together when interpreting clouds.

            2)   IR pictures are available both day and night.

C. **Schedule** -- Operationally, pictures are received once every 30 min.

    1.   From these pictures, the development and dissipation of weather can be seen and followed over the entire area.

D. **Inoperative Satellite** -- At the time of this book, GOES East (75° west) has become inoperative and GOES West has been repositioned at 100°W longitude in order to cover as much of the U.S. as possible.

    1.   As a result, from the west coast out into the Pacific Ocean and from the east coast out into the Atlantic Ocean, the most useful satellite imagery will now be provided by the two NOAA satellites currently in use.

    2.   The next GOES will replace GOES East sometime in the mid 1990s.

END OF CHAPTER

CHAPTER SIX
TERMINAL FORECASTS (FT AND TAF)

> Please take a few minutes to study each of the concepts listed above and anticipate/imagine what they are and how they relate to each other.

The U.S. currently prepares International Terminal Aerodrome Forecasts (TAF) for approximately 80 international U.S. airports. Complete conversion from the domestic Terminal Forecast (FT) to the TAF in the U.S. is expected in 1996. See Appendix E which begins on page 427.

A. **Domestic Terminal Forecasts (FT)** -- A terminal forecast is a description of the surface weather expected to occur at an airport. The forecast includes cloud heights and amounts, visibility, weather, and winds related to flight operations within 5 NM of the center of the runway complex.

1. The term vicinity (VCNTY) covers the area from 5 NM beyond the center of the runway complex out to 25 NM.

   a. EXAMPLE: "TRW VCNTY" means thunderstorms are expected to occur between 5 and 25 NM from the station.

2. Scheduled forecasts are issued by WSFOs (Weather Service Forecast Offices) for their respective areas (Figures 1-5, 1-5A, 1-5B, and 1-5C on pages 256 through 259) three times daily (four times daily for Alaska and Hawaii) and are valid for a 24-hour period. Issue and valid times are according to time zones (see Part III, Chapter 26, Tables and Conversion Graphs, beginning on page 373 for issuance times).

3. The format of the FT is similar to the SAO (Surface Aviation Observation) report.

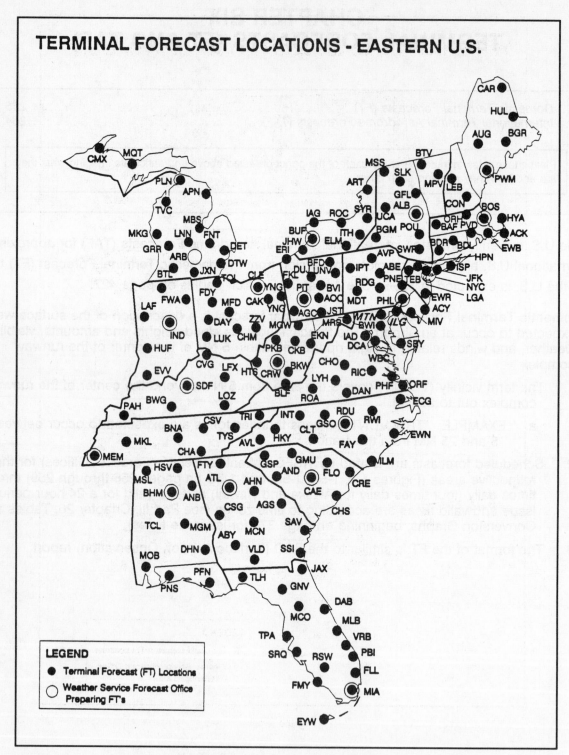

Figure 1-5. Locations of WSFOs, Their Areas of Responsibility and Airports for Which Each Prepares Terminal Forecasts

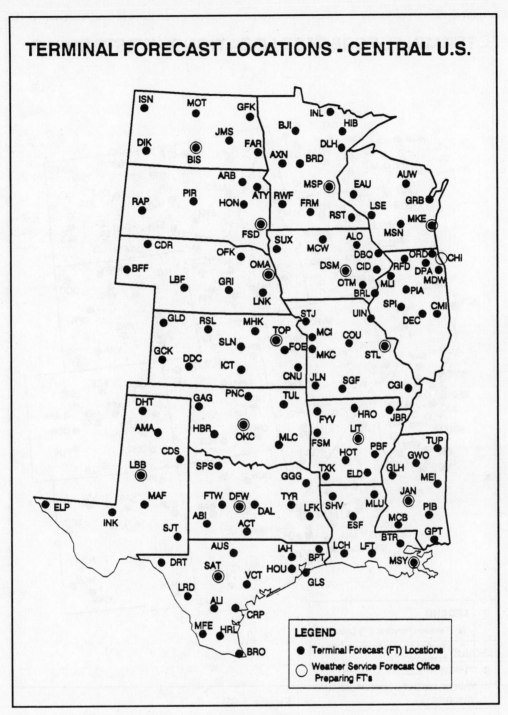

Figure 1-5A.

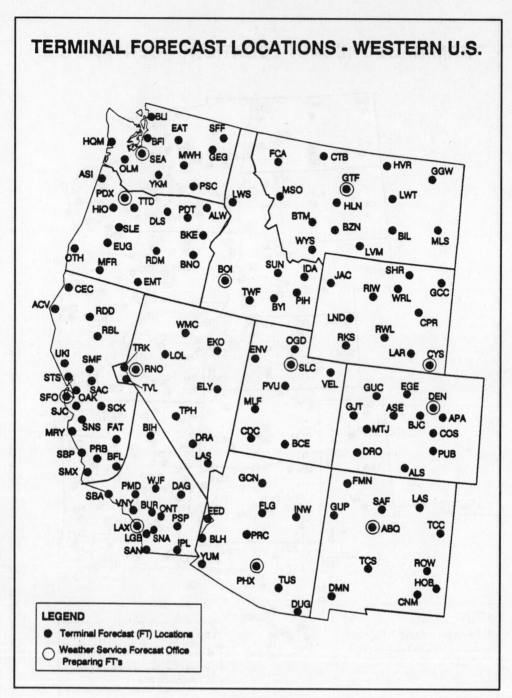

Figure 1-5B.

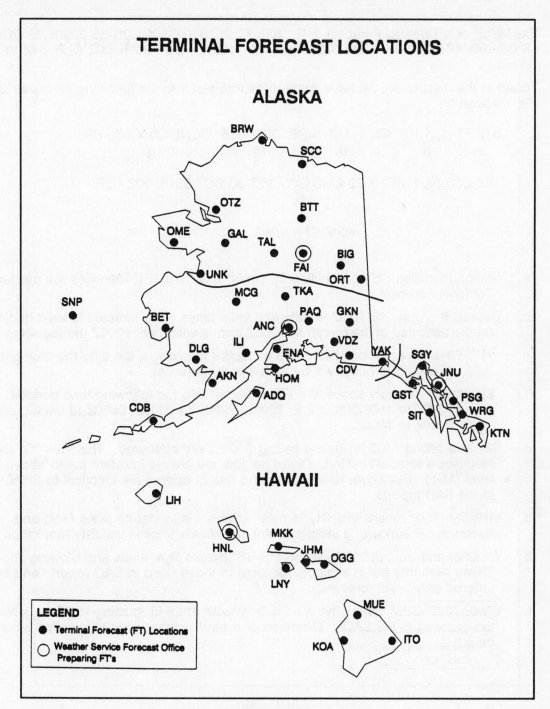

Figure 1-5C.

4.   EXAMPLE of a Terminal Forecast (FT): STL FT 251010 C5 X 1/2S-BS 3325G35 OCNL C0 X 0S+BS. 16Z C30 BKN 3BS 3320 CHC SW-. 22Z 30 SCT 3315. 00Z CLR. 04Z VFR WND..

5.   To aid in the discussion, we have divided the forecast into the following elements lettered "a" through "i".

STL FT 251010 C5 X 1/2 S-BS 3325G35 OCNL C0 X 0S+BS
　a.　　　b.　　　c.　　d.　　e.　　f.　　　　　　　　g.

16Z C30 BKN 3BS 3320 CHC SW-. 22Z 30 SCT 3315. 00Z CLR.
　　　　　　　　　　　　　　h.

04Z VFR WND..
　　　i.

a.   Station identifier. "STL" identifies St. Louis, MO. The "FT" identifies the product as a Terminal Forecast.

b.   Date-time group. "251010" is date and valid times. The forecast is valid beginning on the 25th day of the month at 1000Z and is valid until 1000Z the following day.

   1)   When forecast conditions are expected to change, the time the changes are expected to occur will be stated in the forecast.

   2)   In the example above, the FT is divided into the following time periods: 1000Z to 1600Z, 1600Z to 2200Z, 2200Z to 0000Z, 0000Z to 0400Z, and 0400Z to 1000Z.

c.   Sky and ceiling. "C5 X" means ceiling 500 ft., sky obscured. The letter "C" always identifies a forecast ceiling. Cloud heights are always in reference to above ground level (AGL). Sky cover designators and height coding are identical to those used in the SAO reports.

d.   Visibility. "1/2" means visibility ½ mile. Visibility is in statute miles (SM) and fractions. Absence of a visibility entry specifically implies visibility more than 6 SM.

e.   Weather and obstruction to vision. "S-BS" means light snow and blowing snow. These elements are in symbols identical to those used in SAO reports and are entered only when forecast.

f.   Wind. "3325G35" means the wind is from 330° at 25 kt. gusting to 35 kt., which is the same as in the SAOs. Omission of a wind entry specifically implies wind less than 6 kt.

g. Remarks. "OCNL C0 X 0S+BS" means occasional conditions of ceiling zero, sky obscured, visibility zero, heavy snow and blowing snow. Remarks may be added to more completely describe forecast weather by indicating variations from prevailing conditions.

   1) See Table 4-9 below for the definitions of the variability terms used in remarks.

      a) In this case the occasional conditions described above are expected to occur with a greater than 50% probability but for less than half of the time period from 1000Z to 1600Z.

      b) Thus c. through f. in the example are the prevailing conditions from 1000Z to 1600Z with g. specifying variations from the prevailing conditions during the same time period.

Term	Description
OCNL	Greater than 50% probability of the phenomenon occurring but for less than ½ of the forecast period
CHC	30 to 50% probability (precipitation only)
SLGT CHC	10 to 20% probability (precipitation only)

Table 4-9. Variable Terms

   2) Also, "LLWS" will be included in the remarks section if low-level wind shear is forecast.

   3) If a phenomenon in the remarks is forecast for only a portion of the time period of the forecast group, the time period for the phenomenon is indicated immediately after the phenomenon.

      a) EXAMPLE: 00Z C30 OVC OCNL C12 OVC 02Z-04Z CHC TRW-. 06Z, etc. This forecast says prevailing conditions will be ceiling 3,000 ft. from 0000Z to 0600Z. A ceiling of 1,200 ft. is expected between 0200Z and 0400Z but for less than half of the 2 hours. The chance of thunder-storms with light rain showers is for the period 0000Z to 0600Z.

h. Expected changes. When changes are expected, preceding conditions are followed by a period before the time conditions of the expected change.

   1) "16Z C30 BKN 3BS 3320 CHC SW-. 22Z 30 SCT 3315. 00Z CLR." means by 1600Z the prevailing conditions will be ceiling 3,000 ft. broken, visibility 3 SM, blowing snow, wind 330° at 20 kt., and a 30 to 50% chance of light snow showers.

      a) By 2200Z, the prevailing conditions will change to 3,000 ft. scattered, visibility more than 6 SM (implied), and wind 330° at 15 kt.

      b) By 0000Z, prevailing conditions will become sky clear, visibility more than 6 SM, and wind less than 6 kt. (implied).

i. 6-hour categorical outlook. The last 6 hours of the forecast is a categorical outlook. "04Z VFR WND.." means that from 0400Z until 1000Z (the end of the forecast period), the weather will be no ceiling or ceiling higher than 3,000 ft. and visibility greater than 5 SM (VFR) with wind 25 kt. or stronger. (The term "WND" is included if the sustained winds or gusts are forecast to be 25 kt. or greater.) The double period (..) signifies the end of the forecast for the specific terminal.

1)   Table 4-1 lists applicable categories and remarks that may be appended to the category "VFR" to better describe the expected conditions.

Category	Definition
LIFR	Low IFR-ceiling less than 500 ft. and/or visibility less than 1 mile
IFR	Ceiling 500 to less than 1,000 ft. and/or visibility 1 to less than 3 miles
MVFR	Marginal VFR-ceiling 1,000 to 3,000 ft. and/or visibility 3 to 5 miles inclusive
VFR	No ceiling or ceiling greater than 3,000 ft. and visibility greater than 5 miles
Remarks that may be appended to VFR at forecaster's discretion	**Definition**
VFR CIG ABV 100	Ceiling greater than 10,000 ft. and visibility greater than 5 miles
VFR NO CIG	Cloud coverage less than 6/10 or thin clouds and visibility greater than 5 miles
VFR CLR	Cloud coverage less than 1/10 and visibility greater than 5 miles

Table 4-1. Categories

2)   The cause(s) of below VFR categorical outlooks must be stated.

a)   Below VFR categories (Table 4-2) can be due to ceiling only (CIG), restrictions to visibility only (TRW, F, etc.) or a combination of both (CIG TRW F, etc.).

Example	Definition
LIFR CIG	Low IFR due to low ceiling *only*.
IFR F	IFR due to visibility restricted by fog *only*.
MVFR CIG H K	Marginal VFR due *both* to low ceiling and to visibility restricted by haze and smoke.
IFR CIG R WND	IFR due *both* to low ceiling and to visibility restricted by rain, wind expected to be 25 kt. or greater.

Table 4-2. Examples of Categorical Groupings

6.   The heading of an FT collective identifies the message as an FT along with a 6-digit date-time group giving the transmission time.

   a.   EXAMPLE: "FT130945" means a collective transmitted on the 13th at 0945Z.

   b.   A collective FT message will usually be broken down into states, i.e., "TX 130945" would be followed by a group of FTs for terminals in the state of Texas.

7.   A delayed (RTD), corrected (COR), or amended (AMD) FT is identified in the message rather than in the heading. The following are a delayed FT for Binghamton, NY; a corrected FT for Memphis, TN; and an amended FT for Lufkin, TX.

   BGM FT RTD 131009 1020Z 100 SCT 250 SCT 1810. 12Z 50 SCT
   100 SCT 1913 CHC C30 BKN 3TRW AFT 14Z. 21Z 100 SCT C250 BKN.
   03Z VFR..

   MEM FT COR 132222 2230Z 40 SCT 300 SCT CHC TRW. 02Z CLR.
   16Z VFR..

   LFK FT AMD 1 132218 2225Z C8 OVC 4F OVC OCNL BKN. 23Z 20 SCT
   250 -BKN. 03Z 40 SCT 120 SCT CHC C30 BKN 3TRW. 12Z MVFR
   CIG F..

   a.   Note that in each forecast a time group follows the valid period. This time group is the forecast issue time.

      1)   EXAMPLE: The BGM delayed forecast was issued at 1020Z and not at the scheduled issuance time 0845Z.

         a)   This changes the beginning of the valid forecast period from 0900Z to 1000Z.

   b.   An FT will not be issued unless two consecutive observations are received from the station.

      1)   A routine delayed FT (RTD) is usually due to a station that is not on a 24-hour observing schedule.

         a)   The first two observations of the day are received after the regularly scheduled FT issuance time.

      2)   If the time is known when observations usually end for the day, the phrase "NO AMDTS AFT (TIME) Z" will be appended to the last scheduled FT of the day.

   c.   A corrected FT is necessary when a typographical error is in the FT.

   d.   An amended FT is necessary for a situation in which the forecast has to be revised due to significant changes in the weather.

      1)   Note also that the amended forecast for LFK has the entry "AMD 1." Amended FTs for each terminal are numbered sequentially starting after each scheduled forecast.

B. **International Terminal Aerodrome Forecasts (TAF)** -- The old TAF, the FT and new requirements with adjustments were merged into a new TAF. These new TAF codes have been in use since July 1993. TAF are also discussed in Appendix E which begins on page 427.

1. They are scheduled four times daily for 24-hour periods beginning at 0000Z, 0600Z, 1200Z, and 1800Z.

2. EXAMPLE TAF:

> KPIT 091720Z 1818 22020KT 3SM –SHRA BKN020 FM20 30015G25KT 3SM SHRA OVC015 PROB40 2022 1/2SM TSRA OVC008CB FM23 27008KT 5SM –SHRA BKN020 OVC040 TEMPO 0407 00000KT 1SM –RA FG FM10 22010KT 5SM –SHRA OVC020 BECMG 1315 20010KT P6SM NSW SKC

3. To aid in the discussion, we have divided the forecast into the following elements lettered "a" through "m".

> KPIT   091720Z   1818   22020KT   3SM   –SHRA
> a.   b.   c.   d.   e.   f.
>
> BKN020   FM20 30015G25KT 3SM SHRA OVC015
> g.   h.
>
> PROB40 2022 1/2SM TSRA OVC008CB
> i.
>
> FM23 27008KT 5SM –SHRA BKN020 OVC040
> j.
>
> TEMPO 0407 00000KT 1SM –RA FG
> k.
>
> FM10 22010KT 5SM –SHRA OVC020
> l.
>
> BECMG 1315 20010KT P6SM NSW SKC
> m.

a. Station identifier. The TAF code uses the ICAO 4-letter station identifiers.

> 1) In the contiguous 48 states, the 3-letter identifier is prefixed with a "K"; i.e., the 3-letter identifier for Pittsburgh, PA is PIT while the ICAO identifier is KPIT.

> 2) Elsewhere, the first two letters of the ICAO identifier tell what region the station is in.

>> a) "PA" is Alaska (PACD is Cold Bay); and "PH" is Hawaii (PHTO is Hilo).

b. Issuance time. "091720Z" means the TAF was issued on the 9th day at 1720Z.

c. Valid period. "1818" means the forecast is valid from 1800Z on the 9th day to 1800Z the following days.

d. **Wind.** "22020KT" means the wind is from 220° True North at an average (mean) speed of 20 kt.

    1)   As needed <u>G</u>ust and a 2-digit maximum speed (i.e., 22020G30KT).

    2)   Calm wind is coded "00000KT."

e. **Visibility.** "3SM" means 3 <u>S</u>tatute <u>M</u>iles. It is coded in whole numbers and fractions.

    1)   If visibility is forecast greater than 6 SM it will be coded P6SM (<u>P</u>lus 6 SM).

    2)   Some countries will code a 4-digit group in meters and as required, lowest value with direction. A conversion (meters – SM) is shown in Table 4-3 below.

Meters	Miles	Meters	Miles	Meters	Miles
0000	0	1200	3/4	3000	17/8
0100	1/16	1400	7/8	3200	2
0200	1/8	1600	1	3600	21/4
0300	3/16	1800	11/8	4000	21/2
0400	1/4	2000	11/4	4800	3
0500	5/16	2200	13/8	6000	4
0600	3/8	2400	11/2	8000	5
0800	1/2	2600	15/8	9000	6
1000	5/8	2800	13/4	9999	>6

Table 4-3. Visibility Conversion -- Meters to Miles (SM)

f. Significant present, forecast, and recent weather. Grouped into categories and used in the order listed below.

QUALIFIER

Intensity or Proximity
  – Light   "no sign" Moderate   + Heavy
  VC Vicinity: but not at aerodrome; in U.S., 5-10SM from center of runway complex
     (elsewhere within 8000m)

Descriptor

MI Shallow	BC Patches	DR Drifting	TS Thunderstorm
BL Blowing	SH Showers	FZ Supercooled/freezing	

WEATHER PHENOMENA

Precipitation

DZ Drizzle	RA Rain	SN Snow	SG Snow grains
IC Diamond dust	PE Ice pellets	GR Hail	GS Small hail/snow pellets

UP Unknown precipitation in automated observations

Obscuration

BR Mist	FG Fog	FU Smoke	VA Volcanic ash
SA Sand	HZ Haze	PY Spray	DU Widespread dust

Other

SQ Squall	SS Sandstorm	DS Duststorm	PO Well developed
FC Funnel cloud/tornado/waterspout			dust/sand whirls

---

    1)   If no significant weather is forecast, the code NSW (<u>N</u>o <u>S</u>ignificant <u>W</u>eather) is used.

    2)   "–SHRA" means light showers, rain.

g.   Cloud amount, height, and type.

　　1)   Cloud amount: SKC (SKy Clear), SCT, BKN, and OVC or VV (Vertical Visibility) for obscured sky.

　　　　a)   More than one layer may be reported.

　　2)   Cloud height: 3-digit height in hundreds of feet or vertical visibility for obscured sky.

　　　　a)   An unknown vertical visibility is coded "VV///."

　　3)   Cloud type: either TCU (Towering CUmulus) or CB for CumulonimBus.

　　4)   "BKN020" means broken cloud layer at 2,000 ft.

　　5)   Although not used in the U.S., the code CAVOK (Ceiling And Visibility OK) replaces visibility, weather, and clouds if:

　　　　a)   Visibility is 10 kilometers or more;

　　　　b)   No cloud below 1,500 meters (5,000 ft.) or below the highest minimum sector altitude, whichever is greater and no cumulonimbus; and

　　　　c)   No precipitation, thunderstorm, duststorm, sandstorm, shallow fog, or low drifting dust, sand, or snow.

h.   Significant change is coded FM (FroM) and a 2-digit hour.

　　1)   "FM20 30015G25KT 3SM SHRA OVC015" means from 2000Z the conditions will be wind 300° at 15 kt. gusting to 25 kt., 3 SM visibility, rain showers, and an overcast cloud layer at 1,500 ft.

i.   Probable condition is coded PROB (PROBability) and a 2-digit percent during a 2-digit beginning and 2-digit ending time period.

　　1)   "PROB40 2022 1/2SM TSRA OVC008CB" means a 40% probability from 2000Z to 2200Z of ½ SM visibility, thunderstorm and rain, overcast cloud layer at 800 ft. with cumulonimbus.

j.   Significant change in conditions as indicated by the code "FM."

　　1)   Note more than one cloud layer forecast. A broken layer at 2,000 ft. and an overcast layer at 4,000 ft.

k.   Temporary changes. Changes in forecast conditions that are expected to last for less than a 1-hr. duration and less than half of 2-digit beginning and 2-digit ending time period.

　　1)   "TEMPO 0407 00000KT 1SM −RA FG" means temporary (TEMPOrary) from 0400Z to 0700Z wind calm, 1 SM visibility, light rain and fog.

l.   Significant change in conditions occurs at 1000Z (FM 10).

m.   Expected change. A change expected during 2-digit beginning and 2-digit ending time period.

　　1)   "BECMG 1315 20010KT P6SM NSW SKC" means conditions becoming (BECoMinG) from 1300Z until 1500Z, wind 200° at 10 kt., visibility greater than 6 SM, no significant weather, sky clear.

4.   TAFs exclude temperature, turbulence, and icing forecasts.

END OF CHAPTER

CHAPTER SEVEN
AREA FORECAST (FA)

A. **Definition** -- An Area Forecast (FA) is a forecast of general weather conditions over an area the size of several states. It is used to determine forecast en route weather and to interpolate conditions at airports which do not have FTs issued. Figures 1-4 and 1-4A on pages 268 and 269 map the FA areas.

   1. FAs are issued three times a day by the National Aviation Weather Advisory Unit (NAWAU) in Kansas City, MO, for each of the six areas in the contiguous 48 states. See Part III, Chapter 26, Tables and Conversion Graphs, beginning on page 373 for issuance times.

      a. In Alaska, FAs are issued by the WSFOs in Anchorage, Fairbanks, and Juneau for their respective area.

      b. The WSFO in Honolulu issues FAs for Hawaii.

      c. A specialized FA for the Gulf of Mexico is issued by the National Hurricane Center in Miami, FL.

         1) This product combines both aviation and marine information and is intended to support offshore helicopter operations.

         2) The Gulf of Mexico FA focuses on an area which includes the coastal plains and coastal waters from Apalachicola, FL to Brownsville, TX, and the Gulf of Mexico, west of 85°W longitude and north of 27°N latitude.

   2. The FA is comprised of two sections: HAZARDS/FLIGHT PRECAUTIONS (H), SYNOPSIS AND VFR CLOUDS/WEATHER (C). Each section has a unique communications header which allows replacement of individual sections, due to amendments or corrections, instead of replacing the entire FA.

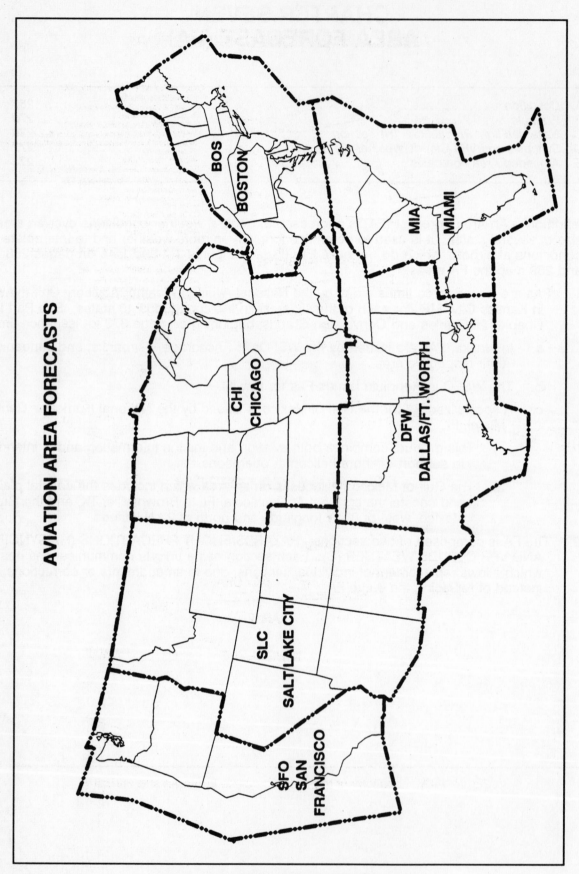

Figure 1-4. Locations of the Area Forecasts

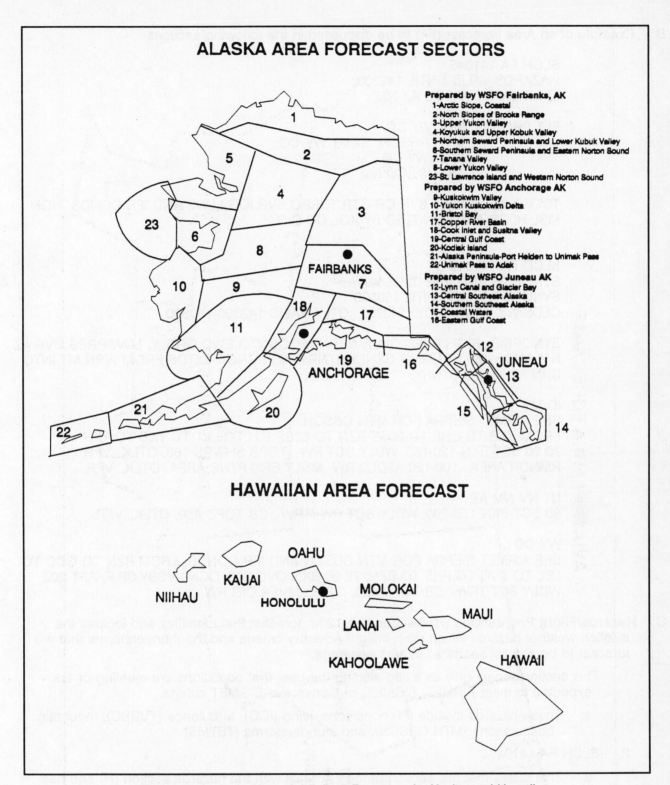

Figure 1-4A. Locations of the Area Forecasts in Alaska and Hawaii

B. **Example** of an Area Forecast (FA) to be discussed in the following sections:

> SLCH FA 141045
> HAZARDS VALID UNTIL 142300
> ID MT NV UT WY CO AZ NM
>
> FLT PRCTNS...IFR...WY CO
>   ...MTN OBSCN...ID MT WY CO
>   ...ICG...WY CO
>   ...TSTMS...CO NM
>
> TSTMS IMPLY PLBL SVR OR GTR TURBC SVR ICG LLWS AND IFR CONDS. NON
> MSL HGTS ARE DENOTED BY AGL OR CIG.
> ...
>
> SLCC FA 141045
> SYNOPSIS AND VFR CLDS/WX
> SYNOPSIS VALID UNTIL 150500
> CLDS/WX VALID UNTIL 142300...OTLK VALID 142300-150500
> .
> SYNOPSIS...HIGH PRES OVER NERN MT CONTG EWD GRDLY. LOW PRES OVR AZ
> NM AND WRN TX RMNG GENLY STNRY. ALF...TROF EXTDS FROM WRN MT INTO
> SRN AZ RMNG STNRY.
> .
> ID MT
> SEE AIRMET SIERRA FOR MTN OBSCN.
> FROM YXH TO SHR TO 30SE BZN TO 60SE PIH TO LKT TO YXC TO YXH.
> 70-90 SCT-BKN 120-150. WDLY SCT RW-. TOPS SHWRS 180. OTLK...VFR
> RMNDR AREA...100-120. ISOLD RW- MNLY ERN PTNS AREA. OTLK...VFR.
> .
> UT NV NM AZ
> 80 SCT-BKN 150-200. WDLY SCT RW-/TRW-. CB TOPS 450. OTLK...VFR.
> .
> WY CO
> SEE AIRMET SIERRA FOR MTN OBSCN AND IFR CONDS. FROM BZN TO GCC TO
> LBL TO DVC TO RKS TO BZN. 70-90 BKN-OVC 200. OCNL VSBY 3R-F. AFT 20Z
> WDLY SCT TRW-. CB TOPS 450. OTLK...MVFR CIG RW.

C. **Hazards/Flight Precautions (H) Section** is a 12-hr. forecast that identifies and locates the aviation weather hazards which meet Inflight Advisory criteria and the thunderstorms that are forecast to be at least scattered in area coverage.

   1. This section serves only as a flag alerting the user that conditions are meeting or are expected to meet AIRMET, SIGMET, or Convective SIGMET criteria.

      a. These hazards include IFR conditions, icing (ICG), turbulence (TURBC), mountain obscurations (MTN OBSCN), and thunderstorms (TSTMS).

   2. SLCH FA 141045

      a. This states that this section of the FA deals with the hazards section (H) and has been issued on the 14th day of the month at 1045Z for the Salt Lake City (SLC) forecast area.

   3. HAZARDS VALID UNTIL 042300

      a. This states that the hazards listed may be valid for the 12-hr. forecast time period 1100Z to 2300Z or for only a portion of the time period.

4.   ID MT NV UT WY CO AZ NM

   a.   This identifies the states and geographical areas that make up the entire SLC forecast area, not only the hazard areas.

5.   FLT PRCTNS...IFR...WY CO
                  ...MTN OBSCN...ID MT WY CO
                  ...ICG...WY CO
                  ...TSTMS...CO NM

   a.   This paragraph states that IFR conditions, mountain obscurations, icing, and thunderstorms are forecast within the 12-hr. period for the listed states within the designated FA boundary.

   b.   If **no** hazards are expected, "NONE XPCD" will be written. Note that specific phenomena which are not expected to occur or are not expected to meet flight precaution criteria will not be listed.

6.   TSTM IMPLY PSBL SVR OR GTR TURBC SVR ICG LLWS AND IFR CONDS

   a.   The above statement is found in all FAs as a reminder of the hazards existing in all thunderstorms.

      1)   Thunderstorms imply possible severe or greater turbulence, severe icing, low-level wind shear, and IFR conditions.

   b.   Thus, these thunderstorm-associated hazards are not spelled out within the body of the FA.

7.   NON MSL HGTS NOTED BY AGL OR CIG

   a.   This statement is contained in every FA. Its purpose is to alert the user that heights, for the most part, are *above mean sea level*. All heights are in hundreds of feet.

      1)   EXAMPLE: 30 BKN 100 HIR TRRN OBSCD means bases of the broken clouds are 3,000 ft. MSL with tops 10,000 ft. MSL. Terrain above 3,000 ft. MSL will be obscured.

   b.   The tops of the clouds, turbulence, icing, and freezing level heights are **always** MSL.

   c.   Heights *above ground level* will be denoted in either of two ways:

      1)   Ceilings by definition are above ground. Therefore, the contraction "CIG" indicates above ground.

         a)   EXAMPLE: "CIGS GENLY BLO 10" means that ceilings are expected to be generally below 1,000 ft. AGL.

      2)   "AGL 20 SCT" means scattered clouds with bases 2,000 ft. above ground level.

   d.   Thus, if the contraction "AGL" or "CIG" is **not** denoted, height is automatically above MSL.

D.   Synopsis and VFR Clouds/Weather (C) Section

1.   SLCC FA 141045
SYNOPSIS AND VFR CLDS/WX

a.   This states that this section of the FA deals with the synopsis and VFR clouds/ weather section (C) and has been issued on the 14th day of the month at 1045Z for the Salt Lake City (SLC) forecast area.

2.   SYNOPSIS VALID UNTIL 150500
CLDS/WX VALID UNTIL 142300...OTLK VALID 142300-150500

a.   This states that the synopsis given is valid for an 18-hr. forecast period 1100Z to 0500Z the following day.

1)   The clouds/weather forecast is valid for a 12-hr. period 1100Z to 2300Z, followed by a 6-hr. categorical outlook period 2300Z to 0500Z.

3.   The synopsis gives a brief summary of the location and movements of fronts, pressure systems, and circulation patterns. References to low ceilings and/or visibilities, strong winds or any other phenomena that the forecaster considers useful may also be included.

a.   SYNOPSIS...HIGH PRES OVER NERN MT CONTG EWD GRDLY. LOW PRES OVR AZ NM AND WRN TX RMNG GENLY STNRY. ALF...TROF EXTDS FROM WRN MT INTO SRN AZ RMNG STNRY.

1)   This paragraph states that high pressure over northeastern Montana will continue to move gradually eastward. Low pressure over Arizona, New Mexico, and western Texas will remain stationary. Aloft, a low pressure trough extending from western Montana into southern Arizona will remain stationary.

4.   The VFR CLDS/WX section contains a 12-hr. specific forecast, followed by a 6-hr. (18-hr. in Alaska) categorical outlook giving a total forecast period of 18 hr. (30 hr. in Alaska).

a.   The VFR CLDS/WX section is usually several paragraphs long. The breakdown may be by states or by well-known geographical areas.

1)   Two Flight Precautions from the hazards section, IFR and MTN OBSCN, will be indicated by a lead sentence that refers you to AIRMET SIERRA (see Part III, Chapter 9, In-Flight Aviation Weather Advisories, beginning on page 279).

b.   The specific forecast section gives a general description of clouds and weather that is significant to flight operations.

c.   Surface visibility and obstructions to vision are included when the forecast is 6 SM or less. Precipitation, thunderstorms, and sustained winds of 20 kt. or greater are always included when forecast.

1)   Expected coverage of showers and thunderstorms is indicated as follows:

Adjective	Coverage
Isolated	Single cells (no percentage)
Widely scattered	Less than 25% of area affected
Scattered	25 to 54% of area affected
Numerous	55% or more of area affected

Table 4-8. Area Coverage of Showers and Thunderstorms

2)   Variability terms used are as follows:

Term	Description
OCNL	Greater than 50% probability of the phenomenon occurring but for less than ½ of the forecast period
CHC	30 to 50% probability (precipitation only)
SLGT CHC	10 to 20% probability (precipitation only)

Table 4-9. Variable Terms

d.   WY CO
SEE AIRMET SIERRA FOR MTN OBSCNS AND IFR CONDS. FROM BZN TO GCC TO LBL TO DVC TO RKS TO BZN. 70-90 BKN-OVC 200. OCNL VSBY 3R-F. AFT 20Z WDLY SCT TRW-. CB TOPS 450.

1)   This part of the VFR CLDS/WX section is the forecast for the states of Wyoming and Colorado. The second line alerts the user to the AIRMET SIERRA for Mountain Obscuration and IFR conditions. The exact area coverage of the forecast is from Bozeman, MT to Gillette, WY to Liberal, KS to Dove Creek, WY to Rock Springs, WY and back to Bozeman, MT.

2)   The base of the broken to overcast layer of clouds is 7,000 to 9,000 ft. MSL with the tops of the clouds at 20,000 ft. MSL. The visibility is expected to be greater than 6 SM and winds less than 20 kt., both implied by omission. The visibility is forecast to be occasionally 3 SM in light rain and fog. After 2000Z, widely scattered thunderstorms and light rain showers are expected with tops to 45,000 ft. MSL.

5.   A categorical outlook, identified by "OTLK," is included for each area breakdown. A categorical outlook of IFR and MVFR can be due to ceilings only (CIG), restriction to visibility only (TRW F, etc.), or a combination of both (CIG TRW F, etc.).

a.   EXAMPLE: OTLK... VFR BCMG MVFR CIG F AFT 09Z means the weather is expected to be VFR, becoming MVFR due to low ceiling and visibilities restricted by fog after 0900Z. "WND" is included in the outlook if winds, sustained or gusty, are expected to be 20 kt. or greater.

E.   Amended Area Forecasts

1.   Amendments to the FA are issued as needed. Only that section of the FA being revised is transmitted as an amendment.

2.   Area forecasts are also amended and updated by in-flight advisories (AIRMETs, SIGMETs, and Convective SIGMETs).

3.   An amended FA is identified by "AMD," a corrected FA is identified by "COR" and a delayed FA is identified by "RTD."

END OF CHAPTER

CHAPTER EIGHT
TWEB ROUTE FORECASTS AND SYNOPSIS

A. **Introduction** -- The Transcribed Weather Broadcast (TWEB) Route Forecast is similar to the Area Forecast (FA) except information is contained in a route format.

1. Forecast sky cover (height and amount of cloud bases), cloud tops, visibility (including vertical visibility), weather, and obstructions to vision are described for a corridor 25 mi. on either side of the route.

2. Cloud bases and tops are always MSL unless noted. Ceilings are always above ground level (AGL).

B. **Synopsis**

1. The Synopsis is a brief statement of frontal and pressure systems affecting the route during the forecast valid period.

C. **Purpose** -- The TWEB Route Forecast is prepared by WSFOs for more than 300 selected routes over the contiguous U.S. (see Figure 1-6 on page 276).

1. These forecasts go into the TWEBs and Pilot's Automatic Telephone Weather Answering Service (PATWAS) transcriptions described in Part III, Chapter 1, The Aviation Weather Service Program, beginning on page 212.

2. Individual route forecasts and synopses are also available by request through any FSS or WSO.

D. **Schedule** -- The TWEB Route Forecasts and Synopses are issued by the WSFOs three times per day according to time zone (see Part III, Chapter 26, Tables and Conversion Graphs, on page 373 for issuance times).

1. The TWEB forecast is valid for a 15-hr. period.

2. This schedule provides 24-hr. coverage with most frequent updating during the hours of greatest general aviation activity.

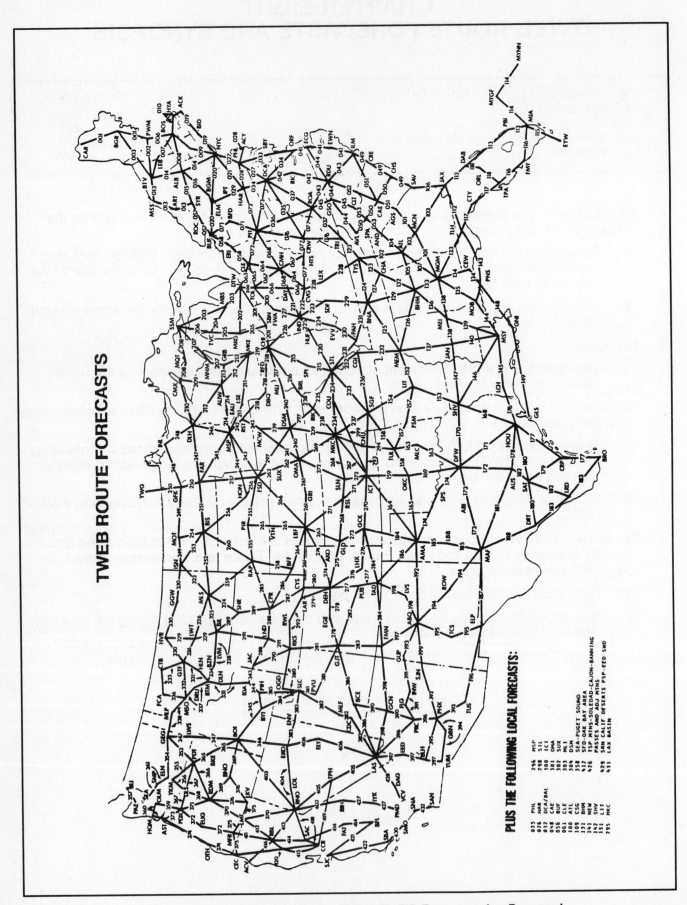

Figure 1-6. Numbered Routes for Which TWEB Forecasts Are Prepared

E.   **Examples**

1.   An example of a TWEB Synopsis:

BIS SYNS 250924. LO PRES TROF MVG ACRS ND TDA AND TNGT. HI PRES MVG
SEWD FM CANADA INTO NWRN ND BY TNGT AND OVR MST OF ND BY WED
MRNG.

BIS	--	Bismark, ND. WSFO issuing the synopsis Route Forecasts
SYNS	--	Synopsis for the area covered by the Route Forecasts
25	--	25th day of the month
0924	--	Valid from 09Z on the 25th to 00Z on the 26th (15 hr.)

(Rest of message)

   Low pressure trough moving across North Dakota today and tonight. High
pressure moving southeastward from Canada into northwestern North Dakota by
tonight and over most of North Dakota by Wednesday morning.

2.   An example of a TWEB Route Forecast:

249 TWEB 250924 GFK-MOT-ISN. ALL HGTS MSL XCP CIGS. GFK VCNTY CIGS
AOA 5 THSD TILL 12Z OTRW OVER RTE CIGS 1 TO 3 THSD VSBY 3 TO 5 MI IN
LGT SNW WITH CONDS BRFLY LWR IN HVYR SNW SHWRS.

249	--	Route number (see Figure 1-6)
TWEB	--	TWEB Route Forecast
25	--	25th day of the month
0924	--	Valid from 09Z on the 25th to 00Z on the 26th (15 hr.)
GFK-MOT-ISN Route	--	Grand Forks to Minot to Williston, ND

(Rest of message)

   All heights above mean sea level except ceilings - Grand Forks vicinity...Ceilings at
or above 5,000 ft. until 1200Z. Otherwise over route...ceilings 1,000 to 3,000 ft....
visibility 3 to 5 SM in light snow with conditions briefly lower in heavier snow
showers.

a.   When visibility is **not** stated, it is implied to be greater than 6 SM.

b.   Because of their varied accessibility and route format, these forecasts are important
and useful weather information available to the pilot for flight operations and
planning. You should become familiar with them and use them regularly.

END OF CHAPTER

CHAPTER NINE
IN-FLIGHT AVIATION WEATHER ADVISORIES
(WST, WS, WA)

> Please take a few minutes to study each of the concepts listed above and anticipate/imagine what they are and how they relate to the other listed concepts.

A. **Definition** -- In-flight Aviation Weather Advisories are forecasts that advise en route aircraft of the development of potentially hazardous weather.

   1. All in-flight advisories in the contiguous 48 states are issued by the National Aviation Weather Advisory Unit (NAWAU) in Kansas City, MO.

      a. In Alaska, the three WSFOs (Anchorage, Fairbanks, and Juneau) issue in-flight advisories for their respective areas.

      b. The WSFO in Honolulu issues advisories for Hawaii.

   2. All heights are referenced to MSL, except in the case of ceilings (CIG), which indicate above ground level.

      a. The advisories are of three types -- Convective SIGMET (WST), SIGMET (WS), and AIRMET (WA).

      b. All in-flight advisories use the same location identifiers (either VORs or well-known geographic areas) to describe the hazardous weather areas.

B. **Convective SIGMET (WST)**

   1. Convective SIGMETs are issued in the contiguous 48 states (i.e., none for Alaska and Hawaii) for any of the following:

      a. Severe thunderstorm due to

         1) Surface winds greater than or equal to 50 kt.
         2) Hail at the surface greater than or equal to ¾ in. in diameter
         3) Tornadoes

      b. Embedded thunderstorms

      c. A line of thunderstorms

      d. Thunderstorms greater than or equal to intensity level 4 affecting 40% or more of an area of at least 3,000 square mi.

   2. Any Convective SIGMET implies severe or greater turbulence, severe icing, and low level wind shear.

      a. A Convective SIGMET may be issued for any convective situation which the forecaster feels is hazardous to all categories of aircraft.

3.   Convective SIGMET bulletins are issued for the Eastern (E), Central (C), and Western (W) United States.

    a.   The areas are separated at 87° and 107°W longitude with sufficient overlap to cover most cases when the phenomenon crosses the boundaries.

        1)   Thus, a bulletin will usually be issued only for the area where the bulk of observations and forecast conditions are located.

    b.   Bulletins are issued hourly at H+55.

        1)   Special bulletins are issued at any time as required and updated at H+55.

    c.   If no criteria meeting a Convective SIGMET are observed or forecast, the message "CONVECTIVE SIGMET...NONE" will be issued for each area at H+55.

    d.   Individual Convective SIGMETs for each area are numbered sequentially (01-99) each day, beginning at 00Z.

        1)   A continuing Convective SIGMET phenomenon will be reissued every hour at H+55 with a new number.

        2)   The text of the bulletin consists of either an observation and a forecast or just a forecast.

        3)   The forecast is valid for up to 2 hr.

4.   The following is an example of a complete WST bulletin, including the outlook, for the Central United States. These are the 20th and 21st Convective SIGMETs of the day in this area. For the Western United States, they would be numbered 20W and 21W. For the Eastern United States, they would be numbered 20E and 21E.

```
MKCC WST 221855
CONVECTIVE SIGMET 20C
VALID UNTIL 2055Z
ND SD
FROM 90W MOT-GFK-ABR-90 MOT
INTSFYG AREA SVR TSTMS MOVG FROM 2445.  TOP ABV 450.  WIND GUSTS TO
60 KTS RPRTD.  TORNADOES...HAIL TO 2 IN...WIND GUSTS TO 65 KTS PSBL ND
PTN

CONVECTIVE SIGMET 21C
VALID UNTIL 2055Z
50SE CDS
ISOLD SVR TSTM D30 MOVG FROM 2420.  TOPS ABV 450.  HAIL TO 2 IN...WIND
GUSTS TO 65 KTS PSBL.

OUTLOOK VALID 222055-230055
AREA 1...FROM INL-MSP-ABR-MOT-INL
SVR TSTMS CONT TO DVLP IN AREA OVR ND.  AREA IS XPCD TO RMN SVR AND
SPRD INTO MN AS STG PVA MOVS OVR VRY UNSTBL AMS CHARACTERIZED BY
-12 LIFTED INDEX.
AREA 2...FROM CDS-DFW-LRD-ELP-CDS
ISOLD STG TSTMS WILL DVLP OVR SWRN AND WRN TX THRUT FCST PD AS UPR
LVL TROF MOVS NEWD OVER VERY UNSTBL AMS.  LIFTED INDEX RMS IN THE -8
TO -10 RANGE.  DRY LINE WILL BE THE FOCUS OF TSTM DVLPMT.
```

    a.   The first message indicates an area of severe thunderstorms in North and South Dakota. These storms have produced wind gusts to 60 kt. with the possibility of tornadoes, hail up to 2 in., and 65 kt. winds over North Dakota.

b.   The second message is about a single severe thunderstorm southeast of Childress, TX.

c.   The outlook section focuses on North and South Dakota in Area 1 and Texas in Area 2.

C. **Area Covered**

1.   SIGMETs and AIRMETs are issued for the six areas corresponding to the FA areas (Figure 1-4 on page 268). The maximum forecast period is 4 hr. for SIGMETs and 6 hr. for AIRMETs.

a.   Both advisories are considered "widespread" because they must be either affecting or be forecast to affect an area of at least 3,000 square mi. at any one time.

b.   At times, the total area to be affected during the forecast period (as outlined by VORs) is very large.

1)   It could be that only a small portion of this total area would be affected at any one time.

2)   An example would be a 3,000 square mi. phenomenon forecast to move across an area totaling 25,000 square mi. during a forecast period.

D. **SIGMET (WS)**

1.   A SIGMET advises of non-convective weather that is potentially hazardous to all aircraft.

a.   In the conterminous U.S., items covered are

1)   Severe icing not associated with thunderstorms.

2)   Severe or extreme turbulence not associated with thunderstorms.

3)   Duststorms, sandstorms, or volcanic ash lowering surface or in-flight visibilities to below 3 SM.

4)   Volcanic eruption.

5)   Tropical storms or hurricanes.

b.   In Alaska and Hawaii, there are no Convective SIGMETs. These criteria are added:

1)   Tornadoes
2)   Lines of thunderstorms
3)   Embedded thunderstorms
4)   Hail equal to or greater than ¾ in. in diameter

2.   SIGMETs are identified by alphabetic designators which include NOVEMBER through YANKEE but exclude SIERRA and TANGO.

a.   The first issuance of a SIGMET will be labeled UWS (Urgent Weather SIGMET), and subsequent issuances at the forecaster's discretion.

b.   Issuances for the same phenomenon will be sequentially numbered, using the original designator until the phenomenon ends.

1)   For example, the first issuance in the CHI area for a phenomenon moving from the SLC area will be SIGMET PAPA 3, if the previous two issuances, PAPA 1 and PAPA 2 had been in the SLC area.

2)   Note that no two different phenomena across the country can have the same alphabetic designator at the same time.

3.   EXAMPLE:

> DFWP UWS 051710
> SIGMET PAPA 1 VALID UNTIL 052110
> AR LA MS
> FROM STL TO 30N MEI TO BTR TO MLU TO STL
> OCNL SVR ICING ABV FRZLVL EXPCD.
> FRZLVL 80 E TO 120 W.
> CONDS CONTG BYD 2100Z.

a.   This is a SIGMET bulletin issued for the DFW area at 1710Z on the 5th day of the month and is valid until 2110Z (maximum valid period for a SIGMET is 4 hr.).

1)   The designator PAPA identifies the phenomenon, in this case severe icing. This is the first issuance of the SIGMET as indicated by "UWS" and "PAPA 1."

2)   The affected states *within* the DFW area are Arkansas, Louisiana, and Mississippi. Freezing level data and a notation that conditions are expected to continue beyond 4 hr. are included.

4.   Several NWS offices have been designated by the International Civil Aviation Organization (ICAO) as Meteorological Watch Offices (MWOs). These offices are responsible for issuing International SIGMETs.

a.   MWOs Anchorage, AK; Fairbanks, AK; Juneau, AK: Honolulu, HI; National Hurricane Center; and the National Meteorological Center have been assigned the use of alphabetic designators ALFA through MIKE within each of their areas.

b.   MWO Guam is assigned the use of NOVEMBER, OSCAR, and PAPA as its alphanumeric designators.

E.   AIRMET (WA)

1.   AIRMETs are advisories of significant weather phenomena but describe conditions at intensities lower than those which trigger SIGMETs. AIRMETs are intended for dissemination to all pilots in the preflight and en route phase of flight to enhance safety.

a.   AIRMETs are issued on a scheduled basis every 6 hr., with the valid period beginning at 0200 UTC each day.

1)   Unscheduled amendments and corrections will be issued as necessary.

b.   AIRMETs are valid for 6 hr. and will contain details of conditions when one or more of the following occur or are forecast to occur:

1)   Moderate icing

2)   Moderate turbulence

3)   Sustained surface winds of 30 kt. or more

4)   Ceiling less than 1,000 ft. and/or visibility less than 3 SM affecting over 50% of the area at one time

5)   Extensive mountain obscurement

2.   AIRMETs have fixed alphanumeric designators of ZULU for icing and freezing level;
       TANGO for turbulence, strong surface winds, and low-level wind shear; and SIERRA for
       Instrument Flight Rules (IFR) and mountain obscuration.

    a.   In addition AIRMET SIERRA will be referenced in the SYNOPSIS AND VFR CLDS/WX
            of this Area Forecast section.

    b.   EXAMPLE:

                CHIC WA 300745
                AIRMET SIERRA FOR IFR VALID UNTIL 301400

                AIRMET IFR...IA MO
                FROM FOD TO CID TO COU TO MKC TO FOD
                CIG BLO 10 AND VSBY BLO 3S–. CONDS CONTG BYD 1400Z AND SPRDG
                OVER ERN IA ERN MO AND IL.

    1)   This AIRMET is part of the Chicago (CHI) Area Forecast and was issued at
            0745Z. The valid time is from 0800Z to 1400Z (6 hr.).

    2)   The AIRMET was issued for IFR conditions in Iowa (IA) and Missouri (MO), and
            reads as follows:

            From Fort Dodge, IA (FOD) to Cedar Rapids, IA (CID) to Columbia, MO
            (COU) to Kansas City, MO (MKC) to FOD. Ceilings below 1,000 ft. and
            visibility below 3 SM, light snow (S–). Conditions continuing beyond 1400Z
            and spreading over eastern IA, eastern MO, and Illinois (IL).

END OF CHAPTER

CHAPTER TEN
WINDS AND TEMPERATURES ALOFT FORECAST (FD)

A. Type and Time of Forecast

1. Winds and temperatures aloft are forecast for specific locations in the contiguous U.S., as shown in Figure 1-3 on page 286. FD forecasts are also prepared for a network of locations in Alaska and Hawaii, as shown in Figure 1-3A on page 287.

   a. Forecasts are made twice daily based on 00Z and 12Z data for use during specific time intervals.

2. Note that charts are available for winds and temperatures aloft for both (1) forecasts and (2) observations. See Section III, Chapter 21, Winds and Temperatures Aloft Charts, beginning on page 331.

B. Example

1. Below is a sample FD message containing a heading and six FD locations. The heading always includes the time during which the FD may be used (1700-2100Z in the example) and a notation "TEMPS NEG ABV 24000," which means that since temperatures above 24,000 ft. are always negative, the minus sign is omitted.

```
FD KWBC 151640

BASED ON 151200Z DATA

VALID 151800Z FOR USE 1700-2100Z TEMPS NEG ABV 24000
```

FT	3000	6000	9000	12000	18000	24000	30000	34000	39000
ALA			2420	2635−08	2535−18	2444−30	245945	246755	246862
AWA		2714	2725+00	2625−04	2531−15	2542−27	265842	256352	256762
DEN			2321−04	2532−08	2434−19	2441−31	235347	236056	236262
HLC		1707−01	2113−03	2219−07	2330−17	2435−30	244145	244854	245561
MKC	0507	2006+03	2215−01	2322−06	2338−17	2348−29	236143	237252	238160
STL	2113	2325+07	2332+02	2339−04	2356−16	2373−27	239440	730649	731960

a. The line labelled "FT" (Forecast Levels) shows 9 to 11 standard FD levels.

   1) The 45,000 ft. and 53,000 ft. levels are not transmitted on teletypewriter circuits but are available in the communications system.

      a) The pilot may request these levels from the FSS briefer or NWS meteorologist.

   2) Levels through 12,000 ft. are in true altitude, and levels 18,000 ft. and above are in pressure altitude.

   3) The FD locations are transmitted in alphabetical order.

b. Note that some lower-level wind groups are omitted.

   1) No winds are forecast within 1,500 ft. of station elevation.

   2) Also, no temperatures are forecast for the 3,000 ft. level or for any level within 2,500 ft. of station elevation.

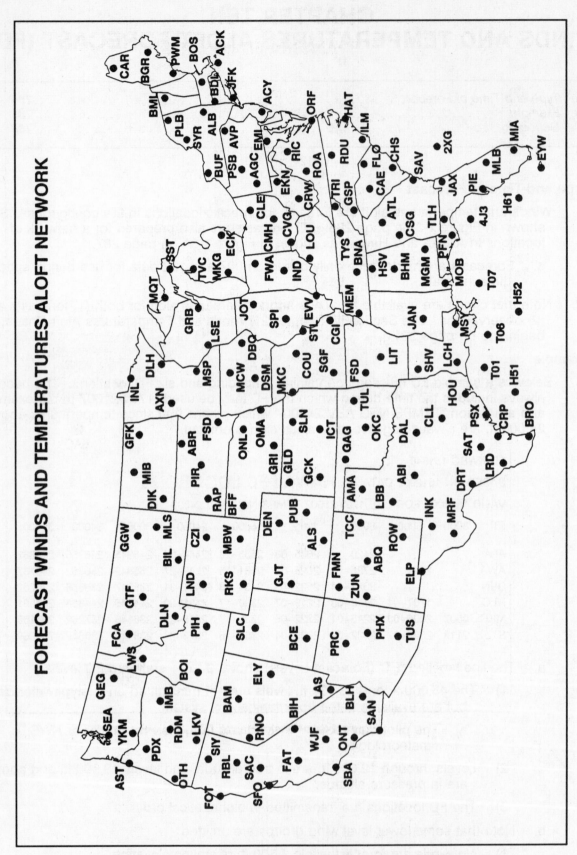

Figure 1-3. Forecast Wind and Temperature Aloft Network

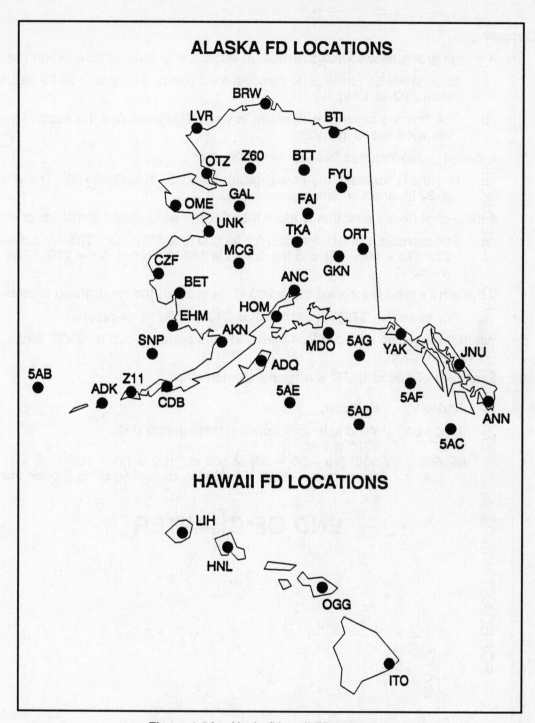

Figure 1-3A. Alaska/Hawaii FD Locations

C. Decoding

1. A 4-digit group shows wind direction, in reference to true north, and wind speed.

    a. Look at the St. Louis (STL) forecast for 3,000 ft. The group 2113 means the wind is from 210° at 13 kt.

    b. The first two digits give direction in tens of degrees and the second two digits are the wind speed in knots.

2. A 6-digit group includes forecast temperatures.

    a. In the STL forecast, the coded group for 9,000 ft. is 2332+02. The wind is from 230° at 32 kt. and the temperature is plus 2°C.

3. If the 2-digit coded direction is more than 36, the wind speed is 100 kt. or more.

    a. For example, the STL forecast for 39,000 ft. is "731960." The wind direction is from 230° (73 – 50 = 23) and the speed is 119 kt. (100 + 19 = 119). The temperature is –60°C.

4. If the wind speed is forecast to be 200 kt. or greater, the wind group is coded at 199 kt.

    a. For example, "7799" is decoded as 270° at 199 kt. or greater.

5. When the forecast speed is less than 5 kt., the coded group is "9900" and is read "LIGHT AND VARIABLE."

6. Examples of decoding FD winds and temperatures:

Coded	Decoded
9900+00	Wind light and variable, temperature 0°C.
2707	270° at 7 kt.
850552	350° (85 – 50 = 35) at 105 kt. (100 + 05 = 105), temperature –52°C if 24,000 ft. or higher; 52°C if lower than 24,000 ft.

END OF CHAPTER

CHAPTER ELEVEN
SPECIAL FLIGHT FORECAST

A. **Purpose** -- When planning a special category flight and scheduled forecasts are insufficient to meet the need, the pilot may request a special flight forecast through any FSS or WSO.

1. Special category flights are hospital or rescue flights; experimental, photographic, or test flights; records attempts; and mass flights (such as air tours, air races, and fly-aways from special events).

B. **Requirements** -- Pilots should make requests far enough in advance to allow ample time for preparing and transmitting the forecast.

1. Advance notice of 6 hr. is desirable.
2. In making a request, the pilot should give the

   a. Aircraft mission.
   b. Number and type of aircraft.
   c. Point of departure.
   d. Route of flight (including intermediate stops, destination, alternates).
   e. Estimated time of departure.
   f. Time en route.
   g. Flight restrictions (such as VFR, below certain altitudes, etc.).
   h. Time forecast is needed.

C. **Example**

1. The forecast is written in plain language contractions as in the following example:

   SPL FLT FCST ABQ-PHOTO MISSION-ABQ 121500Z. THIN CI CLDS AVGG LESS THAN TWO TENTHS CVR. VSBY MORE THAN 30. WNDS AND TEMPS ALF AT FLT ALTITUDE 2320+03. ABQ WSFO 052300Z.

END OF CHAPTER

CHAPTER TWELVE
CENTER WEATHER SERVICE UNIT (CWSU) PRODUCTS

A. **Origin** -- Center Weather Service Unit products are issued by the CWSU meteorologist located in the ARTCCs.

B. **Meteorological Impact Statement (MIS)** -- an unscheduled traffic/flight operations planning forecast of conditions expected to begin generally 4 to 12 hr. after issuance. This enables the impact of expected weather conditions to be included in traffic control related decisions of the near future.

1. An MIS will be issued when the following three conditions are met:

   a. If any one of the following conditions is forecast

      1) Convective SIGMET criteria
      2) Moderate or greater icing and/or turbulence
      3) Heavy or freezing precipitation
      4) Low IFR conditions
      5) Surface winds/gusts 30 kt. or greater
      6) Low-level wind shear within 2,000 ft. of the surface
      7) Volcanic ash, dust, or sandstorm

   b. If the impact occurs on air traffic flow within the ARTCC area of responsibility

   c. If the forecast lead time (the time between issuance and onset of a phenomenon), in the forecaster's judgment, is sufficient to make issuance of a Center Weather Advisory (CWA) unnecessary

2. An example of an MIS:

   ZMP MIS 01 041200-050100
   ...FOR ATC PLANNING PURPOSE ONLY...

   ISOLD LVL 2-4 TSTMS TOPS 350-450 MOVG FM 2925 OVR NW ND. TSTMS MOVG INTO E ND SD W MN AFT 18Z AND TRMG CLSTRS AFT 22Z TOPS 400-500.

   a. This MIS from the Minneapolis, MN, ARTCC is the first issuance of the day. It was issued at 1200Z on the 4th and is valid until 0100Z on the 5th.

C. **Center Weather Advisory (CWA)** -- an unscheduled in-flight flow control, air traffic, and air crew *advisory* for use in anticipating and avoiding adverse weather conditions in the en route and terminal areas.

   1. The CWA is *not* a flight planning forecast but a *nowcast* for conditions beginning within the next 2 hr.

      a. Maximum valid time of a CWA is 2 hr. from the time of issuance.

      b. If conditions are expected to continue beyond the valid period, a statement will be included in the advisory.

   2. A CWA may be issued for the following three situations:

      a. As a supplement to an *existing* in-flight advisory or area forecast (FA) section for the purpose of improving or updating the definition of the phenomenon in terms of location, movement, extent, or intensity *relevant* to the ARTCC area of responsibility. This is important for the following reason.

         1) A SIGMET for severe turbulence was issued by the NAWAU unit and the outline covered the entire ARTCC area for the total 4-hr. valid time period. However, the advisory may only cover a relatively small portion of the ARTCC area at any one time during the 4-hr. period.

      b. When an in-flight advisory has not yet been issued but conditions meet in-flight advisory criteria based on current pilot reports and the information must be disseminated sooner than the NAWAU unit can issue the in-flight advisory.

         1) In this case of an impending SIGMET, the CWA will be issued as an urgent "UCWA" to allow the fastest possible dissemination.

      c. When in-flight advisory criteria are not met but conditions are or will shortly be adversely affecting the safe flow of air traffic within the ARTCC area of responsibility.

   3. An example of a CWA:

   ZKC3 CWA 03 032140-2340
   ISOLD SVR TSTM OVR COU MOVG SWWD 10 KTS. TOP 610. WND GUSTS TO
   55 KTS. HAIL TO 1 INCH RPRTD AT COU. SVR TSTM CONTG BYND 2340.

      a. This CWA from the Kansas City, MO, ARTCC is the third issuance on this phenomenon. Valid time is from the third day of the month at 2140Z to 2340Z.

END OF CHAPTER

CHAPTER THIRTEEN
HURRICANE ADVISORY (WH)

A. **Purpose** -- When a hurricane threatens a coast line, but is located at least 300 NM offshore, an abbreviated hurricane advisory (WH) is issued to alert aviation interests.

   1. The advisory gives the location of the storm center, its expected movement, and the maximum winds in and near the storm center.

   2. It does not contain details of associated weather, such as specific ceilings, visibilities, weather, and hazards that are found in the area, terminal forecasts, and/or in-flight advisories.

B. **Example** of an abbreviated aviation hurricane advisory:

MIA WH 181010
HURCN BOB AT 1000Z CNTRD 29.4N 74.2W OR 400 NMI E OF JACKSONVILLE FL EXPCTED
TO MOV N ABT 12 KT. MAX WNDS 110 KT OVR SML AREA NEAR CNTR AND HURCN WNDS
WITHIN 55-75 NMI.

END OF CHAPTER

CHAPTER FOURTEEN
CONVECTIVE OUTLOOK (AC)

A. **Purpose** -- A convective outlook (AC) describes the prospects for general thunderstorm activity during the next 24 hr.

   1. Areas in which there is a high, moderate, or slight risk of severe thunderstorms are included, as well as areas where thunderstorms may approach severe limits.

      a. Approaching is defined as winds greater than or equal to 35 kt. but less than 50 kt. and/or hail equal to or greater than ½ in. in diameter.

      b. Refer to the Severe Weather Outlook Chart (see Part III, Chapter 23, Severe Weather Outlook Chart, beginning on page 351) for the definitions of high, moderate, and slight risk.

   2. Forecast reasoning is also included in all ACs.

B. **Schedule** -- Outlooks are transmitted by the National Severe Storm Forecast Center (NSSFC) in Kansas City, MO, at 0700Z and 1500Z.

   1. Forecasts in each AC are valid until 1200Z the next day and are used to prepare and update the Severe Weather Outlook Chart.

   2. Use the outlook primarily for planning flights later in the day.

C. **Severe Thunderstorm Criteria**

   1. Winds equal to or greater than 50 kt. at the surface
   2. Hail equal to or greater than 3/4 in. diameter at the surface
   3. Tornadoes

D.   **Example** of a convective outlook:

MKC AC 280700

CONVECTIVE OUTLOOK...REF AFOS NMCGPH940

VALID 281200 - 291200Z

THERE IS A MDT RISK OF SVR TSTMS OVER PTNS ERN AL..MUCH OF GA..SC..NC.. THE SE HALF OF VA..AND EXTRM SRN MD..TO RGT OF LN FM SAV MGR CEW SEM RMG HSS ROA CHO WAL.

THERE IS A SLGT RISK OF SVR TSTMS TO RGT OF LN FM MOB JAN MEM BNA LOZ IAD ACY..CONT..WAL CHO ROA HSS RMG SEM CEW MGR SAV..CONT VRB FMY.

GEN TSTMS ARE FCST TO RGT OF LN FM BLI YKM 4BW WMC TVL SFO..CONT..LCH SNV HOT PAH JST NEL.

STG SFC LO OVR WRN IN EXPCD TO MOV ENE AND DPN DURG NEXT 24 HRS. STLT INDCS UPR LVL IMPULSE FM WRN TN INTO SW AL ROTATING ARND TROF ABT 45 KTS. STG TSTMS CURRENTLY MOVG ACRS AL AHD OF UPR VORT MAX EXPCD TO CONT MOVG E. AMS OVR SE U.S. RMNS MOIST AND UNSTBL BNTH FVRBL UPR LVL DIFLUENCE. SGFNT SVR TSTMS EXPCD FM GA NE INTO SE VA WHERE LO LVL WRM ADVCTN AND CNVRGNC WL BE CONCENTRATED.

ISOLD TSTMS EXPCD TO DVLP PTNS N PAC CST SPRDG INLAND AFT 00Z AS CD AIR ALF ASSOCD WITH NEXT UPR SYS APCHS WRN U.S.
...KLOTH...

END OF CHAPTER

CHAPTER FIFTEEN
SEVERE WEATHER WATCH BULLETIN (WW)

Please take a few minutes to study each of the concepts listed above and anticipate/imagine what they are and how they relate to the other listed concepts.

A. **Purpose** -- A Severe Weather Watch Bulletin (WW) defines areas of possible severe thunderstorms or tornado activity.

 1. The bulletins are issued by the National Severe Storm Forecast Center at Kansas City, MO.

 2. WWs are unscheduled and are issued as required.

B. **Scope** -- A severe thunderstorm watch describes expected severe thunderstorms, and a tornado watch states that the additional threat of tornadoes exists in the designated watch area.

C. **Alert Severe Weather Watch (AWW)** -- In order to alert the WSFOs, WSOs, CWSUs, FSSs, and other users, a preliminary message called the Alert Severe Weather Watch message (AWW) is sent before the main bulletin.

 1. An example of a preliminary message:

    MKC AWW 281909
    WW 56 TORNADO GA SC NC VA AND ADJ CSTL WTRS 282000Z - 290300Z
    AXIS..70 STATUTE MILES EITHER SIDE OF LINE..30 W AGS/AUGUSTA GA/--30 NE
    ECG/ELIZABETH CITY NC/HAIL SURFACE AND ALOFT..3 INCHES. WIND GUSTS..70
    KNOTS. MAX TOPS TO 500. MEAN WIND VECTOR 250/60.

D. **Severe Weather Watch Bulletin Format**

 1. Type of severe weather watch, watch area, valid time period, type of severe weather possible, watch axis, meaning of a watch, and a statement that persons should be on the lookout for severe weather

 2. Other watch information, i.e., references to previous watches

 3. Phenomena, intensities, hail size, wind speeds (knots), maximum CB tops, and estimated cell movement (mean wind vector)

 4. Cause of severe weather

 5. Information on updating ACs

E.   **Example** of a Severe Weather Watch Bulletin:

MKC WW 281914

BULLETIN - IMMEDIATE BROADCAST REQUESTED
TORNADO WATCH NUMBER 56
NATIONAL WEATHER SERVICE KANSAS CITY MO
214 PM EST WED MAR 28 1984

A..THE NATIONAL SEVERE STORMS FORECAST CENTER HAS ISSUED A TORNADO WATCH
FOR

MOST OF SOUTH CAROLINA
MOST OF CENTRAL AND EASTERN NORTH CAROLINA
PARTS OF SOUTHEAST VIRGINIA
PARTS OF EASTERN GEORGIA
ADJOINING COASTAL WATERS

FROM 3 PM EST UNTIL 10 PM EST TODAY.

TORNADOES..LARGE HAIL..DANGEROUS LIGHTNING AND DAMAGING THUNDERSTORM
WINDS ARE POSSIBLE IN THESE AREAS.

THE TORNADO WATCH AREA IS ALONG AND 70 STATUTE MILES EITHER SIDE OF A LINE
FROM 30 MILES WEST OF AUGUSTA GEORGIA TO 30 MILES NORTHEAST OF ELIZABETH
NORTH CAROLINA.

REMEMBER..A TORNADO WATCH MEANS CONDITIONS ARE FAVORABLE FOR TORNADOES
AND SEVERE THUNDERSTORMS IN AND CLOSE TO THE WATCH AREA. PEOPLE IN THESE
AREAS SHOULD BE ON THE LOOKOUT FOR THREATENING WEATHER CONDITIONS AND
LISTEN FOR LATER STATEMENTS AND POSSIBLE WARNINGS.

C..TORNADOES AND A FEW SVR TSTMS WITH HAIL SFC AND ALF TO 3 IN. EXTRM TURBC
AND SFC WND GUSTS TO 70 KT. A FEW CBS WITH MAX TOPS TO 500. MEAN WIND
VECTOR 250/60.

D..PARAMETERS IN CAROLINAS VERY STG WITH STG LOW LVL FLOW AND CNVRGNC ALG
WITH UNSTABLE AMS. MESO LOW MVG RPDLY THRU ERN AL WILL CONT ENE INTO
CAROLINAS THIS EVENG. MEANWHILE SVR TSTMS LIKELY DVLPG IN ADVANCE THIS AFTN
DUE TO ABV MENTIONED CONDS.

F.   **Status Reports** are issued as needed to show progress of storms and to delineate areas no
longer under the threat of severe storm activity.

1.   Cancellation bulletins are issued when it becomes evident that no severe weather will
develop or that storms have subsided and are no longer severe.

G.   **Local Warnings** -- When tornadoes or severe thunderstorms have developed, local WSOs and
WSFOs issue local warnings.

END OF CHAPTER

CHAPTER SIXTEEN
SURFACE ANALYSIS CHART

Please take a few minutes to study each of the concepts listed above and anticipate/imagine what they are and how they relate to the other listed concepts.

A. Introduction

1. A surface analysis is commonly referred to as a surface weather analysis chart. The surface analysis chart is a computer-prepared chart that covers the contiguous 48 states and adjacent areas.

   a. The chart is transmitted every 3 hr.

   b. Figure 5-1, on page 298, is a section of a surface weather chart and Figure 5-2, on page 299, illustrates the symbols depicting fronts and pressure centers.

B. Valid Time of the chart corresponds to the time of the plotted observations. A date-time group in Universal Coordinated Time (UTC) tells the user when conditions portrayed on the chart were occurring.

C. Isobars are solid lines depicting the sea level pressure pattern and are usually spaced at 4-mb intervals.

1. When the pressure gradient is weak, dashed isobars are sometimes inserted at 2-mb intervals to more clearly define the pressure pattern.

2. Each isobar is labelled by a 2-digit number.

   a. For example, 32 signifies 1032.0 mb, 00 signifies 1000.0 mb, 92 signifies 992.0 mb, and 88 signifies 988.0 mb.

D. Pressure Systems -- The letter "L" denotes a low pressure center and the letter "H" denotes a high pressure center.

1. The pressure at each center is indicated by a 2-digit underlined number which is interpreted the same as the isobar labels.

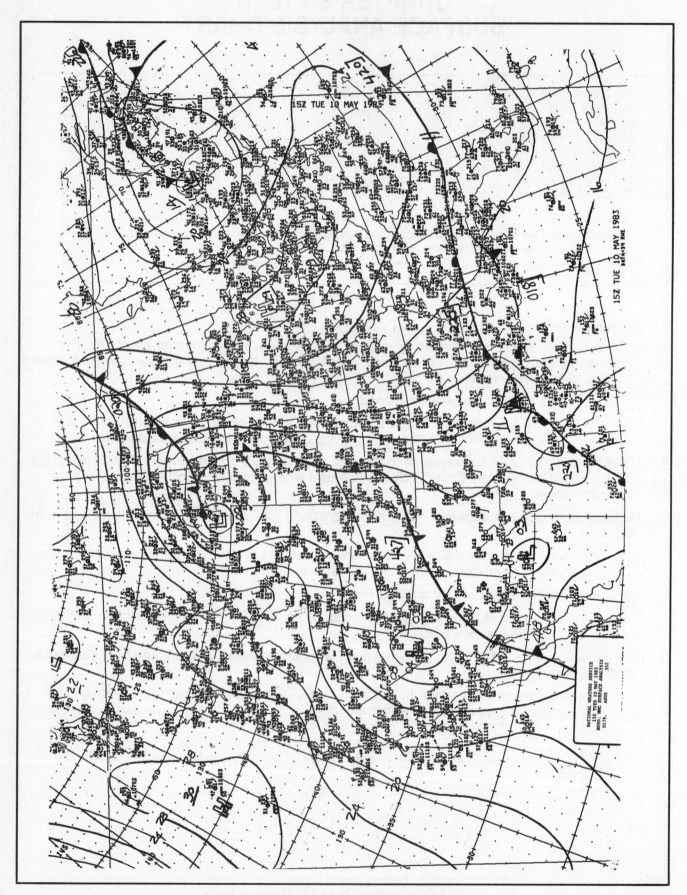

Figure 5-1. Surface Weather Analysis Chart

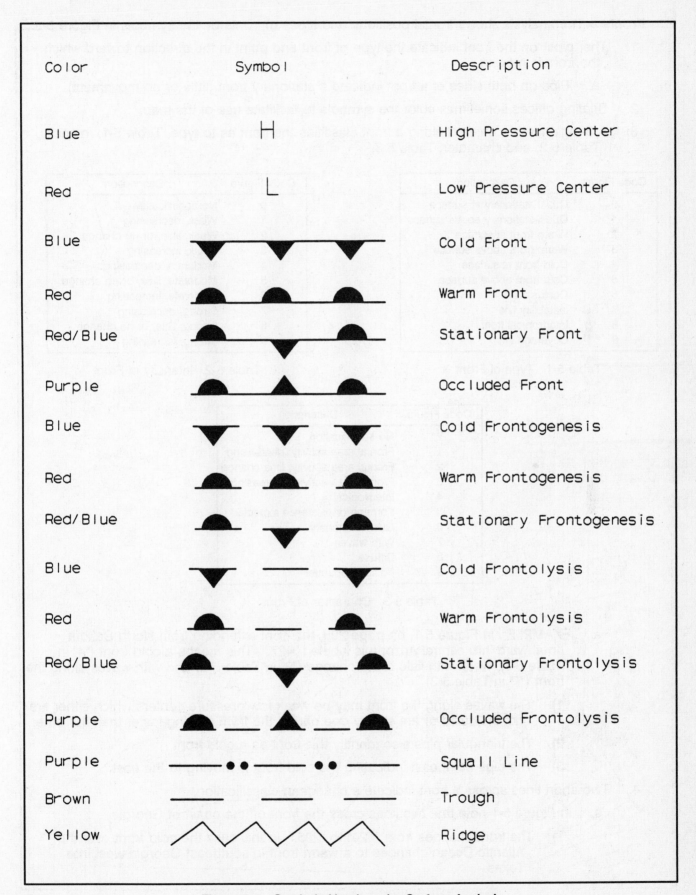

Color	Symbol	Description
Blue	H	High Pressure Center
Red	L	Low Pressure Center
Blue		Cold Front
Red		Warm Front
Red/Blue		Stationary Front
Purple		Occluded Front
Blue		Cold Frontogenesis
Red		Warm Frontogenesis
Red/Blue		Stationary Frontogenesis
Blue		Cold Frontolysis
Red		Warm Frontolysis
Red/Blue		Stationary Frontolysis
Purple		Occluded Frontolysis
Purple		Squall Line
Brown		Trough
Yellow		Ridge

Figure 5-2. Symbols Used on the Surface Analysis

E. **Fronts** -- The analysis shows frontal positions and types of fronts by the symbols in Figure 5-2.

1. The "pips" on the front indicate the type of front and point in the direction toward which the front is moving.

    a. Pips on both sides of a front indicate a stationary front (little or no movement).

2. Briefing offices sometimes color the symbols to facilitate use of the map.

3. A 3-digit number entered along a front classifies the front as to type, Table 5-1; intensity, Table 5-2; and character, Table 5-3.

Code Figure	Description
0	Quasi-stationary at surface
1	Quasi-stationary above surface
2	Warm front at surface
3	Warm front above surface
4	Cold front at surface
5	Cold front above surface
6	Occlusion
7	Instability line
8	Intertropical front
9	Coverage line

Table 5-1. Type of Front

Code Figure	Description
0	No specification
1	Weak, decreasing
2	Weak, little, or no change
3	Weak, increasing
4	Moderate, decreasing
5	Moderate, little, or no change
6	Moderate, increasing
7	Strong, decreasing
8	Strong, little, or no change
9	Strong, increasing

Table 5-2. Intensity of Front

Code Figure	Description
0	No specification
1	Frontal area activity, decreasing
2	Frontal area activity, little change
3	Frontal area activity, increasing
4	Intertropical
5	Forming or existence expected
6	Quasi-stationary
7	With waves
8	Diffuse
9	Position doubtful

Table 5-3. Character of Front

a. EXAMPLE: In Figure 5-1, on page 298, the front extending from North Dakota southward into central Arizona is labeled "427." This means a cold front ("4" in Table 5-1), weak with little or no change ("2" in Table 5-2) and with waves along the front ("7" in Table 5-3).

1) The waves along the front may be weak low pressure centers which either are not indicated or are simply one part of the front moving faster than the other.

2) The triangular pips also identify this front as a cold front.

3) The pips point east indicating the cold front is moving to the east.

4. Two short lines across a front indicate a change in classification.

a. In Figure 5-1 note that two lines cross the front off the coast of Georgia.

1) The front changes from "420" to "225." In this case the cold front over the Atlantic Ocean changes to a warm front in southeast Georgia west into Texas.

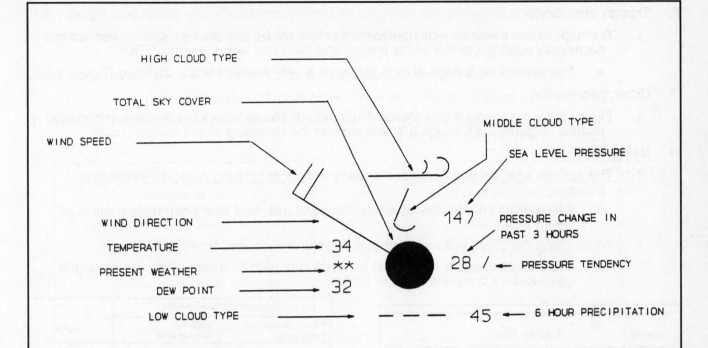

1. Total sky cover: OVERCAST (Figure 5-4).
2. Temperature: 34 DEGREES F, Dew Point: 32 DEGREES F.
3. Wind: FROM THE NORTHWEST AT 20 KNOTS (relative to True North).

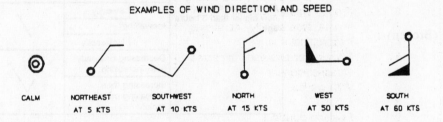

4. Present Weather: CONTINUOUS LIGHT SNOW (Figure 5-6).
5. Predominant low, middle, high cloud reported: STRATO FRACTUS OR CUMULUS FRACTUS OF BAD WEATHER, ALTOCUMULUS IN PATCHES, AND DENSE CIRRUS (Figure 5-7).
6. Sea Level Pressure: 1014.7 MILLIBARS (mbs).
   <u>NOTE</u>: Pressure is always shown in 3 digits to the nearest tenth of a millibar.
   For 1000 mbs or greater, prefix a "10" to the 3 digits. For less than
   1000 mbs, prefix a "9" to the 3 digits.
7. Pressure change in past 3 hours: INCREASED STEADILY OR UNSTEADILY BY 2.8 mbs.
   The actual change is in tenths of a millibar. (Figure 5-5)
8. 6-hour precipitation: 45 hundredths of an inch.
   The amount is given to the nearest hundredth of an inch.

Figure 5-3. Station Model and Explanation

F.   Trough and Ridge

1.   A trough of low pressure with significant weather will be depicted as a thick, dashed line running through the center of the trough and identified with the word "TROF."

   a.   The symbol for a ridge of high pressure is very rarely, if at all, depicted (Figure 5-2).

G.   Other Information

1.   Figure 5-3 is an example of a station model which shows where the weather information is plotted. Figures 5-4 through 5-7 help explain the decoding of the station model.

H.   Using the Chart

1.   The surface analysis chart provides a ready means of locating pressure systems and fronts.

   a.   It also gives an overview of winds, temperatures, and dew point temperatures at chart time.

2.   When using the chart, keep in mind that weather moves and conditions change.

   a.   The surface analysis chart must be used in conjunction with other information to give a more complete weather picture.

Symbol	Total Sky Cover
◯	Sky Clear (less than 1/10)
◑	1/10 to 5/10 inclusive (Scattered)
◕	6/10 to 9/10 inclusive (Broken)
●	10/10 (Overcast)
⊗	Sky obscured or partially obscured

Figure 5-4. Sky Cover Symbols

Description of Characteristic		
Primary Unqualified Requirement	Additional Requirements	Graphic
HIGHER Atmospheric pressure now higher than 3 hours ago	Increasing then decreasing	∧
	Increasing then steady; or	⌐
	increasing then increasing more slowly	
	Increasing Steadily / Unsteadily	/
	Decreasing or steady then increasing; or	√
	increasing then increasing more rapidly	
THE SAME Atmospheric pressure now same as 3 hours ago	Increasing then decreasing	∧
	Steady	—
	Decreasing then increasing	∨
LOWER Atmospheric pressure now lower than 3 hours ago	Decreasing then increasing	∨
	Decreasing then steady; or	⌐
	decreasing then decreasing more slowly	
	Decreasing Steadily / Unsteadily	\
	Steady or increasing then decreasing; or	∧
	decreasing then decreasing more rapidly	

Figure 5-5. Barometer Tendencies

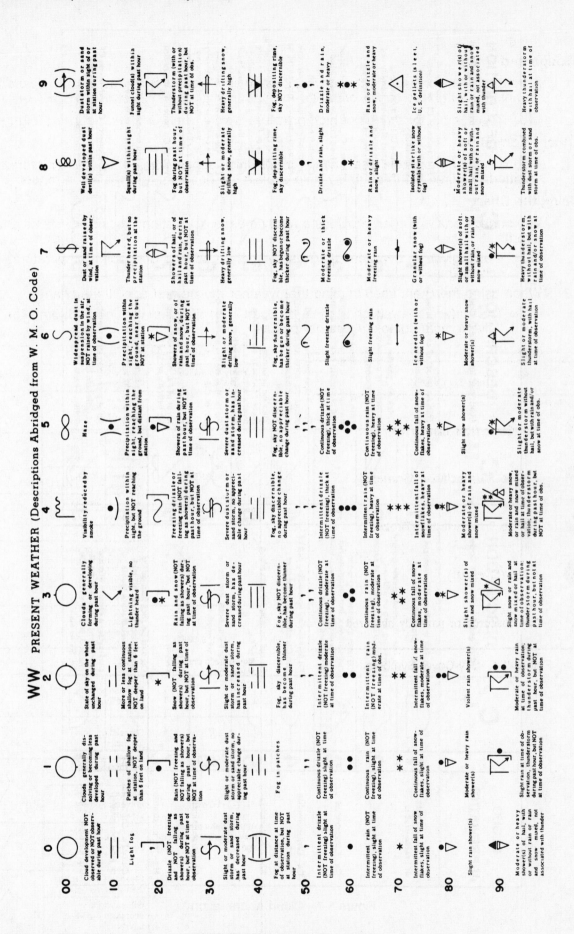

Figure 5-6. Present Weather

CLOUD ABBREVIATION	C_L	DESCRIPTION (Abridged from W.M.O. Code)	C_M	DESCRIPTION (Abridged from W.M.O. Code)	C_H	DESCRIPTION (Abridged from W.M.O. Code)
St or Fs - Stratus or Fractostratus	1	Cu, fair weather, little vertical development & flattened	1	Thin As (most of cloud layer semitransparent)	1	Filaments of Ci, or "mares tails", scattered and not increasing
Ci - Cirrus	2	Cu, considerable development, towering with or without other Cu or SC bases at same level	2	Thick As, greater part sufficiently dense to hide sun (or moon), or Ns	2	Dense Ci in patches or twisted sheaves, usually not increasing, sometimes like remains of CB; or towers tufts
Cs - Cirrostratus	3	Cb with tops lacking clear-cut outlines, but distinctly not cirriform or anvil shaped; with or without Cu, Sc, St	3	Thin Ac, mostly semi-transparent; cloud elements not changing much at a single level	3	Dense Ci, often anvil-shaped derived from or associated Cb
Cc - Cirrocumulus	4	Sc formed by spreading out of Cu; Cu often present also	4	Thin AC in patches; cloud elements continually changing and/or occurring at more than one level	4	Ci, often hook-shaped gradually spreading over the sky and usually thickening as a whole
Ac - Altocumulus	5	Sc not formed by spreading out of Cu	5	Thin Ac in bands or in a layer gradually spreading over sky and usually thickening as a whole	5	Ci and Cs, often in converging bands or Cs alone; generally overspreading and growing denser; the continuous layer not reaching 45 altitude
As - Altostratus	6	St or Fs or both, but no Fs of bad weather	6	Ac formed by the spreading out of Cu	6	Ci & Cs often in converging bands or Cs alone; generally overspreading and growing denser the continuous layer exceeding 45 altitude
Sc - Stratocumulus	7	Fs and/or Fc of bad weather (scud)	7	Double-layered Ac, or a thick layer of Ac, not increasing; or Ac with As and/or Ns	7	Veil of Cs covering the entire sky
Ns - Nimbostratus	8	Cu and Sc (not formed by spreading out of Cu) with bases at different levels	8	Ac in the form of Cu-shaped tufts or Ac with turrets	8	Cs not increasing and not covering entire sky
Cu or Fc - Cumulus or Fractocumulus	9	Cb having a clearly fibrous (cirriform) top, often anvil-shaped, with or without Cu Sc, ST or scud	9	Ac of a chaotic sky, usually at different levels; patches of dense Ci are usually present	9	Cc alone or Cc with some Ci or Cs but the Cc being the main cirriform cloud
Cb - Cumulonimbus						

Figure 5-7. Cloud Abbreviation

END OF CHAPTER

CHAPTER SEVENTEEN
WEATHER DEPICTION CHART

Please take a few minutes to study each of the concepts listed above and anticipate/imagine what they are and how they relate to the other listed concepts.

A. **Introduction**

1. The weather depiction chart, Figure 6-1 on page 307, is computer-prepared from Surface Aviation Observations (SAO).

   a. The weather depiction chart gives a broad overview of the observed flying category conditions at the valid time of the chart.

   b. This chart begins at 01Z each day, is transmitted at 3 hr. intervals, and is valid at the time of the plotted data.

B. **Plotted Data** for each station are:

1. Total sky cover. The amount of sky cover is shown by the station circle shaded as in Table 6-1.

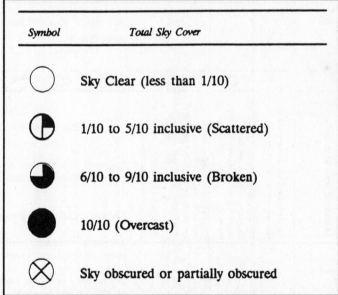

Symbol	Total Sky Cover
○	Sky Clear (less than 1/10)
◑	1/10 to 5/10 inclusive (Scattered)
◕	6/10 to 9/10 inclusive (Broken)
●	10/10 (Overcast)
⊗	Sky obscured or partially obscured

Table 6-1. Total Sky Cover

2. Cloud height or ceiling. Cloud height above ground level is entered under the station circle in hundreds of feet, the same as coded in an SAO report.

   a. If total sky cover is scattered, the cloud height entered is the base of the lowest layer.

   b. If total sky cover is broken or greater, the cloud height entered is the ceiling.

      1) Broken or greater total sky cover without a height entry indicates thin sky cover.

   c. A partially or totally obscured sky is shown by the sky cover symbol "X."

      1) However, a partially obscured sky without a cloud layer above is denoted by the absence of a height entry.

2)   A partially obscured sky with clouds above will have a cloud layer or ceiling height entry.

3)   A totally obscured sky always has a height entry of the ceiling (vertical visibility into the obscuration).

3.   Weather and obstructions to vision. Weather and obstructions to vision symbols are entered to the left of the station circle. Figure 5-6 on page 303 explains most of the symbols used.

a.   When an SAO reports clouds topping ridges, a symbol unique to the weather depiction chart is entered to the left of the station circle:

b.   When several types of weather and/or obstructions to visibility are reported at a station, only the most significant one is entered (i.e., the highest coded number in Figure 5-6 on page 303).

4.   Visibility. When visibility is 6 SM or less, it is entered to the left of the weather or obstructions to vision symbol.

a.   Visibility is entered in statute miles and fractions of a mile.
b.   Table 6-2 shows examples of plotted data.

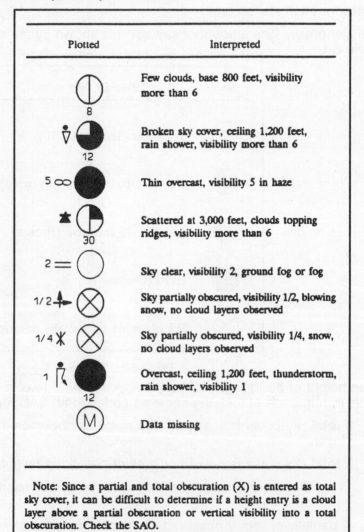

Plotted	Interpreted
	Few clouds, base 800 feet, visibility more than 6
	Broken sky cover, ceiling 1,200 feet, rain shower, visibility more than 6
	Thin overcast, visibility 5 in haze
	Scattered at 3,000 feet, clouds topping ridges, visibility more than 6
	Sky clear, visibility 2, ground fog or fog
	Sky partially obscured, visibility 1/2, blowing snow, no cloud layers observed
	Sky partially obscured, visibility 1/4, snow, no cloud layers observed
	Overcast, ceiling 1,200 feet, thunderstorm, rain shower, visibility 1
	Data missing

Note: Since a partial and total obscuration (X) is entered as total sky cover, it can be difficult to determine if a height entry is a cloud layer above a partial obscuration or vertical visibility into a total obscuration. Check the SAO.

Table 6-2. Examples of Plotting on the Weather Depiction Chart

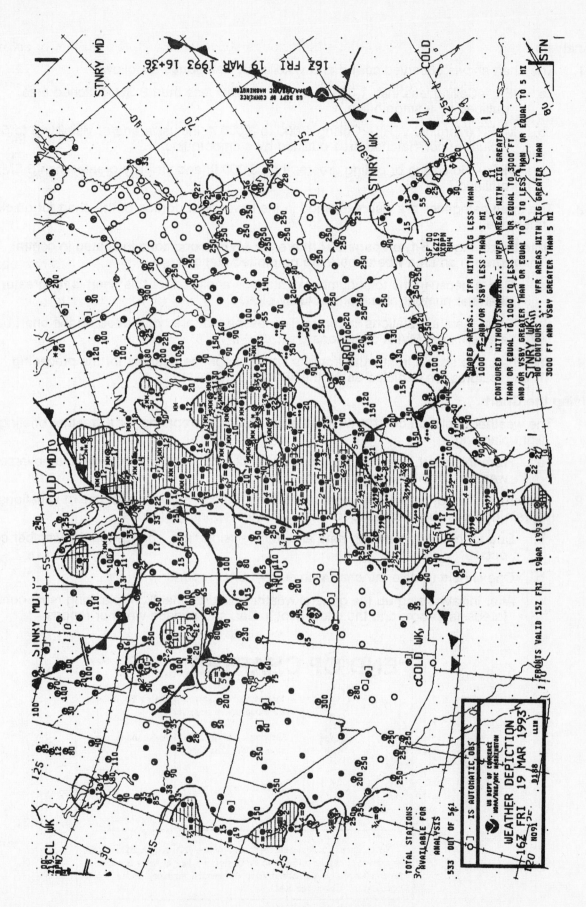

Figure 6-1. A Weather Depiction Chart

C. **Analysis**

1. The chart shows observed ceiling and visibility by categories as follows:

   a. IFR -- Ceiling less than 1,000 ft. and/or visibility less than 3 SM; hatched area outlined by a smooth line.

   b. MVFR (Marginal VFR) -- Ceiling 1,000 to 3,000 ft. inclusive and/or visibility 3 to 5 SM inclusive; non-hatched area outlined by a smooth line.

   c. VFR -- No ceiling or ceiling greater than 3,000 ft. and visibility greater than 5 SM; not outlined.

2. The three categories are also explained in the lower right portion of the chart for quick reference.

3. Note that in Figure 6-1, on page 307, there are MVFR conditions indicated in central Tennessee in an area where there are no plotted stations.

   a. This is because the total number of stations analyzed for this chart is far greater than the number of stations actually plotted.

   b. Thus, there are stations in central Tennessee that are not plotted on the chart but are reporting MVFR conditions.

4. The chart also shows fronts and troughs from the surface analysis for the preceding hour. These features are depicted as on the surface chart.

D. **Using the Chart**

1. The weather depiction chart is an ideal place to begin preparing for a weather briefing and flight planning.

   a. From this chart, one can get a bird's-eye view of areas of favorable and adverse weather conditions for chart time.

2. This chart may not completely represent the en route conditions because of variations in terrain and possible weather occurring between reporting stations.

   a. Due to the delay between data and transmission time, changes in the weather could occur.

   b. One should update the chart with current SAO reports.

   c. After initially sizing up the general weather picture, final flight planning must consider forecasts, progs, and the latest pilot, radar, and surface weather reports.

END OF CHAPTER

CHAPTER EIGHTEEN
RADAR SUMMARY CHART

Please take a few minutes to study each of the concepts listed above and anticipate/imagine what they are and how they relate to the other listed concepts.

A. Type and Time of Report

1. A radar summary chart, Figure 7-1 on page 310, graphically displays a collection of radar weather reports (SDs).

   a. This computer-generated chart is constructed from regularly scheduled radar observations and is valid at the time of these reports (H+35).

      1) A current radar summary chart is available between 16 and 24 hr. daily depending on the system being used.

   b. The chart displays the type of precipitation echoes, their intensity, intensity trend, configuration, coverage, echo tops and bases, and movement.

      1) Severe weather watches are plotted if they are in effect when the chart is valid.

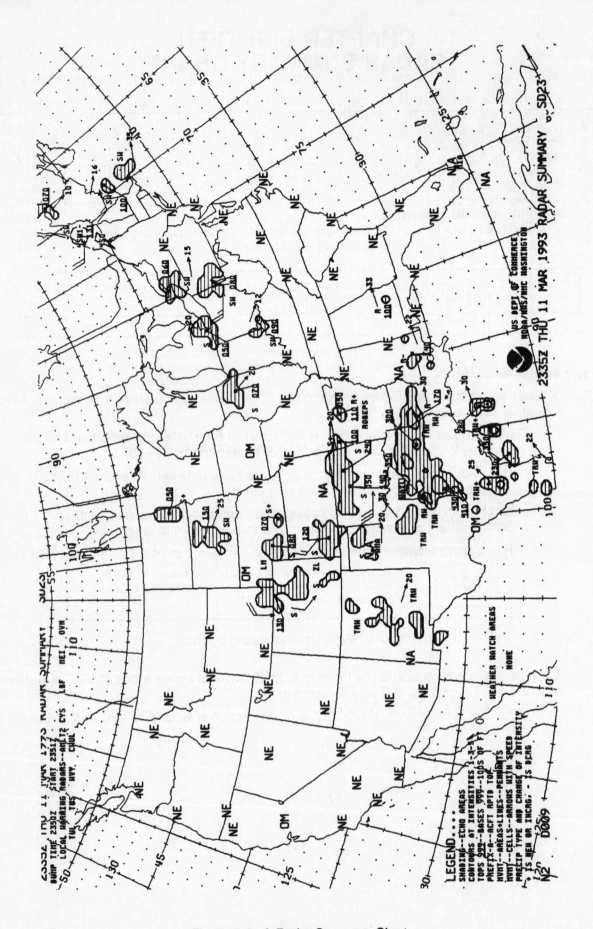

Figure 7-1. A Radar Summary Chart

B. **Echo Type**

1. Radar primarily detects particles of precipitation size within a cloud or falling from a cloud.

2. The type of precipitation can be determined by the radar operator from the scope presentation in combination with other sources.

    a. Table 7-1, below, lists the boxed symbols used to denote types of echoes.

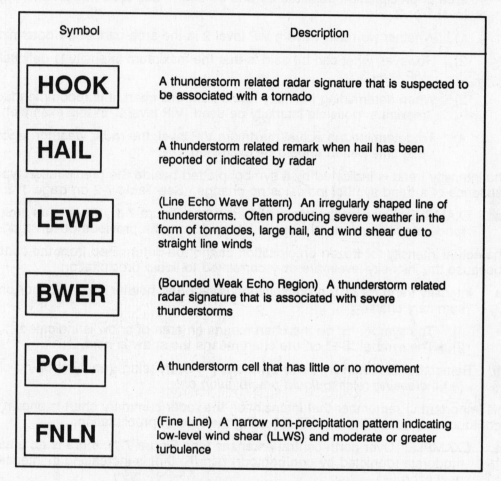

Symbol	Description
HOOK	A thunderstorm related radar signature that is suspected to be associated with a tornado
HAIL	A thunderstorm related remark when hail has been reported or indicated by radar
LEWP	(Line Echo Wave Pattern) An irregularly shaped line of thunderstorms. Often producing severe weather in the form of tornadoes, large hail, and wind shear due to straight line winds
BWER	(Bounded Weak Echo Region) A thunderstorm related radar signature that is associated with severe thunderstorms
PCLL	A thunderstorm cell that has little or no movement
FNLN	(Fine Line) A narrow non-precipitation pattern indicating low-level wind shear (LLWS) and moderate or greater turbulence

Table 7-1. Explanation of Boxed Symbols

3. Note that in north-central Texas there is a "HAIL" symbol inside a box ("HAIL" and other symbols are explained in Table 7-1 above).

    a. The actual location of the "HAIL" is drawn from the symbol to that part of the echo.

C.   Intensity and Intensity Trend

1.   The intensity is obtained from the Video Integrator Processor (VIP) and is indicated on the chart by *contours*. The six VIP levels are combined into three contours as indicated in Table 7-2 on page 313.

    a.   EXAMPLE: Over southeastern Alabama (see Figure 7-1 on page 310), there is an area of precipitation depicted by one contour. This area has an intensity of VIP level 1 or 2.

        1)   Whether we really have a VIP level 2 in the area cannot be determined.

        2)   However, what can be said is that the maximum intensity is definitely below VIP level 3.

        3)   When determining intensity levels from this chart, it is recommended that the maximum possible intensity be used (VIP level 2, in this example).

        4)   To determine the actual maximum VIP level, the radar weather report (SD) for that time period should be examined.

2.   The intensity trend is indicated by a symbol plotted beside the precipitation type. The absence of a trend symbol indicates no change. See Table 7-2 on page 313.

    a.   EXAMPLE: Over southeastern Alabama (see Figure 7-1), there is an area of light to moderate rain with no change in intensity from the previous observation.

3.   The actual intensity for frozen precipitation cannot be determined from the contours because the intensity levels are only correlated to liquid precipitation.

    a.   Intensity trend for frozen precipitation is reported neither on an SD nor on the radar summary chart.

        1)   The symbol "S" on the chart means an area of snow is indicated.
        2)   The symbol "S+" on the chart means the snow is *new*.

    b.   Remember the intensity trend symbols; (–) decreasing, (no symbol) no change, and (+) increasing refer to *liquid precipitation only*.

4.   It is important to remember that intensity on the radar summary chart is shown by contours and not by the symbol following the type of precipitation.

    a.   EXAMPLE: Over north-central Arkansas (see Figure 7-1), there is an area of light to moderate (depicted by one contour) rain (R) that is increasing in intensity or is a new echo (+).

D.   Echo Configuration and Coverage

1.   The configuration is the arrangement of echoes. There are three designated arrangements: a LINE of echoes, an AREA of echoes, and an isolated CELL.

2.   Coverage is simply the area covered by echoes. All of the hatched area inside of the contours on the chart is considered to be covered by echoes.

    a.   When the echoes are reported as a LINE, a line will be drawn through them on the chart.

    b.   Where there is 8/10 coverage or more, the line is labeled as solid (SLD) at both ends.

        1)   In the absence of this label it can be assumed that there is less than 8/10 coverage.

VIP LEVEL	ECHO INTENSITY	PRECIPITATION INTENSITY	RAINFALL RATE in/hr STRATIFORM	RAINFALL RATE in/hr CONVECTIVE
1	WEAK	LIGHT	LESS THAN 0.1	LESS THAN 0.2
2	MODERATE	MODERATE	0.1 - 0.5	0.2 - 1.1
3	STRONG	HEAVY	0.5 - 1.0	1.1 - 2.2
4	VERY STRONG	VERY HEAVY	1.0 - 2.0	2.2 - 4.5
5	INTENSE	INTENSE	2.0 - 5.0	4.5 - 7.1
6	EXTREME	EXTREME	MORE THAN 5.0	MORE THAN 7.1

* The numbers representing the intensity level do not appear on the chart. Beginning from the first contour line, bordering the area, the intensity level is 1-2, second contour is 3-4, and the third contour is 5-6.

450
Highest precipitation top in area in hundreds of feet MSL
(45,000 FEET MSL)

SYMBOL MEANING

R	RAIN
RW	RAIN SHOWER
HAIL	HAIL
S	SNOW
IP	ICE PELLETS
SW	SNOW SHOWER
L	DRIZZLE
T	THUNDERSTORM
ZR, ZL	FREEZING PRECIPITATION
NE	NO ECHOES OBSERVED
NA	OBSERVATIONS UNAVAILABLE
OM	OUT FOR MAINTENANCE
STC	STC ON - all precipitation may not be seen
ROBEPS	RADAR OPERATING BELOW PERFORMANCE STANDARDS
RHINO	RANGE HEIGHT INDICATOR NOT OPERATING

SYMBOLS USED ON CHARTS
SYMBOL MEANING

+	INTENSITY INCREASING OR NEW ECHO
-	INTENSITY DECREASING
NO SYMBOL	NO CHANGE IN INTENSITY
35	CELL MOVEMENT TO NE AT 35 KNOTS
	LINE OR AREA MOVEMENT TO EAST AT 20 KNOTS
LM	LITTLE MOVEMENT
MA	ECHOES MOSTLY ALOFT
PA	ECHOES PARTLY ALOFT

SYMBOL MEANING

	LINE OF ECHOES
SLD	8/10 OR GREATER COVERAGE IN A LINE
WS999	SEVERE THUNDERSTORM WATCH
WT999	TORNADO WATCH
LEWP	LINE ECHO WAVE PATTERN
HOOK	HOOK ECHO
BWER	BOUNDED WEAK ECHO REGION
PCLL	PERSISTENT CELL
FNLN	FINE LINE

Table 7-2. Key to Radar Summary Chart

E. Echo Heights

1. Echo heights in locations with radars designed for weather detection are obtained by use of range height indicators and are *PRECIPITATION* tops and bases.

    a. In those areas not served by National Weather Service radars, the tops are obtained from pilot reports and are actual *CLOUD* tops.

    b. Usually, echo height will be missing in the western mountain regions because ARTCC radars are used.

2. Heights are displayed in hundreds of feet MSL and should be considered only as approximations because of radar limitations.

    a. Tops are entered above a short line, and any available bases are entered below.

        1) The top height displayed is the highest in the indicated area.

    b. EXAMPLE:

        $\frac{220}{080}$ Bases 8,000 ft., Maximum top 22,000 ft.

        $\underline{500}$ Bases at surface, Maximum top 50,000 ft.

        $\overline{020}$ Bases 2,000 ft., Maximum top either missing or reported in another place

3. Absence of a figure below the line indicates that the echo base is at the surface. Radar detects tops more readily than bases because precipitation usually reaches the ground.

    a. EXAMPLE: Over north-central Arkansas (see Figure 7-1), the maximum precipitation top in the area is 11,000 ft. MSL.

        1) The location is indicated by a line drawn to the symbol.

        2) The base of the rain associated with the maximum top and probably over most of the area is 5,000 ft. MSL.

F. Echo Movement

1. Individual cell movement is indicated by an arrow with the speed in knots entered as a number at the top of the arrowhead.

    a. Little movement is identified by "LM."

2. Line or area movement is indicated by a shaft and barb combination with the shaft indicating the direction and the barbs the speed.

    a. A half barb (ı) is 5 kt., a whole barb (ı) is 10 kt., and a pennant (▲) is 50 kt.

3. EXAMPLE: Over southeastern Wyoming (see Figure 7-1), no cell movement is given, but the area movement is toward the south at 20 kt.

    a. Over southeastern Alabama, no area movement is given, but the cell movement of the rain is toward the east at 33 kt.

G. **Severe Weather Watch Areas**

1. Severe weather watch areas are outlined by heavy dashed lines, usually in the form of a large rectangular box.

    a. There are two types, tornado watches and severe thunderstorm watches.

2. The type of watch and the watch number are enclosed in a small rectangle (see Table 7-2 on page 313) and positioned as closely as possible to the northeast corner of the watch box.

    a. Weather watch areas are also listed by type and number at the bottom of the chart (over Mexico) together with the issuance time and valid until time.

        1) If there are no weather watch areas, it will so indicate by the word "NONE," as in Figure 7-1.

H. **Canadian Data**

1. Radar data from Canadian radar stations are plotted when available. The stations are Halifax, Holyrood, Mechanics Settlement, Broadview, Elbow, Vivian, Mont Apica, Britt, Carp, Exeter, Montreal, River Harbour, Upsala, Villerpy, King City, Carvel, Vulcan, Cold Lake, and McGill.

    a. The data is displayed in AZRAN (azimuth-range) format with echo areas outlined by solid lines.

    b. Area, line, and cell movements are shown in the same manner as U.S. data.

    c. An alphanumeric code associated with each echo shows, in order, area coverage, precipitation type, intensity, and intensity trend.

        1) For area coverage, a blank designator represents cells, 1 equals less than 1/10 coverage, 4 equals 1/10 to 5/10 coverage, 7 equals 6/10 to 9/10 coverage, and 10 equals 10/10 coverage.

        2) For intensity levels, 0 is very weak, 1 is weak, 2 is moderate, 3 is strong, and 4 is very strong. Levels 1 through 4 are comparable to the U.S. VIP levels 1 through 4.

        3) Precipitation type and intensity trend are the same as U.S. data.

2. Figure 7-1 shows a Canadian report in southeast Canada that also covers a small part of northwestern Vermont and northeastern New York.

    a. The report "4SO" is decoded as an area of very weak snow with no change in intensity.

        1) 1/10 to 5/10 of the area is covered with snow with no movement reported.
        2) The maximum top within the area is not reported.

    b. Note that Canadian echo top reports are converted from meters to feet and are plotted to the nearest hundreds of feet MSL.

        1) For example, 197 is 19,700 ft. MSL.

3. It is sometimes difficult to interpret the data when both U.S. and Canadian reports are plotted. Do not confuse a Canadian report with a severe weather watch box.

    a. A Canadian report is a box outlined by a solid line, not a dashed line as required for a severe weather watch area.

I.   Using the Chart

1. The radar summary chart helps preflight planning by identifying general areas and movement of precipitation and/or thunderstorms.

   a. Radar detects ONLY drops or ice particles of precipitation size; it DOES NOT detect clouds and fog.

   b. Thus, the absence of echoes does not guarantee clear weather, and cloud tops may be higher than the tops detected by radar.

   c. The chart must be used in conjunction with other charts, reports, and forecasts.

2. Examine chart notations carefully. Always determine location and movement of echoes.

   a. If echoes are anticipated near the planned route, take special note of echo intensity and trend.

   b. Be sure to examine the chart for missing radar reports (NA, OM) before assuming that no echoes are present.

   c. EXAMPLE: In Figure 7-1, the Miami (MIA) radar report in southeast Florida is shown as not available (NA). There could be echoes in the southern half of Florida but too far away to be detected by the other surrounding radars.

3. Suppose the planned flight route goes through an area of widely scattered thunderstorms in which no increase is anticipated. If these storms are separated by good VFR weather, they can be visually detected and circumnavigated.

   a. However, widespread cloudiness may conceal the thunderstorms.

   b. To avoid these embedded thunderstorms, it would be necessary either to use airborne radar or to detour the area.

   c. Details on avoiding hazards of thunderstorms are given in Part I, Chapter 11, Thunderstorms, beginning on page 115.

4. Remember that the radar summary chart is for preflight planning only and should be updated by hourly radar reports.

   a. Once airborne, the pilot must avoid individual storms by in-flight observations either by visual detection, by using airborne radar, or by requesting radar echo information from FSS Flight Watch.

   b. FSS Flight Watch has access to Radar Remote Weather Displays (RRWDS).

5. There can be an interpretation problem concerning an area of precipitation that is reported by more than one radar site.

   a. EXAMPLE: Station A may report RW− with cell movement toward the northeast at 10 kt. For the same area, station B may be reporting TRW+ with cell movement toward the northeast at 30 kt.

   b. This difference in reports may be due to a different perspective and distance of the radar site from the area of echoes. The area may be moving away from station A and approaching station B.

      1) The rule of thumb is to use the plotted data associated with the area which presents the greatest hazard to aviation. In this case, the station B report should be used.

END OF CHAPTER

CHAPTER NINETEEN
U.S. LOW-LEVEL SIGNIFICANT WEATHER PROG

Please take a few minutes to study each of the concepts listed above and anticipate/imagine what they are and how they relate to the other listed concepts.

A. Type and Time of Forecast

1. The low-level prog is a 4-panel chart as shown in Figure 8-1, on page 318.

    a. The two lower panels are 12- and 24-hr. surface progs (SFC PROG).

    b. The two upper panels are 12- and 24-hr. progs of significant weather (SIG WX) from the surface to 400 mb (24,000 ft. MSL).

    c. The charts show conditions as they are forecast to be at the valid time (VT) of the chart.

    d. This chart is issued four times daily with the 12- and 24-hr. forecasts based on the 00Z, 06Z, 12Z, and 18Z synoptic data.

       1) EXAMPLE: The prog in Figure 8-1 is based on the 06Z, THU 03 JUN 1993 initial data.

B. Surface Prog

1. The two surface prog panels use standard symbols for fronts, significant troughs, and pressure centers.

    a. High and low pressure centers are indicated by a 2-digit number. These 2-digit numbers are underlined on the manually prepared chart, but not on the computer prepared chart.

    b. Isobars depicting forecast pressure patterns are included on some 24-hr. surface progs.

2. The surface prog also outlines areas of forecast precipitation and/or thunderstorms as shown in the lower panels of Figure 8-1.

    a. Smooth lines enclose areas of expected continuous or intermittent (stable air) precipitation and dash-dot lines enclose areas of showers and thunderstorms (unstable air precipitation).

    b. Areas of continuous or intermittent precipitation with embedded showers and thunderstorms are also enclosed by dash-dot lines.

3. Note that the symbols indicate precipitation type and character (Tables 8-1 and 8-2, on page 319).

    a. If precipitation affects half or more of an area, that area is shaded.

       1) The absence of shading denotes more sparse precipitation, specifically coverage of less than half of the area.

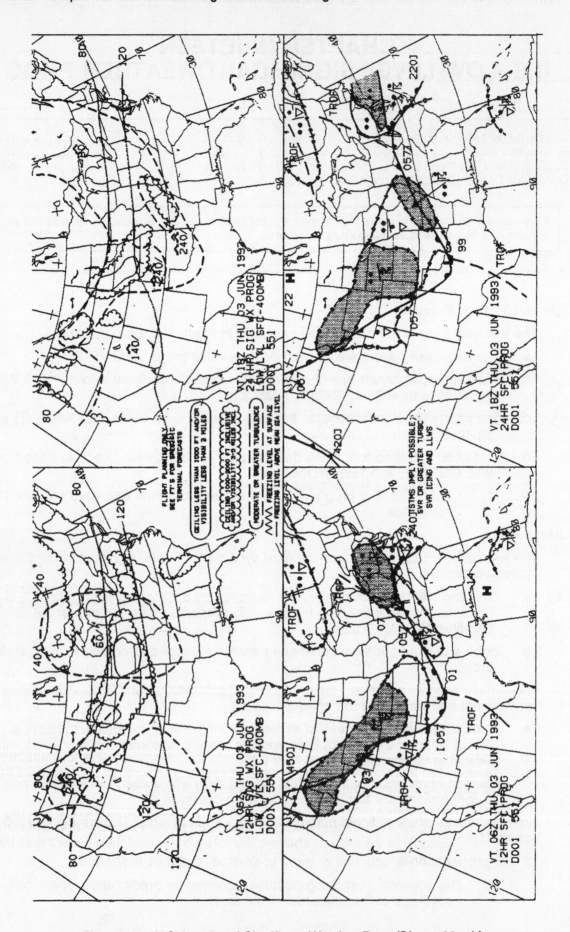

Figure 8-1. U.S. Low-Level Significant Weather Prog (Sfc. —400 mb)

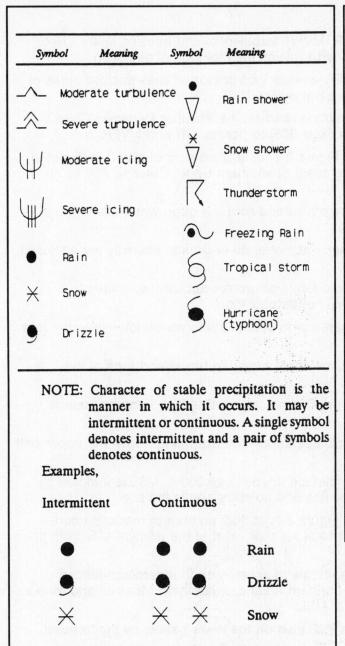

NOTE: Character of stable precipitation is the manner in which it occurs. It may be intermittent or continuous. A single symbol denotes intermittent and a pair of symbols denotes continuous.

Examples,

Intermittent	Continuous	
●	● ●	Rain
●	● ●	Drizzle
✳	✳ ✳	Snow

Table 8-1. Some Standard Weather Symbols

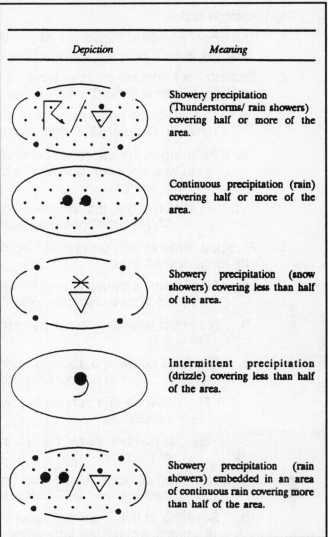

Table 8-2. Significant Weather Prognostic Symbols

b.   Look at the lower left panel of Figure 8-1. At 06Z, the forecast is for thunderstorms embedded in an area of continuous rain covering more than half the area that is shaded from northwest Kansas to southeast Washington.

1)   On the same prog, showery precipitation (rain showers/thunderstorms) are forecast from central Arkansas northeast to the Kentucky and Ohio border. Since this area is not shaded, the forecast precipitation covers less than half the area.

C.   Significant Weather

1.   The upper panels of Figure 8-1 depict IFR, MVFR, turbulence, and freezing levels. Note that the legend near the center of the chart explains the methods of depiction.

2.   Smooth lines enclose areas of forecast IFR weather and scalloped lines enclose areas of marginal weather (MVFR). VFR areas are not outlined.

   a.   This is not the same manner of depiction used on the Weather Depiction Chart (Part III, Chapter 17, beginning on page 305) to portray IFR and MVFR.

   b.   Referring to the upper left panel of Figure 8-1, at 06Z, an area of MVFR is depicted from western North Carolina to the coast of southern South Carolina and south-eastern Georgia.

      1)   Notice that the depiction does not extend over the open water even though MVFR conditions may exist.

3.   Forecast areas of non-convective turbulence of moderate or greater intensity are enclosed by long, dashed lines.

   a.   Since thunderstorms *always* imply moderate or greater turbulence, areas of thunderstorm-related turbulence will not be outlined.

   b.   A symbol entered within a general area of forecast turbulence denotes intensity (see Table 8-1).

   c.   Numbers below and above a short line show expected bases and tops of the turbulent layer in hundreds of feet MSL.

      1)   Absence of a number below the line indicates turbulence from the surface upward.

      2)   No number above the line indicates turbulence extending above the upper limit of the chart.

      3)   Turbulence forecast from the surface to above 24,000 ft. MSL is denoted by the notation "SFC" below the line and no entry above the line.

   d.   Referring to the upper left panel of Figure 8-1, at 06Z, an area of moderate non-thunderstorm-related turbulence is forecast over most of the western U.S. from the surface to 12,000 ft. MSL.

      1)   At 18Z (upper right panel), moderate to severe non-thunderstorm-related turbulence is forecast over northern Arkansas, southern Missouri and Illinois from the surface to 24,000 ft. MSL.

   e.   Thunderstorm-related turbulence is indicated on the lower panels by the forecast areas of thunderstorms.

4.   Freezing level height contours for the *highest* freezing level are drawn at 4,000 ft. intervals.

   a.   Contours are labeled in hundreds of feet MSL.

   b.   The zig-zag line shows where the freezing level is forecast to be at the surface and is labeled "SFC."

   c.   An upper freezing level contour crossing the surface/32°F (0°C) line indicates multiple freezing levels.

      1)   Multiple freezing levels indicate layers of warmer air aloft.

      2)   If clouds and precipitation are forecast in this area, icing hazards should be considered.

5.   The low-level significant weather prog does not specifically outline areas of icing. However, icing is implied in clouds and precipitation above the freezing level. Interpolate for freezing levels between the given contours.

   a.   EXAMPLE: In the upper left panel of Figure 8-1 at 06Z, the forecast *highest* freezing level over Washington, DC is approximately 12,000 ft. MSL.

D.   36- and 48-hr. Surface Weather Prog

1.   This prog is an extension of the 12- and 24-hr. surface prog and is based on the 00Z and 12Z initial synoptic data.

   a.   The prog in Figure 8-2, on page 322, is a continuation of the 12- and 24-hr. prog in Figure 8-1.

2.   The depiction of data is about the same as the 12- and 24-hr. surface prog with the following exceptions.

   a.   Freezing precipitation is not forecast.

   b.   Scalloped lines denote areas of *overcast* clouds with *no* reference to the height of the cloud base.

   c.   A prognostic discussion is included to explain the forecaster's reasoning for the 12- through 48-hr. surface progs.

3.   The 36- and 48-hr. surface prog should only be used for outlook purposes. That is, just to get a very general picture of the weather conditions that are in the relatively distant future.

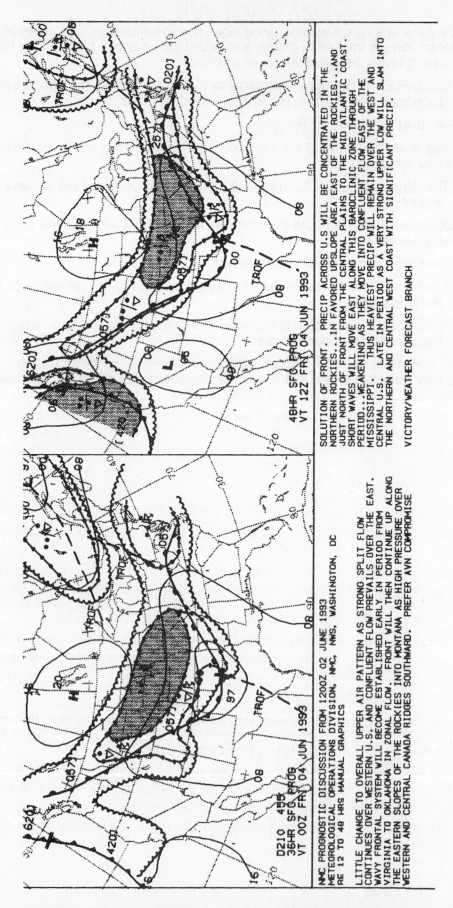

Figure 8-2. U.S. Low-Level 36- and 48-hr. Significant Weather Prog

CHAPTER TWENTY
HIGH-LEVEL SIGNIFICANT WEATHER PROG

Please take a few minutes to study each of the concepts listed above and anticipate/imagine what they are and how they relate to the other listed concepts.

A. Introduction

1. The High-Level Significant Weather Prog is a computer-generated forecast for both domestic and international flights.

   a. The U.S. National Meteorological Center (NMC), near Washington D.C., is a component of the World Area Forecast System (WAFS).

   b. NMC is designated in the WAFS as both a World Area Forecast Center (WAFC) and a Regional Area Forecast Center (RAFC).

   c. The main function of the NMC as a WAFC is to prepare global forecasts of upper winds and upper air temperatures in grid-point form and to supply the forecasts to associated RAFCs.

      1) One important RAFC function is to supply users with forecast winds and temperatures aloft along with a forecast of significant weather.

B. Content -- The following significant weather is depicted on the charts:

1. The abbreviation or symbol "CB" (cumulonimbus clouds) is used to depict thunderstorm activity.

   a. By definition, this symbol refers to either the occurrence or expected occurrence of an area of widespread cumulonimbus clouds along a line with little or no space between individual clouds, or cumulonimbus clouds embedded in cloud layers or concealed by haze or dust.

      1) It does not refer to isolated or scattered (occasional) cumulonimbus clouds that are not embedded in cloud layers or concealed by haze or dust.

   b. The symbol "CB" automatically implies moderate or greater turbulence and icing, and these are not depicted separately.

   c. CB data will normally be identified as ISOL EMBD CB (isolated embedded CB), OCNL EMBD CB (occasional embedded CB), ISOL CB in HAZE (isolated CB in haze), or OCNL CB in HAZE (occasional CB in haze).

      1) In rare instances, CB coverage above FL 240 may exceed 4/8 coverage; in these instances, CB activity will be described as FRQ CB (frequent cumulonimbus with little or no separation).

   d. The meanings of these area coverage terms are: ISOL (less than 1/8), OCNL (1/8 to 4/8), and FRQ (5/8 to 8/8).

    e.   **CB bases that are below FL 240 are shown as XXX.**

        1)   CB tops are to be expressed in hundreds of feet MSL.

        2)   The area to which the forecast applies is shown by scalloped lines.

    f.   **EXAMPLE:**

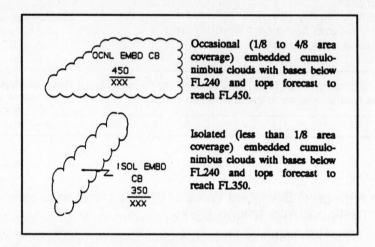

2.   Tropical cyclones are depicted by the symbol shown in the upper left-hand corner of the diagram below.

    a.   Areas of associated cumulonimbus activity, if meeting the previously given criteria (ISOL EMBD CB, OCNL EMBD CB, ISOL CB IN HAZE, OCNL CB IN HAZE, FRQ CB), are enclosed by scalloped lines and labeled with the vertical extent.

    b.   **EXAMPLE:**

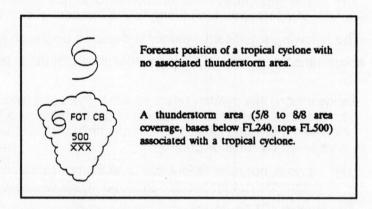

    c.   Notes:

        1)   The names of tropical cyclones, when relevant, are entered adjacent to the symbol.

        2)   A significant weather chart depicting the tropical cyclone symbol will state that the latest tropical cyclone advisory, rather than the tropical cyclone's forecast position on the chart, is to be given public dissemination.

3.   Severe squall lines. The example below shows a forecast severe squall line with associated CBs. Coverage 5/8 to 8/8 with bases below FL 240 and top forecast to reach FL 500.

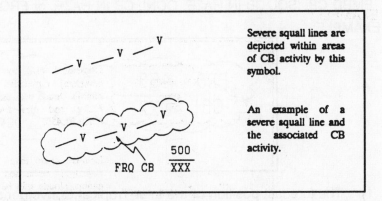

Severe squall lines are depicted within areas of CB activity by this symbol.

An example of a severe squall line and the associated CB activity.

4.   Moderate or severe turbulence (in clouds or clear air)

a.   Areas of forecast moderate or greater clear air turbulence (CAT) are bounded by heavy, dashed lines.

1)   CAT includes all turbulence (wind shear induced and mountain wave induced) not caused by convective activity.

b.   Areas are labeled with the appropriate turbulence symbol (Table 8-1 on page 319) and the vertical extent in hundreds of feet MSL.

1)   EXAMPLE:

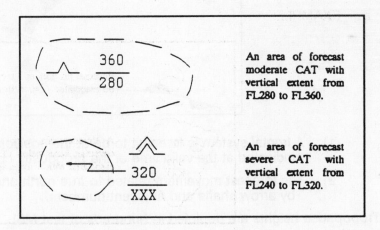

An area of forecast moderate CAT with vertical extent from FL280 to FL360.

An area of forecast severe CAT with vertical extent from FL240 to FL320.

5.   Widespread sandstorm/duststorm

a.   Areas of widespread sandstorms (BN) and duststorms (BD) are enclosed by scalloped lines and labeled by symbol and vertical extent.

b.   EXAMPLE:

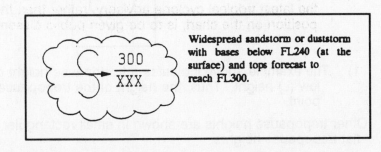

Widespread sandstorm or duststorm with bases below FL240 (at the surface) and tops forecast to reach FL300.

6. Areas of cumulonimbus activity associated with the Inter-Tropical Convergence Zone are enclosed by scalloped lines and labeled with the vertical extent.

    a. Note that the CBs must meet the previous given criteria (ISOL EMBD CB, OCNL EMBD CB, ISOL CB IN HAZE, OCNL CB IN HAZE, or FRQ CB).

    b. EXAMPLE:

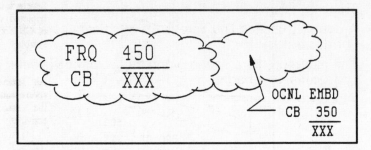

       1) The forecast position of the Inter-Tropical Convergence Zone is shown by the associated thunderstorm areas.

       2) The coverage for the frequent CBs is 5/8 to 8/8 with bases below FL 240 with tops at FL 450.

       3) The coverage for the occasional CBs is 1/8 to 4/8 with bases below FL 240 and tops at FL 350.

7. The forecast surface positions, speed, and direction of movement of frontal systems associated with significant weather are also depicted.

    a. EXAMPLE:

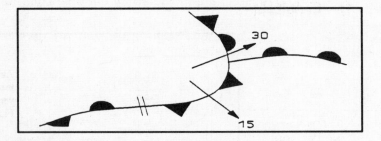

       1) A frontal system is forecast to be at the position and with the orientation indicated at the valid time of the prognostic chart.

       2) The forecast movement related to true north and speed in knots are indicated by arrow shafts and adjacent numbers.

8. Tropopause heights are depicted in hundreds of feet MSL.

    a. A five-sided polygon (as shown below) depicts only an area of the highest (peak) or lowest (depression) height of the tropopause.

       1) The example above indicates a tropopause height of 22,000 ft. MSL which is a low (L) height. Thus, the height of the tropopause is higher around this point.

    b. Other tropopause heights are shown in small rectangular boxes to define areas of flat tropopause heights.

c.   EXAMPLE: In Figure 8-3 below, the height of the tropopause over New Orleans is 45,000 ft. MSL. It slopes down (decreases height) to a low of 32,000 ft. MSL over northeastern North Dakota.

9.   The height and maximum wind speed of jet streams having a core speed of 80 kt. or greater are shown. The height is given as a flight level (FL).

a.   The beginning of the line indicates a core speed of 80 kt.

b.   A double, hatched line across the jet stream core indicates a 20-kt. speed increase or decrease.

c.   The maximum core speed along the jet stream is depicted by arrow shafts, pennants, and barbs.

d.   EXAMPLE:

1)   A jet stream with a forecast maximum speed of 130 kt. (two pennants @ 50 kt. + three barbs @ 10 kt.) at a height of 42,000 ft. MSL.

a)   The extreme left line starts at 80 kt.

b)   The first hatched, double line indicates a speed increase of 20 kt. to 100 kt. and the second double-hatched line shows an increase of 20 kt. to 120 kt.

c)   The double-hatched line to the right of the maximum speed indicates a decrease of 20 kt. to 120 kt.

2)   Wind directions are indicated by the orientation of arrow shafts in relation to true north.

C.   **U.S. High-Level Significant Weather Prog** -- as shown in Figure 8-3, encompasses airspace from 24,000 ft. to 63,000 ft. pressure altitude.

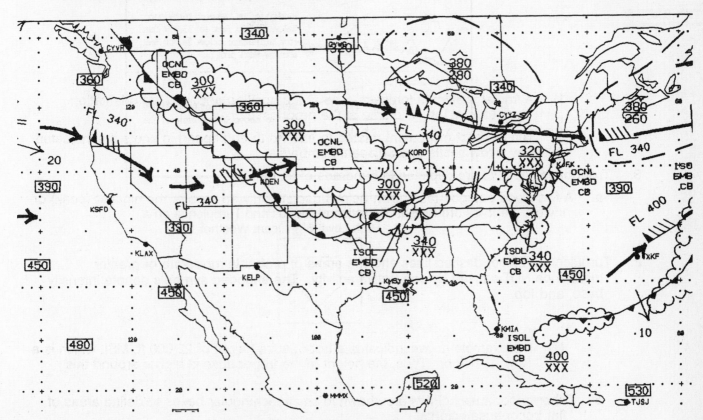

Figure 8-3. U.S. High-Level Significant Prog

1.   Figure 8-3 outlines areas of forecast turbulence and cumulonimbus clouds.

   a.   Table 8-3 below interprets some examples of chart notation.

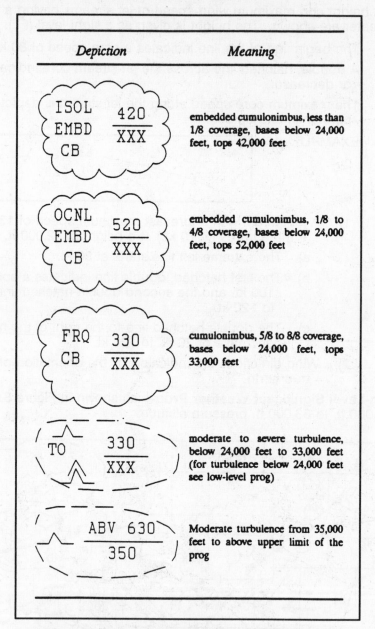

Table 8-3. Depiction of Clouds and Turbulence on a
High-Level Significant Weather Prog

2.   Turbulence. Large, dashed lines enclose areas of probable moderate or greater turbulence NOT caused by convective activity. The encoded symbols denote intensity, base, and top.

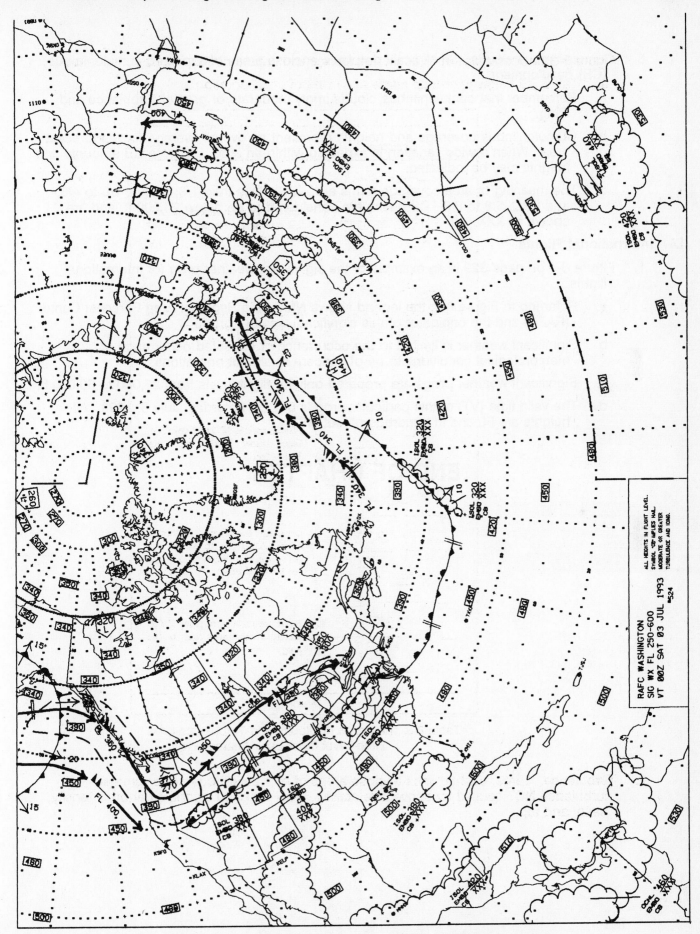

Figure 8-4. International High-Level Significant Weather Prog

3.   Cumulonimbus clouds. Small scalloped lines enclose areas of expected cumulonimbus (CB) development.

a.   Remember that cumulonimbus clouds imply moderate or greater turbulence and icing.

b.   Cumulonimbus coverage and heights represent an overall average for the forecast area. When a wide variation is expected within an area, separate CB amounts and heights may be indicated.

c.   The meaning of area coverage terms are: ISOL, less than 1/8; OCNL, 1/8 to 4/8; and FRQ, 5/8 to 8/8. CB bases are considered to be below 24,000 ft. and are shown as XXX.

D.   International Flights

1.   Figure 8-4 on page 329 is an example of the significant weather prog for international flights.

a.   Referring to Figure 8-4, the legend shows NMC as a Regional Area Forecast Center (RAFC) and the originator of this significant weather prog.

b.   Significant weather is limited to the occurrence or expected occurrence of meteorological conditions to be of concern to aircraft operations.

c.   Significant weather progs are prepared only for flight levels from 25,000 to 60,000 ft.

d.   The valid time (VT) of this particular prog is 00Z on Saturday, July 3, 1993. All heights are FL and in hundreds of feet MSL.

END OF CHAPTER

CHAPTER TWENTY-ONE
WINDS AND TEMPERATURES ALOFT CHARTS

> Please take a few minutes to study each of the concepts listed above and anticipate/imagine what they are and how they relate to the other listed concepts.

This chapter contains two types of charts for winds and temperatures aloft: forecast and observed. Also recall that winds and temperatures aloft forecasts are available in a table format as described and illustrated in Part III, Chapter 10, Winds and Temperatures Aloft Forecast (FD), beginning on page 285.

A. Forecast Winds and Temperatures Aloft (FD)

1. Forecast winds and temperatures aloft are prepared for eight levels on eight separate panels.

   a. The levels are 6,000; 9,000; 12,000; 18,000; 24,000; 30,000; 34,000; and 39,000 ft. MSL.

   b. They are available daily as 12 hr. progs valid at 1200Z and 0000Z.

      1) A legend on each panel shows the valid time and the level of the panel.

   c. Levels through 12,000 ft. are in true altitude and levels at and above 18,000 ft. are in pressure altitude.

   d. Figure 9-1 on page 332, is an example of a winds and temperatures aloft forecast panel for 24,000 ft. MSL.

2. Each station that prepares a winds and temperatures aloft forecast is represented on the panel by a station circle (see Table 9-1 on page 333).

   a. Temperature is in whole degrees Celsius for each forecast point and is entered above and to the right of the station circle.

   b. Arrows with pennants and barbs, similar to those used on the surface map, show wind direction and speed.

      1) Wind speed is indicated by the sum of three types of indicators.

         a) A half barb (ı) is 5 kt., a whole barb (ı) is 10 kt., and a pennant (▲) is 50 kt.

      2) See Figure 9-1 on page 332 for examples of these indicators on the chart.

   c. Wind direction is drawn to the nearest 10° with the second digit of the coded direction entered at the outer end of the arrow.

      1) To determine wind direction, obtain the general direction from the arrow and then use the digit to determine the direction to the nearest 10°.

      2) For example, a wind in the northwest quadrant with a digit of "3," indicates 330°.

      3) A calm or light and variable wind is shown by "99" entered to the lower left of the station circle.

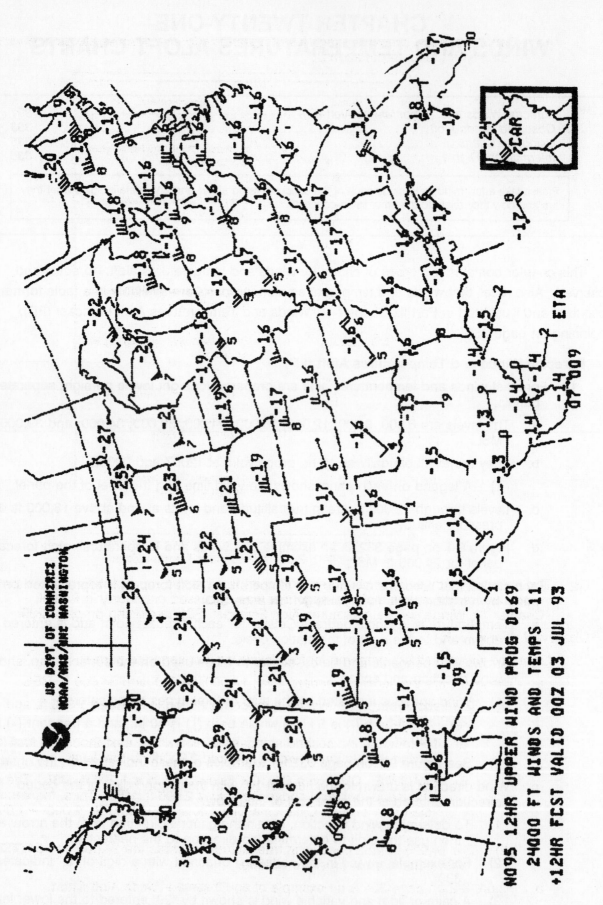

Figure 9-1. A Panel of Forecast Winds and Temperatures Aloft for 24,000 ft.

d. Table 9-1 below presents examples of a station's forecast temperatures and winds aloft with their interpretations.

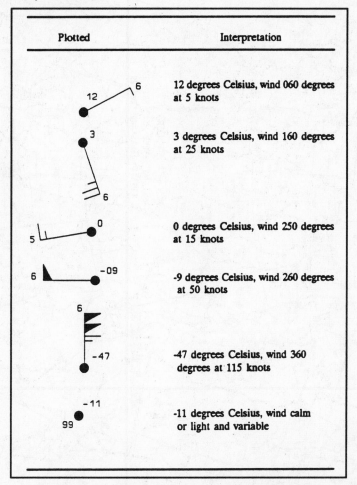

Plotted	Interpretation
	12 degrees Celsius, wind 060 degrees at 5 knots
	3 degrees Celsius, wind 160 degrees at 25 knots
	0 degrees Celsius, wind 250 degrees at 15 knots
	-9 degrees Celsius, wind 260 degrees at 50 knots
	-47 degrees Celsius, wind 360 degrees at 115 knots
	-11 degrees Celsius, wind calm or light and variable

Table 9-1. Plotted Winds and Temperatures

3. This forecast winds and temperatures aloft chart is a graphic representation of the forecast winds and temperatures aloft message that was discussed previously in Part III, Chapter 10, Winds and Temperatures Aloft Forecast (FD), beginning on page 285.

B. Observed Winds Aloft

1. Observed winds aloft are prepared for four levels on four separate panels.

   a. The levels are the second standard level, 14,000, 24,000, and 34,000 ft. MSL.

   1) The second standard level for a reporting station is between 1,000 ft. and 2,000 ft. AGL.

   a) To compute the second standard level, find the next thousand-foot level above the station elevation and add 1,000 ft. to that level.

   b) EXAMPLE: Oklahoma City, OK field elevation is 1,290 ft. MSL. The next thousand-foot level above the field is 2,000 ft. MSL. Thus, the second standard level for Oklahoma City, OK is 3,000 ft. MSL (2,000 + 1,000), or 1,710 ft. AGL.

   2) The 14,000 ft. MSL panel is in true altitude whereas the 24,000 and 34,000 ft. MSL panels are in pressure altitude.

   b. Figure 9-2 on page 334 is an example of an Observed Winds Aloft Chart.

   1) Figure 9-3 on page 335 is an example of a panel of Observed Winds and Temperatures Aloft for 34,000 ft. MSL.

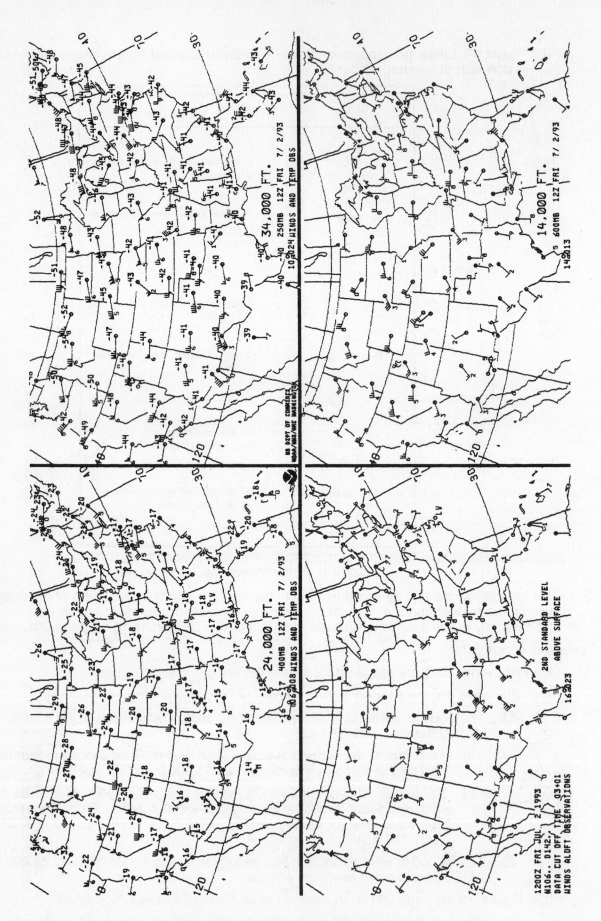

Figure 9-2. An Observed Winds Aloft Chart

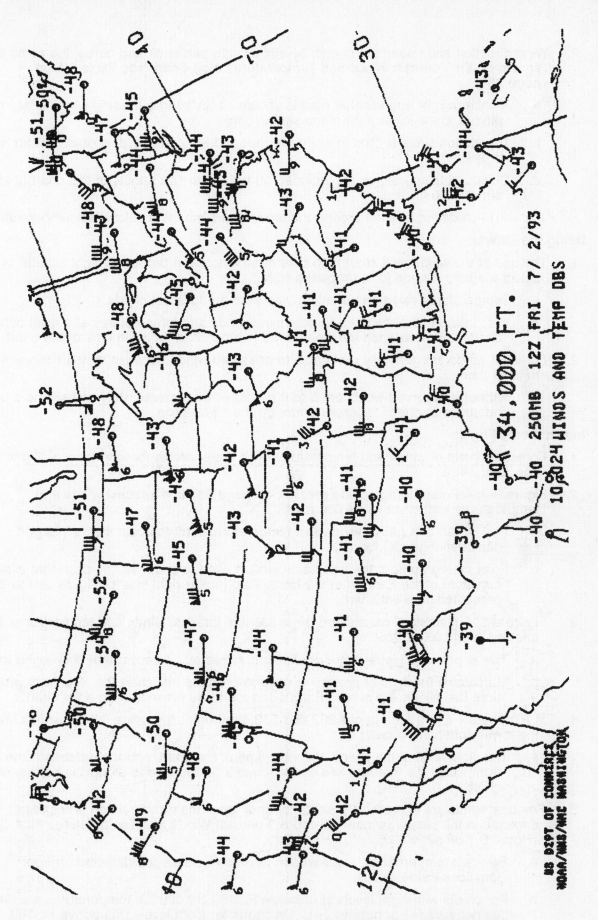

Figure 9-3. A Panel of Observed Winds and Temperatures Aloft for 34,000 ft.

2.   Wind direction and speed are shown by arrows with pennants and barbs, the same as shown on the Forecast Winds and Temperatures Aloft Chart (see Table 9-1 on page 333).

    a.   A calm or light and variable wind is shown as "LV" and a missing wind as "M," both plotted to the lower right of the station circle.

    b.   The station circle is filled in when the reported temperature/dew point spread is 5°C or less.

    c.   Observed temperatures are included on the upper two panels of this chart (24,000 ft. and 34,000 ft.).

        1)   A dotted bracket around the temperature means a calculated temperature.

C. Using the Charts

1.   The use of the winds aloft chart is to determine winds at a proposed flight altitude or to select the best altitude for a proposed flight.

    a.   Temperatures can also be determined from the forecast charts.

    b.   Interpolation must be used to determine winds and temperatures at a level between charts and data when the time period is other than the valid time of the chart.

2.   Forecast winds are generally preferable to observed winds because they are more relevant to flight time.

    a.   Although observed winds are 5 to 8 hr. old when received, they can still be a useful reference to check for gross errors on the 12-hr. prog.

D. International Flights

1.   Forecast charts of winds and temperatures aloft are available for international flights at specified levels.

2.   Figure 9-4, A Polar Sterographic Forecast, on page 337 is a forecast winds and temperatures aloft chart for 34,000 ft. MSL.

    a.   This is part of a global winds and temperatures aloft forecast that is in a grid (latitude/longitude) format.

    b.   Polar sterographic is the type of projection used to produce this chart that allows the curvature of the Earth from the North Pole (upper right side of Figure 9-4) to be presented on a flat chart.

3.   Figure 9-5, A Mercator Forecast, on page 338 is a forecast winds and temperatures aloft forecast for 45,000 ft. MSL.

    a.   This is part of a global winds and temperatures aloft forecast that is in a grid format.

    b.   Mercator is the type of projection used to produce this chart (i.e., Mercator projection) that allows the curved Earth's surface to be presented on a flat chart.

4.   In Figures 9-4 and 9-5 on pages 337 and 338, the originating office (NMC) is indicated in the lower right-hand corner.

    a.   The flight level of the charts, the valid time of the chart, and the database time (data from which the forecast was derived) make up the legend along the bottom of each chart.

5.   Forecast winds are expressed in knots for spot locations with direction and speed depicted in the same manner as the U.S. Forecast Winds and Temperatures Aloft Chart (Figure 9-1 on page 332).

    a.   Forecast temperatures, expressed in degrees Celsius, are depicted for spot locations inside circles.

    b.   For charts with flight levels at or below FL 180 (18,000 ft.), temperatures are depicted as negative (−) or positive (+). On charts for flight levels (FL) above FL 180, temperatures are always negative, and no sign is depicted.

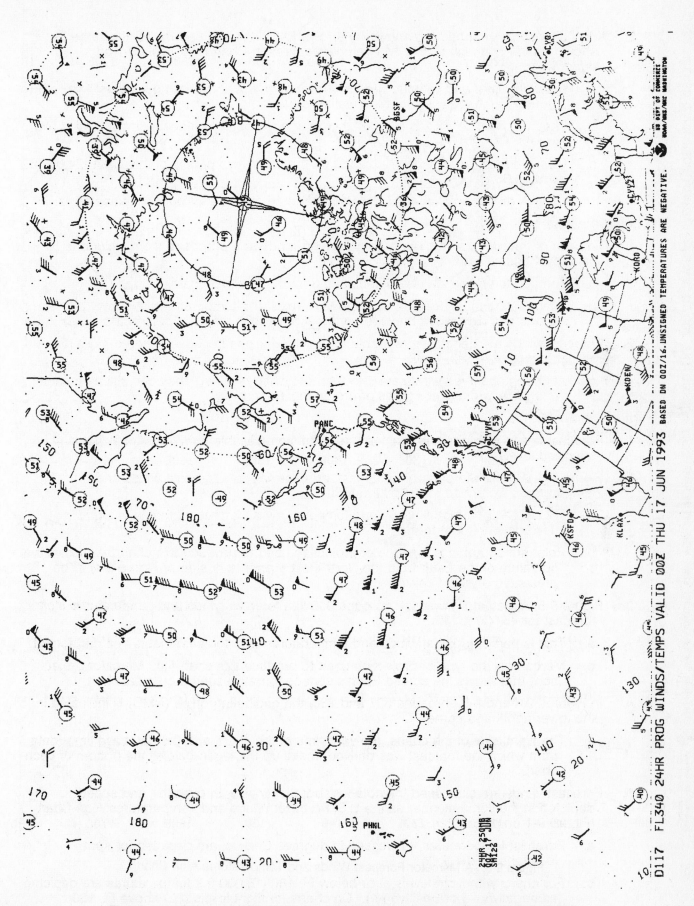

Figure 9-4. A Polar Sterographic Forecast Winds and Temperatures Aloft Chart

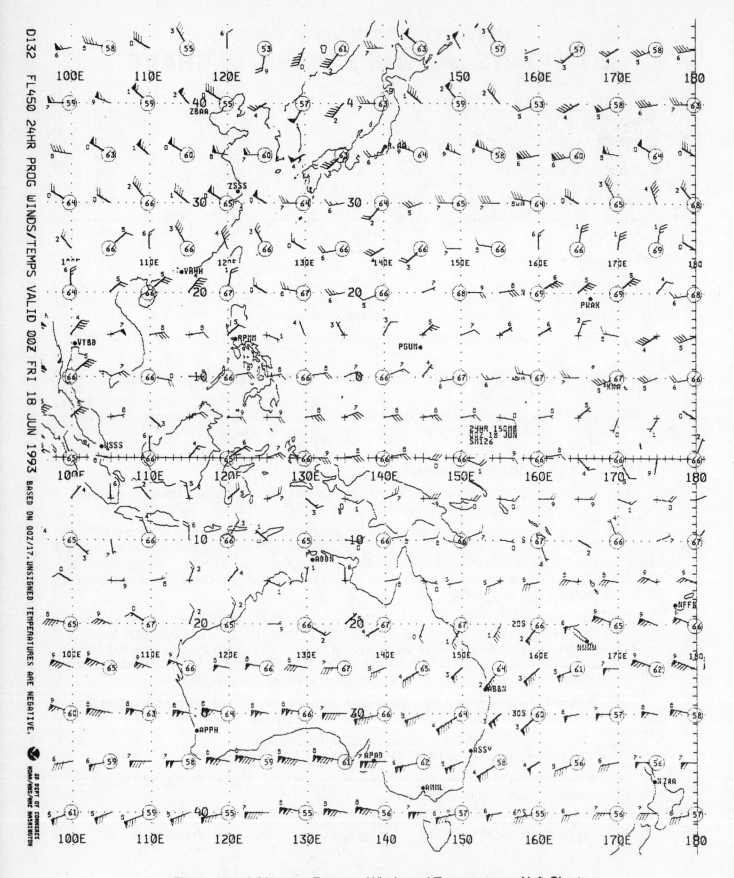

Figure 9-5. A Mercator Forecast Winds and Temperatures Aloft Chart

END OF CHAPTER

CHAPTER TWENTY-TWO
COMPOSITE MOISTURE STABILITY CHART

A. Type and Time of Report

1. The **composite moisture stability chart** (Figure 10-1 on page 340) is an analysis chart using observed upper air data. The chart is composed of four panels, including stability, freezing level, precipitable water, and average relative humidity.

    a. This computer-generated chart is available twice daily with valid times of 12Z and 00Z.

    b. On this chart, the pressure levels used are surface, 1000 mb, 850 mb, 700 mb, and 500 mb.

        1) Significant levels are those where significant changes in temperature and/or moisture occur when compared to higher or lower levels.

    c. The availability of upper air data (on all the panels) for analysis is indicated by the shape of the station model.

        1) The station model legend is shown on the lower left side of the Precipitable Water panel (see upper right panel of Figure 10-1 on page 340).

B. Stability Panel

Stability Panel -- the upper left panel of the composite moisture stability chart outlines areas of stable and unstable air. (See Figure 10-2 on page 341 for an enlarged depiction of this panel.) There are two stability indices that are computed for each upper air station. The top value is the *lifted index*, which is plotted above a short line, and the *K index*, plotted below the line. An "M" indicates the value is missing.

1. The lifted index (LI) is computed as if a parcel of air near the surface were lifted to 500 mb (18,000 ft. MSL). As the air is lifted it cools, at approximately 2°C/1,000 ft., due to expansion. The temperature the parcel would have at 500 mb is then subtracted from the actual (environmental) 500-mb temperature. This difference is the lifted index which is positive, negative, or zero and indicates the stability of the parcel of air.

    a. A **positive index** means that a low-level parcel of air, *if lifted*, would be colder than the surrounding air at 500 mb.

        1) The air is stable and would resist vertical motion.
        2) Large positive values (+8) would indicate very stable air.

    b. A **negative index** means that a low-level parcel of air, *if lifted* to 500 mb, would be warmer than the surrounding air.

        1) The air is unstable and suggests the possibility of convection.
        2) Large negative values (−4 or less) would indicate very unstable air.

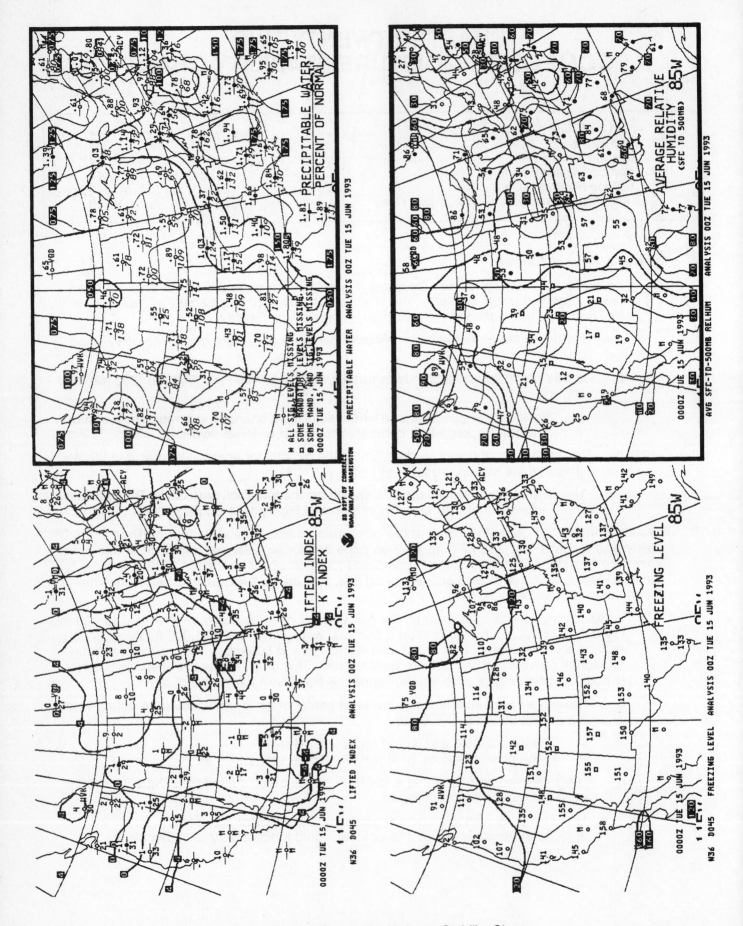

Figure 10-1. A Composite Moisture Stability Chart

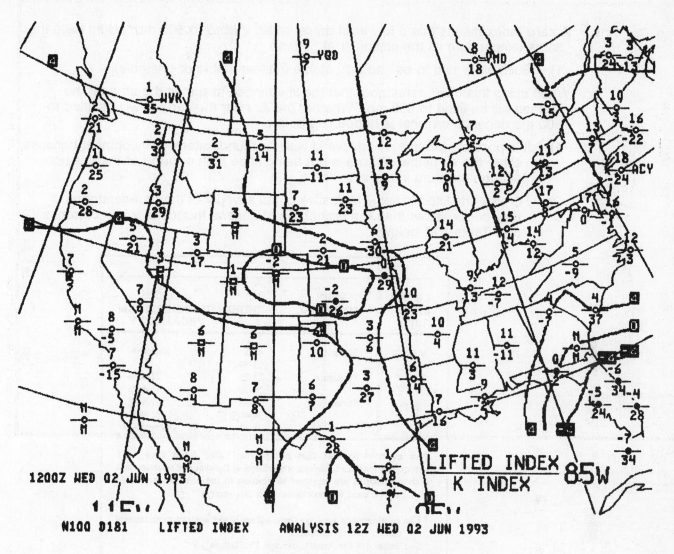

Figure 10-2. A Stability Panel

c. A **zero index** means that a low-level parcel of air, *if lifted* to 500 mb, would have the same temperature as the actual air at 500 mb.

    1) Such air is said to be neutrally stable (neither stable or unstable).

d. When using this chart, remember that the lifted index assumes the air near the surface will be lifted to 500 mb. Whether the air near the surface will be lifted to 500 mb depends on what is happening below.

    1) It is possible to have a negative LI with no thunderstorm development because either the air below 500 mb is not being lifted high enough or there is not enough moisture in the air.

    2) For use, the lifted index is indicative of the severity of the thunderstorms, if they occur, rather than the probability of general thunderstorm occurrence (see Table 10-1 below).

LIFTED INDEX (LI)	SEVERE POTENTIAL	"K" * INDEX	AIRMASS THUNDERSTORM PROBABILITY
0 to -2	Weak	<15	near 0%
		15-20	20%
-3 to -5	Moderate	21-25	21-40%
		26-30	41-60%
≤ -6	Strong	31-35	61-80%
		36-40	81-90%
		>40	near 100%

It is essential to note that an unstable Lifted Index does NOT automatically mean thunderstorms. Look at the synoptic situation and if thunderstorms are expected to develop in the unstable air, Table 10-1 may be used in accordance with this section.

* Use caution when applying these values in the western mountainous terrain due to elevation.

See Table 4-9 for Areal Coverage Definitions

Table 10-1. Thunderstorm Potential

    3) Also note that the LI can change dramatically just by daytime heating and nighttime cooling.

        a) Daytime heating tends to make the LI value decrease (i.e., more unstable) and nighttime cooling tends to make the LI increase (i.e., more stable).

2. The K index is primarily for the meteorologist. It examines the temperature and moisture profile of the environment. The K index is not really a stability index because the parcel of air is not lifted and compared to the environment. The K index is computed using three variables:

K = (850 mb temp − 500 mb temp) + (850 mb dew point) − (700 mb temp/dew point spread)

   a. The first variable (850 mb temp − 500 mb temp) compares the temperature at 850 mb (5,000 ft. MSL) to the temperature at 500 mb (18,000 ft. MSL).

      1) The larger the temperature difference, the more unstable the air and the higher the K index.

   b. The second variable (850 mb dew point) is a measure of low-level moisture.

      1) Note that since the dew point variable is added, high moisture content at 850 mb increases the K index.

   c. The third variable (700 mb temp/dew point spread) is a measure of saturation at 700 mb (10,000 ft. MSL).

      1) The greater the temperature/dew point spread, the drier the air; and since the term is subtracted, it lowers the K index.

      2) The greater the degree of saturation at 700 mb, the larger the K index.

   d. During the thunderstorm season, a large K index indicates conditions favorable for air mass thunderstorms (see Table 10-1, Thunderstorm Potential, on page 342).

      1) However, the K index and meaning can decrease significantly for thunderstorm development associated with steady state thunderstorms.

         a) Remember, steady state thunderstorms are usually associated with weather systems, and these systems will affect the variables used in computing the K index.

   e. In winter, because of cold temperatures and low moisture values, the temperature terms completely dominate the K index computation.

      1) Because of the lack of moisture, even fairly large values do not mean conditions are favorable for thunderstorms.

      2) Be aware that the K index can change significantly over a short time period due to temperature and moisture advection.

3. Analysis

   a. Stability is based on the lifted index only.

   b. Station circles are blackened for LI values of zero or below.

   c. Contour lines are drawn for values of +4 and below at intervals of 4 (+4, 0, −4, −8, etc.).

4. Using the stability panel

   a. When clouds and precipitation are forecast or are occurring, the LI is used to determine the type of clouds and precipitation. That is, stratiform clouds and continuous precipitation occur with stable air, while convective clouds and showery precipitation occur with unstable air.

   b. Stability is also very important when considering the type, extent, and intensity of aviation weather hazards. For example, a quick estimate of areas of probable convective turbulence can be made by associating the areas with unstable air.

      1) An area of extensive icing would be associated with stratiform clouds and steady precipitation which are characterized by stable air.

C. **Freezing Level Panel** -- the lower left panel of the composite moisture stability chart is an analysis of the observed freezing level data from upper air observations (see Figure 10-3, below).

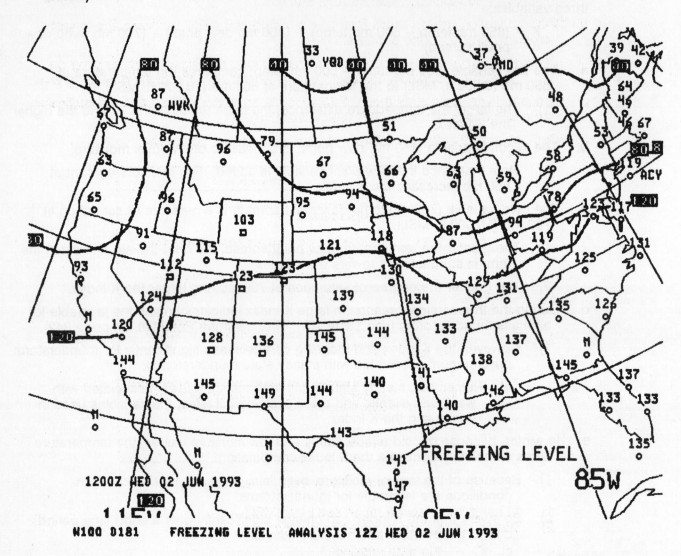

Figure 10-3. A Freezing Level Panel

1. Analysis

   a. Solid lines are contours of the lowest freezing height and are drawn for 4,000-ft. intervals and labeled in hundreds of feet MSL.

   b. When a station reports more than one crossing of the 0°C isotherm, the lowest crossing is used in the analysis.

      1) This convention is in contrast to the low-level significant weather prog on which the depicted forecast freezing level aloft is the highest freezing level.

   c. A dashed line represents the 32°F (0°C) isotherm at the surface and will outline an area of stations reporting "BF" (below freezing).

   d. See Table 10-2 below for interpretation of plotted freezing levels.

Plotted	Interpretation
O BF	Entire observation is below freezing (0 degree Celsius).
28 ✳	Freezing level is at 2,800 feet; Temperatures below freezing above 2,800 feet.
120 □	Freezing level at 12,000 feet; Temperatures above 12,000 feet are below freezing.
110 51 O BF	Temperatures are below freezing from the surface to 5,100 feet; above freezing from 5,100 to 11,000 feet and below freezing above 11,000 feet.
90 34 O 4	Lowest freezing level is at 400 feet; below freezing from 400 feet to 3,400 feet; above freezing from 3,400 to 9,000 feet and below freezing above 9,000 feet.
M O	Data are missing.

Table 10-2. Plotting Freezing Levels

   1) All heights are above mean sea level (MSL).
   2) Station models are depicted four ways:

      a) O   No data missing
      b) ✳   All significant levels missing
      c) □   Some mandatory levels missing
      d) ⊘   Some mandatory and significant levels missing

2.   Using the Freezing Level panel

    a.   The contour analysis shows an overall view of the lowest observed freezing level.

        1)   Always plan for possible icing in clouds or precipitation, especially between the temperatures of 0°C and −10°C.

    b.   Plotted multiple crossings of the 0°C isotherm always show an inversion of warm air above subfreezing temperatures (see Table 10-3 below).

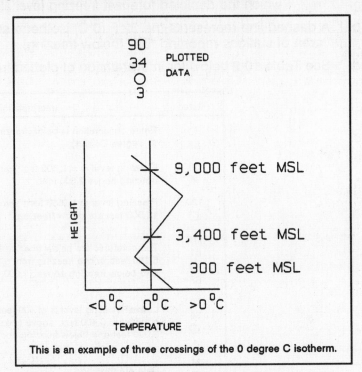

Table 10-3. Vertical Temperature Profile of Freezing
Levels

        1)   This situation can produce very hazardous icing when precipitation is occurring.

            a)   See AIRMET ZULU (for icing and freezing level), which will state the areas of expected icing more specifically.

            b)   The low-level significant weather prog shows anticipated changes in the freezing level.

D.   **Precipitable Water Panel** -- the upper right panel of the chart, is an analysis of the water vapor content from the surface to the 500-mb level (see Figure 10-4 below for an enlarged depiction of this panel). The amount of water vapor observed is shown as precipitation water, which is the amount of liquid precipitation that would result if all the water vapor were condensed.

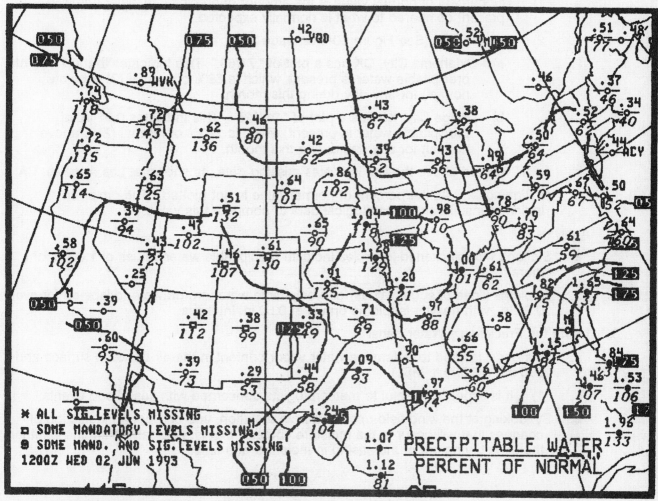

PRECIPITABLE WATER   ANALYSIS 12Z WED 02 JUN 1993

Figure 10-4. A Precipitable Water Panel

1. Plotted data

    a. At each station, precipitable water values to the nearest hundredth of an inch are plotted above a short line and the percent of normal value for the month below the line.

        1) The percent of normal value is the amount of precipitable water actually present compared to what is normally expected.

        2) EXAMPLES: (See Figure 10-4 on page 347)

            a) Oklahoma City, OK has a plot of ".71/69." This indicates that 0.71 in. of precipitable water is present, which is 69% of normal (31% below normal) for any day during this month.

            b) Dodge City, KS has a plot of ".91/125." This indicates 0.91 in. of precipitable water is present, which is 125% of normal (25% above normal) for any day during this month.

    b. An "M" plotted above the line indicates missing data, as shown at Los Angeles, CA.

    c. At Birmingham, AL the percent of normal value is not plotted. The omission indicates insufficient climatological data to compute this value.

2. Analysis

    a. Stations with blackened-in circles indicate precipitable water values of 1.00 in. or more.

    b. Isopleths (lines of equal values) of precipitable water are drawn and labeled for every 0.25 in., with heavier isopleths drawn at 0.50 in. intervals.

3. Using the Precipitable Water panel

    a. This panel is used to determine water vapor content in the air between surface and 500 mb (18,000 ft. MSL).

        1) It is especially useful to meteorologists concerned with flash flood events.

    b. By looking at the wind field upstream from a station, you can get an indication of changes that will occur in the moisture content; that is, you can determine whether the air is drying out or increasing in moisture with time.

E. **Average Relative Humidity Panel** -- the lower right panel of the chart is an analysis of the average relative humidity from the surface to 500 mb (see Figure 10-5 below for an enlarged depiction of this panel). The values are plotted as a percentage for each reporting station. An "M" indicates the value is missing.

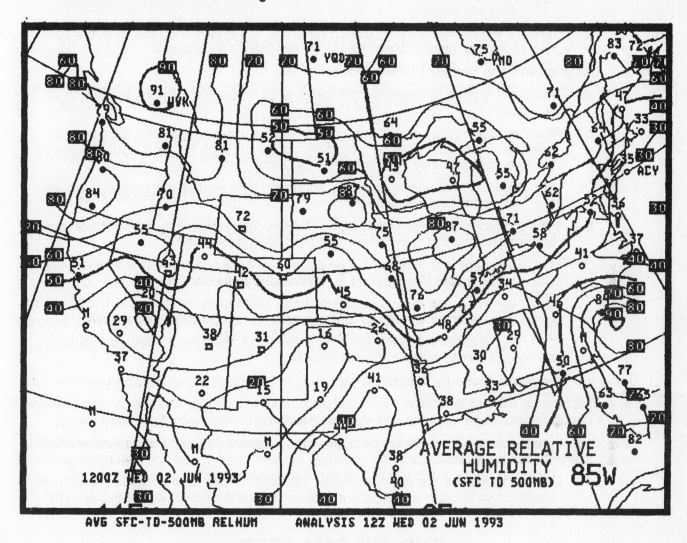

Figure 10-5. An Average Relative Humidity Panel

1.   Analysis

    a.   Station circles are blackened for humidities of 50% and higher.

        1)   Station identifiers depicted by a box or an asterisk indicate that some of the data are missing.

    b.   Isopleths of relative humidity, called isohumes, are drawn and labeled every 10% with heavier isohumes drawn for values of 10%, 50%, and 90%.

2.   Using the Average Relative Humidity Panel

    a.   This panel is used to determine the average air saturation from the surface to 500 mb.

    b.   When average relative humidities of 50% and greater are indicated, clouds and possible precipitation can be assumed, due to the high average relative humidity through approximately 18,000 ft. MSL (500 mb).

        1)   It is likely that a layer or layers will have 100% relative humidity with clouds and possibly precipitation.

    c.   It is important to remember that high values of relative humidity do not necessarily mean high values of water vapor content (precipitable water).

        1)   For example, in Figure 10-4 on page 347, Spokane, WA has less water vapor content than Key West, FL (.72 and 1.96, respectively).

            a)   However, in Figure 10-5 on page 349, the average relative humidities are nearly the same for both stations (81% and 82%, respectively).

            b)   If rain were falling at both stations, the result would likely be lighter precipitation totals for Spokane.

F.   **Using the Composite Moisture Stability Chart**

1.   This chart is used to determine the characteristics of a particular weather system in terms of stability, moisture, and possible aviation hazards.

2.   Even though this chart is several hours old when received, the weather system will tend to move these characteristics with it.

    a.   Caution should be exercised, however, because modification of these characteristics could occur through development, dissipation, or the movement of the system.

END OF CHAPTER

CHAPTER TWENTY-THREE
SEVERE WEATHER OUTLOOK CHART

> Please take a few minutes to study each of the concepts listed above and anticipate/imagine what they are and how they relate to the other listed concepts.

A. Type and Time of Forecast

1. The severe weather outlook chart, Figure 11-1 on page 352, is a 48-hr. outlook for thunderstorm activity.

   a. This chart is presented in two panels.

      1) The left-hand panel covers the first 24-hr. period beginning at 12Z and depicts areas of possible general thunderstorm activity as well as severe thunderstorms.

      2) The right-hand panel covers the following day beginning at 12Z and is an outlook for the possibility of severe thunderstorms only.

   b. This computer-prepared chart is issued once daily in the morning (about 08Z).

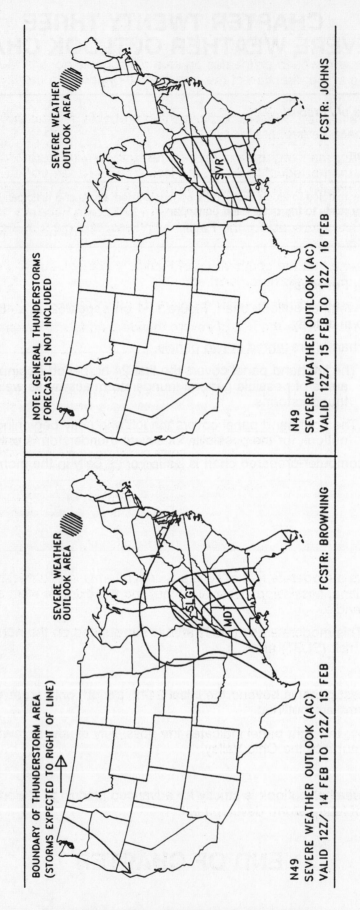

Figure 11-1. A Severe Weather Outlook Chart

B. **Left Panel**

 1. A line with an arrowhead delineates an area of probable *general* thunderstorm activity. When facing in the direction of the arrow, the thunderstorm activity is expected to the right of the line.

    a. An area labeled "APCHG" indicates that probable general thunderstorm activity may approach severe intensity.

       1) "Approaching" means that, at the surface, winds are expected to be greater than or equal to 35 kt., but less than 50 kt., and/or hail is expected to be greater than or equal to ½ in. in diameter, but less than ¾ in. at the surface.

    b. In Figure 11-1, from 12Z on February 14 to 12Z on February 15, general thunderstorms are forecast for the Pacific Northwest and most of the southeast part of the country.

       1) Note that the southern tip of Florida is not included in the general thunderstorm forecast area.

 2. The hatched area indicates possible severe thunderstorms.

    a. Table 11-1 shows the risk of severe thunderstorms and possible coverage.

Notation	Coverage
SLIGHT RISK	2 to 5% coverage or 4 to 10 radar grid boxes containing severe thunderstorms per 100,000 square miles.
MODERATE RISK	6 to 10% coverage or 11 to 21 radar grid boxes containing severe thunderstorms per 100,000 square miles.
HIGH RISK	More than 10% coverage or more than 21 radar grid boxes containing severe thunderstorms per 100,000 square miles.

Table 11-1. Notation of Coverage

    b. There is a moderate risk of severe thunderstorms in southeast Arkansas, most of Louisiana, Mississippi, west Alabama, and the extreme west section of the Florida Panhandle.

       1) This moderate risk (MDT) area is surrounded on the north and east by a slight risk (SLGT) area.

C. **Right Panel**

 1. As this forecast panel is beyond the initial 24-hr. period, **only** areas of possible severe thunderstorms are outlined.

 2. In Figure 11-1, the right panel indicates the possibility of severe thunderstorms from the Gulf Coast north to the Ohio Valley.

D. **Using the Chart**

 1. The severe weather outlook is strictly for advanced planning. It alerts all interests to the possibility of future storm development.

END OF CHAPTER

CHAPTER TWENTY-FOUR
CONSTANT PRESSURE ANALYSIS CHARTS

> Please take a few minutes to study each of the concepts listed above and anticipate/imagine what they are and how they relate to the other listed concepts.

A. Type and Time of Report

1. Any surface of equal pressure in the atmosphere is a constant pressure surface. A constant pressure analysis chart is an upper air weather map where all the information depicted is at the specified pressure of the chart.

   a. Twice daily, six computer-prepared constant pressure charts (850-mb, 700-mb, 500-mb, 300-mb, 250-mb, and 200-mb) are transmitted over the facsimile circuits.

      1) The valid times of these charts are the same as the radiosonde (observation) times, 12Z and 00Z.

      2) Plotted for the specified level at each reporting station is

         a) Observed temperature,

         b) Temperature-dew point spread,

         c) Wind,

         d) Height of the pressure surface, and

         e) The height changes over the previous 12-hr. period. Figures 12-2 through 12-7 are sections of each constant pressure chart.

   b. Pressure altitude (height in the standard atmosphere) for each of the six pressure surfaces is shown in Table 12-1 on page 356.

      1) For example, 700 mb of pressure has a pressure altitude of 10,000 ft. in the standard atmosphere.

         a) In the real atmosphere 700 mb of pressure only approximates 10,000 ft. because the real atmosphere is seldom standard.

      2) For direct use of a constant pressure chart, assume a flight is planned at 10,000 ft.

         a) The 700-mb chart is approximately 10,000 ft. MSL and is the best source for observed temperature, temperature-dew point spread, moisture, and wind for that flight level.

PRESSURE (millibars)	PRESSURE ALTITUDE in feet (flight level)	PRESSURE ALTITUDE in meters	TEMPERATURE DEW/POINT SPREAD	ISOTACHS	CONTROL INTERVAL (meters)	DECODE STATION HEIGHT PLOT PREFIX TO PLOTTED VALUE	DECODE STATION HEIGHT PLOT SUFFIX TO PLOTTED VALUE	EXAMPLES OF STATION HEIGHT PLOTTING PLOTTED	EXAMPLES OF STATION HEIGHT PLOTTING HEIGHT
850	5,000	1,500	yes	no	30	1	—	530	1,530
700	10,000	3,000	yes	no	30	2 or 3*	—	180	3,180
500	18,000	5,500	yes	no	60	——	0	582	5,820
300	30,000	9,000	yes**	yes	120	——	0	948	9,480
250	34,000	10,500	yes**	yes	120	1	0	063	10,630
200	39,000	12,000	yes**	yes	120	1	0	164	11,640

NOTE:
1. The pressure altitudes are rounded off to the nearest thousand for feet and to the nearest 500 for meters.
2. All heights are above Mean Sea Level (MSL).
3. * Prefix a "2" or "3," whichever brings the height closer to 3,000 meters.
4. ** Omitted when the air is too cold (temperature less than -41 degrees).

Table 12-1. Features of the Constant Pressure Charts - U.S.

B. Plotted Data

1. Figure 12-1 on page 357 illustrates and decodes the standard radiosonde data plot. Table 12-2 on page 357 gives a data plot example for each chart level.

   a. Aircraft and satellite observations are used in analysis over areas of sparse data.

      1) A square is used instead of a station circle to signify an aircraft report. The flight level of the aircraft is plotted in hundreds of feet.

      2) Temperature and wind data are also plotted for that flight level.

      3) The time of the report is indicated, to the nearest hour, UTC.

      4) EXAMPLE: Figure 12-5 on page 362 has an aircraft report at about 33°N and 125°W. Decoded, the report indicates the flight level was 33,000 ft., the temperature was −45°C, winds were from 320° at 40 kt., and the time of the report (to the nearest hour) was 2100 UTC (Z).

   b. A star identifies satellite wind estimates made from cloud tops.

      1) Figure 12-4 on page 361 has an example at about 52°N and 162°W.

      2) Decoded, the report indicates that the height was about 20,000 ft., winds were from 330° at 10 kt., and the time of the report (to the nearest hour) was 2200 UTC.

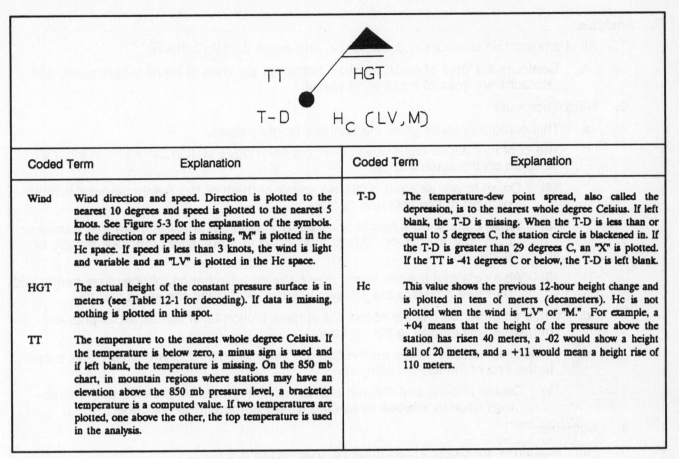

Figure 12-1. Radiosonde Data Station Plot

Coded Term	Explanation	Coded Term	Explanation
Wind	Wind direction and speed. Direction is plotted to the nearest 10 degrees and speed is plotted to the nearest 5 knots. See Figure 5-3 for the explanation of the symbols. If the direction or speed is missing, "M" is plotted in the Hc space. If speed is less than 3 knots, the wind is light and variable and an "LV" is plotted in the Hc space.	T-D	The temperature-dew point spread, also called the depression, is to the nearest whole degree Celsius. If left blank, the T-D is missing. When the T-D is less than or equal to 5 degrees C, the station circle is blackened in. If the T-D is greater than 29 degrees C, an "X" is plotted. If the TT is -41 degrees C or below, the T-D is left blank.
HGT	The actual height of the constant pressure surface is in meters (see Table 12-1 for decoding). If data is missing, nothing is plotted in this spot.	Hc	This value shows the previous 12-hour height change and is plotted in tens of meters (decameters). Hc is not plotted when the wind is "LV" or "M." For example, a +04 means that the height of the pressure above the station has risen 40 meters, a -02 would show a height fall of 20 meters, and a +11 would mean a height rise of 110 meters.
TT	The temperature to the nearest whole degree Celsius. If the temperature is below zero, a minus sign is used and if left blank, the temperature is missing. On the 850 mb chart, in mountain regions where stations may have an elevation above the 850 mb pressure level, a bracketed temperature is a computed value. If two temperatures are plotted, one above the other, the top temperature is used in the analysis.		

	850 MB	700 MB	500 MB	300 MB	250 MB	200 MB
WIND	LIGHT AND VARIABLE	010/20 KTS	210/60 KTS	270/25 KTS	240/30 KTS	MISSING
TT	22 C	9 C	-19 C	-46 C	-55 C	-60 C
T-D	4 C	17 C	>29 C	not plotted	not plotted	not plotted
DEW POINT	18 C	-8 C	DRY	DRY	DRY	DRY
HGT	1,479 m	3,129 m	5,580 m	9,190 m	10,370 m	11,910 m
Hc	not plotted	- 30 m	+ 30 m	+ 100 m	+ 10 m	not plotted

Table 12-2. Examples of Radiosonde Plotted Data

C. Analysis

1. All charts contain contours and isotherms, and some contain isotachs.

    a. Contours are lines of equal heights, isotherms are lines of equal temperature, and isotachs are lines of equal wind speed.

2. Height contours

    a. This contour analysis gives the charts a *height* pattern.

    b. The contours depict highs, lows, troughs, and ridges aloft in the same manner as isobars on the surface chart.

        1) On an upper air chart, then, we speak of "high or low height centers" instead of "high or low pressure centers."

        2) When comparing a height analysis to a pressure analysis, note that a *contour* high, low, trough, or ridge is analogous to a pressure high, low, trough, or ridge.

        3) Also note that the two terms may be used interchangeably because height and pressure analyses are just two ways of describing the same features.

    c. Since an upper air chart is above the surface friction layer, winds for all practical purposes flow parallel to the contours.

    d. To decode contour values on the 850-mb through 300-mb chart, simply affix a zero to the end of the three-digit code.

        1) On the 200-mb and 250-mb charts, a one (1) must be prefixed to the three-digit code in addition to placing a zero at the end of the code.

3. Isotherms

    a. Isotherms are dashed lines drawn at intervals of 5°C.

    b. The isotherm analysis shows horizontal temperature variations at that chart altitude.

    c. Figure 12-2 on page 359 is an example of an 850-mb chart. Note the dashed line extending from North Carolina west through Montana and labeled "+10" in northern Nebraska. This is the +10° isotherm.

        1) North of this isotherm, the temperatures, at approximately 5,000 ft., are below +10°C.

        2) South of the isotherm, the temperatures are above +10°C.

    d. By inspecting the isotherm pattern, one can determine if a flight would be toward colder or warmer air.

        1) Subfreezing temperatures and a temperature-dew point spread of 5°C or less would indicate the possibility of icing.

    e. On the 300-, 250-, and 200-mb charts, the isotherms are the heavy dashed lines.

4. Isotachs

    a. Isotachs, the short, lightly dashed lines, appear only on the 300-, 250-, and 200-mb charts.

    b. Isotachs are drawn at 20-kt. intervals beginning at 10 kt.

    c. To aid in identifying areas of strong winds, hatching denotes wind speeds of 70 to 110 kt.

        1) A clear area within a hatched area indicates that the wind speed is between 110 and 150 kt.

        2) Note the alternating hatched/clear areas (Figure 12-5 on page 362) that extend from west of Oregon in the Pacific Ocean. The clear area within the hatching indicates winds greater than 110 kt., but less than 150 kt.

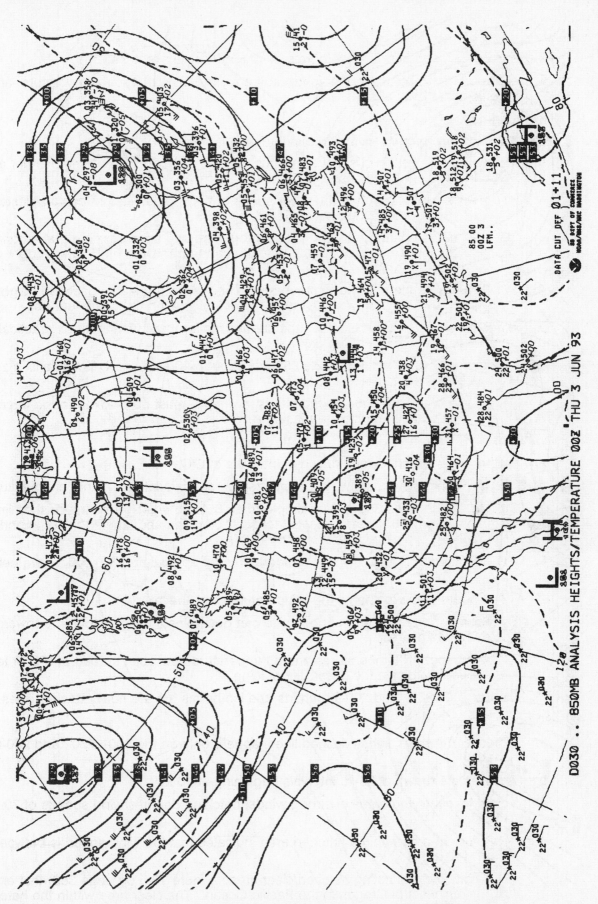

Figure 12-2. An 850-mb Analysis (Pressure Altitude 5,000 ft.)

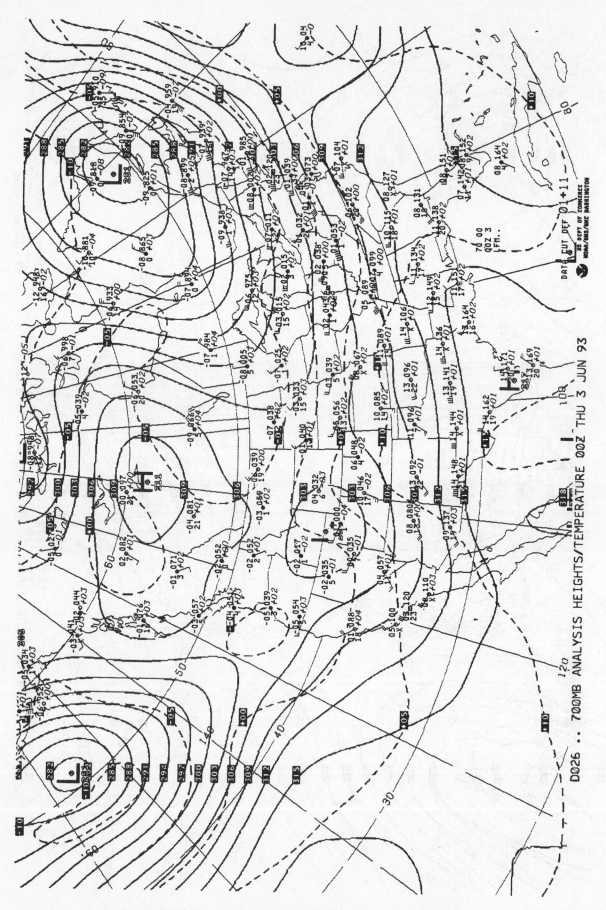

Figure 12-3. A 700-mb Analysis (Pressure Altitude 10,000 ft.)

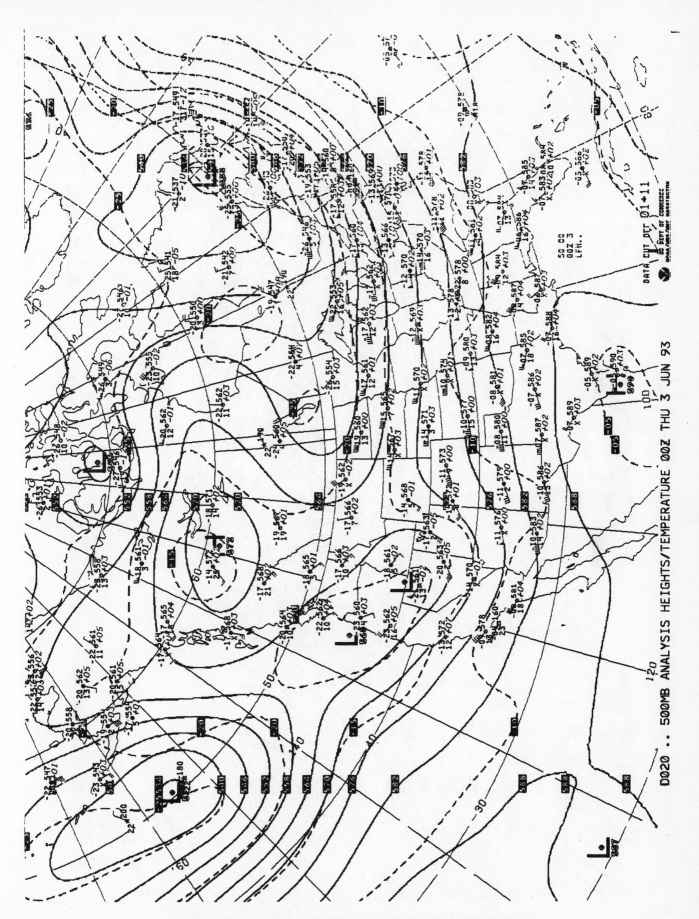

Figure 12-4. A 500-mb Analysis (Pressure Altitude 18,000 ft.)

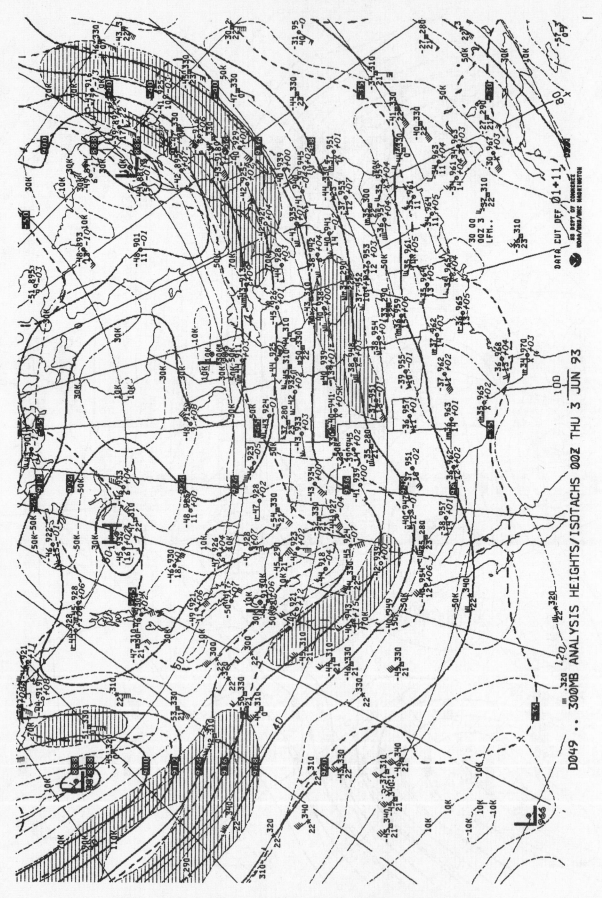

Figure 12-5. A 300-mb Analysis (Pressure Altitude 30,000 ft.)

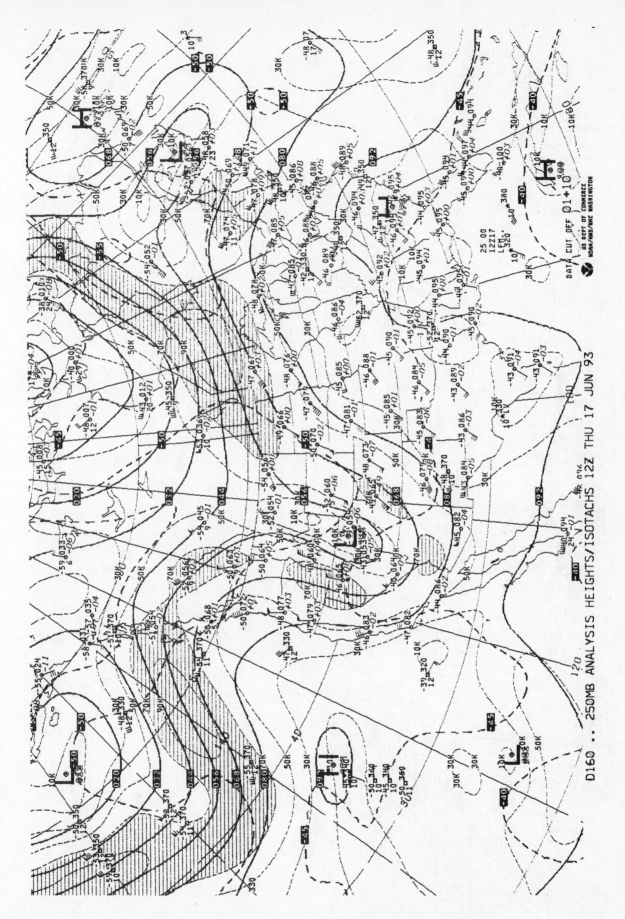

Figure 12-6. A 250-mb Analysis (Pressure Altitude 34,000 ft.)

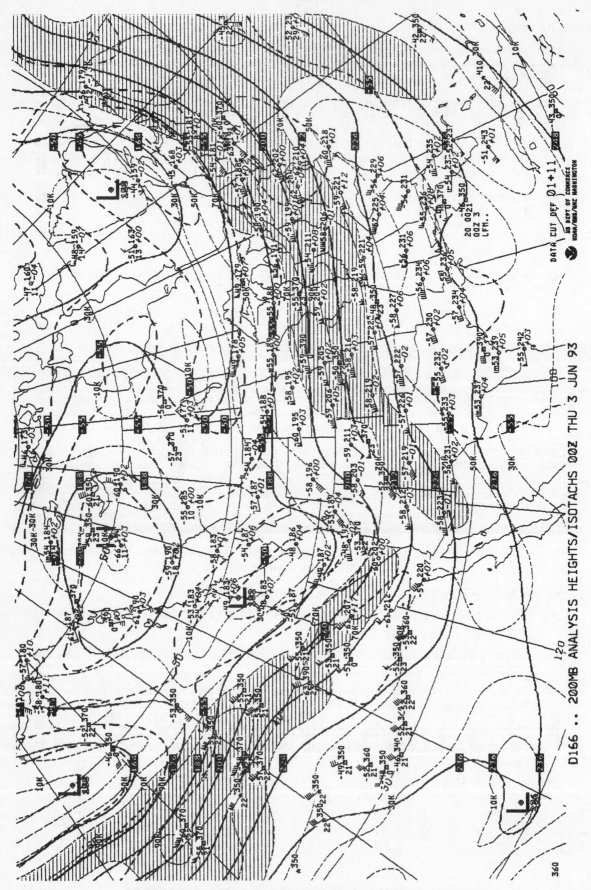

Figure 12-7.   A 200-mb Analysis (Pressure Altitude 39,000 ft.)

D. **Three-Dimensional Aspects**

1. As established earlier, a height analysis may be treated as a pressure analysis. Closely spaced contours indicate strong winds just like closely spaced isobars. Winds blow clockwise around a contour high and counter-clockwise around a low.

2. Features on a synoptic surface chart and the associated upper air charts are generally related.

    a. However, a weak surface system often either loses its identity in a large-scale upper air pattern or another system may be more evident on the upper air charts than on the surface chart.

    b. In fact, many times weather is more closely associated with an upper air pattern than with the features on the surface map.

3. As a general rule, a surface low is a producer of bad weather and a high is a producer of good weather. However, an upper air low or trough usually means bad weather, also.

    a. The area of cloudiness and precipitation found with an upper air low is usually associated with a surface low.

    b. Sometimes an upper level low with clouds and precipitation will move over a shallow surface high with corresponding bad weather in the high.

    c. As with a surface high, an upper air high usually means good weather.

        1) An exception would be an upper air high or ridge that has a stabilizing effect on the layers of the atmosphere below it.

        2) Smoke, haze, dust, low stratus, and fog may persist for an extended period but the surface map shows no cause for the restriction.

4. Lows generally slope to the west, toward colder air, with ascending altitude for developing low pressure systems.

    a. Due to this slope, winds aloft with an upper system often blow across the associated surface system.

5. Surface fronts, lows, and highs tend to move with the upper winds.

    a. For example, strong winds aloft across a front will cause the front to move rapidly, but if upper winds are parallel to a front, it moves slowly, if at all.

6. An old, nondeveloping low pressure system tilts little with height.

    a. The low becomes almost vertical and is clearly evident on both surface and upper air maps.

    b. Upper winds encircle the surface low, rather than blowing across it. This causes the storm to move very slowly.

    c. As a result, extensive and persistent cloudiness, precipitation, and generally adverse flying weather occur.

    d. The term "cold low" describes such a system and is usually identified on the surface chart as an old, occluded low with the warm air having been cut off from the low pressure center.

7. In contrast to the cold low is the "thermal low." A dry, sunny region becomes quite warm from intense surface heating. This results in a surface low pressure area.

    a. The warm air is carried to high levels by convective currents, but very few clouds occur because of the lack of moisture.

    b. This warm surface low often is "capped" by a high aloft.

    c. Unlike the cold low, the thermal low is relatively shallow with weak pressure gradients and no well-defined cyclonic circulation.

        1) However, be alert for high density altitude, light to moderate convective turbulence, and isolated showers and thunderstorms if sufficient moisture is present.

    d. The thermal low is a semipermanent feature of the desert regions in the southwestern United States and northern Mexico during warm weather.

8. These are only a few examples of associating weather with upper air features. They point out the need to view weather in the three dimensions to get a complete picture of the atmosphere. This is the first step in understanding the atmosphere and its associated weather.

E. Using the Charts

1. From these charts, a pilot can approximate the observed temperature, wind, and temperature-dew point spread along a proposed route.

2. A constant pressure chart usually can be selected close to a proposed flight altitude.

    a. For an altitude about midway between two charted surfaces, interpolate between the two charts.

3. Determine temperature from plotted data or the pattern of isotherms.

    a. To find areas of high moisture content, look for reports that have the station circle shaded.

        1) This indicates a temperature-dew point spread of 5°C or less. A small spread indicates the possibility of clouds, precipitation, and icing.

4. Wind speed from the 300-mb, 250-mb, and 200-mb charts can be determined by the isotachs. Below this level, wind speeds can be determined from the plotted data.

5. As stated earlier, constant pressure charts often show the cause of weather and its movement more clearly than does the surface map.

    a. For example, the large-scale wind flow around a low aloft may spread cloudiness, low ceilings, and precipitation far more extensively than indicated by the surface map alone.

6. Note: Keep in mind that constant pressure charts are observed weather.

END OF CHAPTER

CHAPTER TWENTY-FIVE
TROPOPAUSE DATA CHART

> Please take a few minutes to study each of the concepts listed above and anticipate/imagine what they are and how they relate to the other listed concepts.

A. Type and Time of Forecast

1. The tropopause data chart is a two-panel chart containing a maximum wind prog and a vertical wind shear prog.

    a. The chart is prepared for the contiguous 48 states (Figure 13-1 on page 368) and is available once a day with a valid time of 18Z.

    b. Both panels show forecast parameters at the tropopause level. The first panel depicts the forecast winds at the tropopause and the second panel gives the tropopause height and vertical wind shear (VWS).

B. Tropopause Winds

1. The tropopause winds prog depicts wind direction by streamlines. The streamlines are the solid lines that are not labeled. See Figure 13-2 on page 369 for an enlarged depiction of this panel.

    a. Since winds parallel the streamlines and generally flow from west to east, direction can be obtained by following the streamline flow.

    b. A high or low may be encircled by a closed streamline.

    c. To determine whether a closed streamline is a high or a low you must memorize the circulation around these systems.

2. Wind speed is shown by isotachs at 20-kt. intervals. The isotachs are the dashed lines and are labeled in knots.

    a. Areas of wind speeds between 70 and 110 kt. are hatched as are wind speeds between 150 and 190 kt.

    b. Note that the shading criteria are the same as used on the higher level constant pressure analysis and progs.

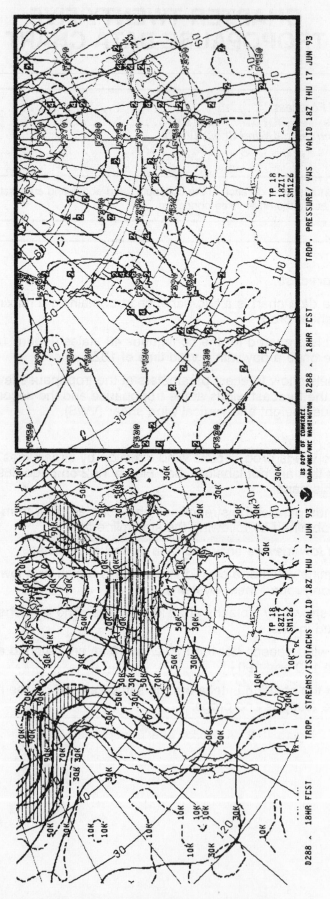

Figure 13-1. A Tropopause Data Chart

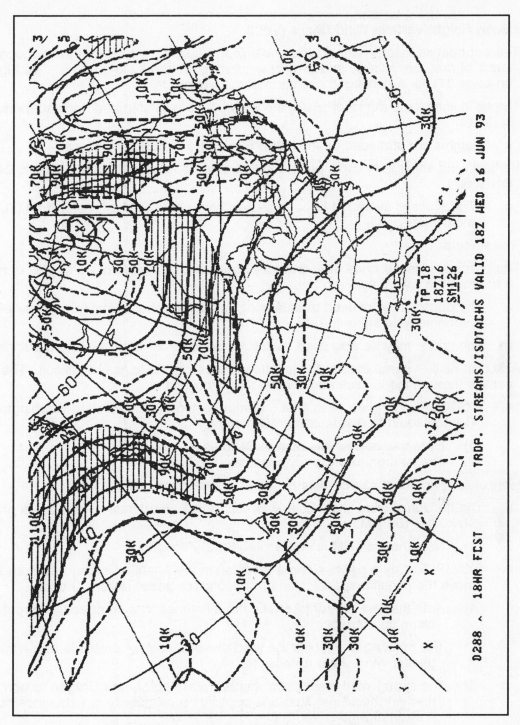

Figure 13-2. A Section of a Tropopause Wind Prog

C. **Tropopause Height/Vertical Wind Shear (VWS)**

1. The tropopause height/vertical wind shear prog depicts the height of the tropopause in terms of pressure altitude and vertical wind shear in knots per 1,000 ft. See Figure 13-3 on page 371 for an enlarged depiction of this panel.

2. The solid lines trace intersections of the tropopause with standard constant pressure surfaces.

    a. Heights are preceded by the letter "F" and are in hundreds of feet.

3. Vertical wind shear is in knots per 1,000 ft. and is depicted by dashed lines at 2-kt. intervals.

    a. Wind shear is averaged through a layer from about 8,000 ft. below to 4,000 ft. above the tropopause.

D. **Using the Panels**

1. The progs are issued once daily and may be used for a period of up to plus or minus 6 hr. from the valid time.

    a. The panels may be used to determine vertical and horizontal wind shear as clues to probable wind shear turbulence.

    b. The charts may be also used to determine winds for high level flight planning.

2. Although neither panel depicts the jet stream, locating the jet is not difficult. The jet passes through the isotach and vertical shear maxima.

    a. EXAMPLE: In Figure 13-2, a jet maximum extends from western Washington and Oregon south through northern California.

        1) It reappears over the eastern Dakotas.
        2) The jet then extends northeast into Canada.

3. Horizontal wind shear can be determined from the spacing of the isotachs.

    a. The horizontal wind shear critical for turbulence (moderate or greater) is greater than 18 kt. per 150 NM.

        1) Note that 150 NM is equal to about 2½° longitude.

    b. EXAMPLE: Lay a pencil along a meridian in the Atlantic Ocean. Measure 2½° and move the pencil perpendicular to the isotach across western Oregon.

        1) Note that the horizontal shear, the difference in wind speed, is about 40 kt. along this distance.

        2) This spacing represents the wind shear critical for probable moderate or greater wind shear turbulence.

        3) The strong wind shear from western Washington and Oregon to northern Utah then northeast into Montana suggests a probability of turbulence due to horizontal wind shear.

4. Vertical wind shear can be determined directly from the dashed lines in Figure 13-3.

    a. The vertical shear critical for probable turbulence is 6 kt. per 1,000 ft.

        1) This critical value can be found in central Oregon and in eastern Canada.

        2) An area of extremely high probability of moderate or greater turbulence is over Oregon and Washington. This is where the horizontal shear is about 60 kt. per 150 NM and the vertical shear is in excess of 6 kt. per 1,000 ft.

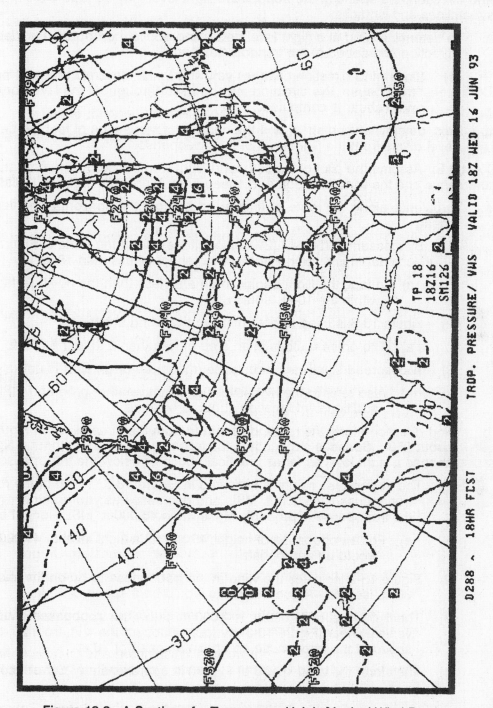

Figure 13-3. A Section of a Tropopause Height/Vertical Wind Prog

5. Wind direction and speed at the tropopause flight level may be read directly from the streamlines and isotachs.

    a. To determine wind at a flight level below and above the tropopause, determine the direction and speed at the tropopause.

        1) Since wind direction changes very little within several thousand feet of the tropopause, this direction may be used throughout the layer for which vertical wind shear is computed.

    b. Next, determine wind shear and the number of thousands of feet the desired flight level differs from the flight level of the tropopause.

6. EXAMPLE: Assume the pilot of a westbound flight wants to know the probability of turbulence and the wind direction and speed from Atlanta, Georgia, to Dallas, Texas.

    a. Turbulence along the route is determined by both the horizontal and vertical wind shear.

        1) By measuring 150 NM (2½° of longitude) along the route in Figure 13-2, you can determine the horizontal wind shear is less than 18 kt. per 150 NM.

            a) In fact, the flight begins in an area less than the 30-kt. isotach and ends within the 10-kt. isotach.

        2) Figure 13-3 is used to determine vertical wind shear.

            a) The entire route is in an area of less than 2 kt. per 1,000 ft.

        3) Widespread significant turbulence (moderate or greater) is unlikely.

            a) Also refer to the high-level significant weather prog and PIREPs for additional information about turbulence.

    b. Wind direction along the route, determined from the streamlines in Figure 13-2, is about 250°. Speed is strongest at the tropopause, so a westbound flight should choose a flight level as far as practical above or below the tropopause.

        1) The tropopause height is indicated in Figure 13-3. North of the route the tropopause is 45,000 ft. MSL, and decreases as you go farther north. Thus, the tropopause is probably higher than 45,000 ft. MSL along the route.

            a) For a tropopause of height 45,000 ft., a flight level of 41,000 ft. MSL would be appropriate.

        2) Figure 13-2 indicates the wind at the tropopause to be on the low side of the 30-kt. isotach. A good estimate would be a speed of 20 kt.

        3) The flight level of 41,000 ft. is 4,000 ft. below the tropopause. Multiply the 1-kt. vertical shear (Figure 13-3) by four. Subtract the 4 kt. from 20, the wind speed at the tropopause, to obtain a speed of 16 kt.

        4) Therefore, the wind speed at FL 410 is approximately 250° at 16 kt.

END OF CHAPTER

CHAPTER TWENTY-SIX
TABLES AND CONVERSION GRAPHS

Please take a few minutes to study each of the concepts listed above and anticipate/imagine what they are and how they relate to the other listed concepts.

A. Introduction

1. This chapter provides graphs and tables that can be used operationally in decoding weather messages during preflight and in-flight planning and in transmitting pilot reports. Information included covers

   a. Icing intensities and reporting.

   b. Turbulence intensities and reporting.

   c. Locations of probable turbulence by intensity as it relates to weather and terrain features.

   d. Density altitude computations.

   e. Selected contractions.

   f. Selected acronyms.

   g. Standard temperature, speed, and pressure conversions.

   h. Scheduled issuance and valid times of forecast products.

B. Icing Intensities

1. The table of icing intensities (Table 14-1 on page 374) classifies each intensity according to its operational effects on aircraft.

C. Turbulence Intensities

1. The table of turbulence intensities (Table 14-2 on page 375) classifies each intensity according to its effects on aircraft control and structural integrity and on articles and occupants within the aircraft.

Intensity	Airframe ice accumulation	Pilot report
Trace	Ice becomes perceptible. Rate of accumulation slightly greater than rate of sublimation. It is not hazardous even though deicing/anti-icing equipment is not used unless encountered for an extended period of time (over one hour).	Aircraft indenification, location, time UTC, intensity and type of icing*, altitude/FL, aircraft type, IAS
Light	The rate of accumulation may create a problem if flight is prolonged in this environment (over one hour). Occasional use of deicing/anti-icing equipment removes/prevents accumulation. It does not present a problem if the deicing/anti-icing equipment is used.	
Moderate	The rate of accumulation is such that even short encounters become potentially hazardous and use of deicing/antiicing equipment or diversion is necessary.	
Severe	The rate of accumulation is such that deicing/anti-icing equipment fails to reduce or control the hazard. Immediate diversion is necessary.	

* Icing may be rime, clear and mixed.

Rime ice:	Rough milky opaque ice formed by the instanteous freezing of small supercooled water droplets.
Clear ice:	A glossy, clear or translucent ice formed by the relatively slow freezing of large supercooled water droplets.
Mixed ice:	A combination of rime and clear ice.

Table 14-1. Icing Intensities, Airframe Ice Accumulation, and Pilot Report

Intensity	Aircraft Reaction	Reaction Inside Aircraft	Reporting Term Definition
Light	Turbulence that momentarily causes slight, erratic changes in altitude and/or attitude (pitch, roll, yaw). Report as *Light Turbulence*.* **or** Turbulence that causes slight, rapid, and somewhat rhythmic bumpiness without appreciable changes in altitude or attitude. Report as *Light Chop*.	Occupants may feel a slight strain against belts or shoulder straps. Unsecured objects may be displaced slightly. Food service may be conducted and little or no difficulty is encountered in walking.	Occasional -- less than 1/3 of the time. Intermittent -- 1/3 to 2/3 of the time. Continuous -- More than 2/3 of the time.
Moderate	Turbulence that is similar to Light Turbulence but of greater intensity. Changes in altitude and/or attitude occur but the aircraft remains in positive control at all times. It usually causes variations in indicated airspeed. Report as *Moderate Turbulence*.* **or** Turbulence that is similar to Light Chop but of greater intensity. It causes rapid bumps or jolts without appreciable changes in aircraft altitude or attitude. Report as *Moderate Chop*.	Occupants feel definite strains against seat belts or shoulder straps. Unsecured objects are dislodged. Food service and walking are difficult.	**NOTE** 1. Pilots should report location(s), time (UTC), intensity, whether in or near clouds, altitude, type of aircraft and, when applicable, duration of turbulence.
Severe	Turbulence that causes large, abrupt changes in altitude and/or attitude. It usually causes large variations in indicated airspeed. Aircraft may be momentarily out of control. Report as *Severe Turbulence*.*	Occupants are forced violently against seat belts or shoulder straps. Unsecured objects are tossed about. Food service and walking are impossible.	2. Duration may be based on time between two locations or over a single location. All locations should be readily identifiable.
Extreme	Turbulence in which the aircraft is violently tossed about and is practically impossible to control. It may cause structural damage. Report as *Extreme Turbulence*.*		

* High level turbulence (normally above 15,000 feet AGL) that is not associated with cumuliform cloudiness, including thunderstorms, should be reported as CAT (clear air turbulence) preceded by the appropriate intensity, or light or moderate chop.

Table 14-2. Turbulence Reporting Criteria

D. **Locations of Probable Turbulence by Intensities** as it relates to weather and terrain features

   1. Light turbulence

      a. In hilly and mountainous areas, even with light winds
      b. In and near small cumulus clouds
      c. In clear-air convective currents over heated surfaces
      d. With weak wind shears in the vicinity of

         1) Troughs aloft
         2) Lows aloft
         3) Jet streams
         4) The tropopause

      e. In the lower 5,000 ft. of the atmosphere

         1) When winds are near 15 kt.
         2) Where the air is colder than the underlying surfaces

2. Moderate turbulence

    a. In mountainous areas with a wind component of 25 to 50 kt. perpendicular to and near the level of the ridge

        1) At all levels from the surface to 5,000 ft. above the tropopause with preference for altitudes

            a) Within 5,000 ft. of the ridge level
            b) At the base of relatively stable layers below the base of the tropopause
            c) Within the tropopause layer

        2) Extending downstream from the lee of the ridge for 150 to 300 mi.

    b. In and near thunderstorms in the dissipating stage

    c. In and near other towering cumuliform clouds

    d. In the lower 5,000 ft. of the troposphere

        1) When surface winds are 30 kt. or more
        2) Where heating of the underlying surface is unusually strong
        3) Where there is an invasion of very cold air

    e. In fronts aloft where:

        1) Vertical wind shear exceeds 6 kt. per 1,000 ft., and/or
        2) Horizontal wind shear exceeds 18 kt. per 150 mi.

3. Severe turbulence

    a. In mountainous areas with a wind component exceeding 50 kt. perpendicular to and near the level of the ridge

        1) In 5,000 ft. layers

             a) At and below the ridge level in rotor clouds or rotor action
            b) At the tropopause
            c) Sometimes at the base of other stable layers below the tropopause

        2) Extending downstream from the lee of the ridge for 50 to 150 mi.

    b. In and near growing and mature thunderstorms

    c. Occasionally in other towering cumuliform clouds

    d. 50 to 100 mi. on the cold side of the center of the jet stream, in troughs aloft, and in lows aloft where:

        1) Vertical wind shear exceeds 10 kt. per 1,000 ft., and
        2) Horizontal wind shear exceed 40 kt. per 150 NM.

4. Extreme turbulence

    a. In mountain wave situations, in and below the level of well-developed rotor clouds. Sometimes it extends to the ground.

    b. In severe thunderstorms (most frequently in organized squall lines) indicated by

        1) Large hailstones (diameter ¾ in. or greater),
        2) Strong radar echoes, or
        3) Almost continuous lightning.

E. Density Altitude Computation

1. Use the graph below to find density altitude either on the ground or aloft.

   a. Set the aircraft's altimeter to 29.92 in.; it now indicates pressure altitude.
   b. Read the outside air temperature.
   c. Enter the graph at the pressure altitude and move horizontally to the temperature.
   d. Read the density altitude from the sloping lines.

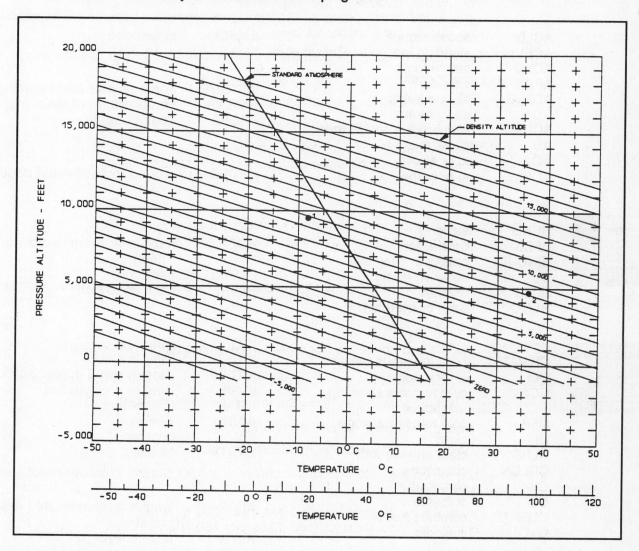

2. EXAMPLES:

   a. Density altitude in flight. Pressure altitude is 9,500 ft. and the temperature is −8°C. Find 9,500 ft. on the left of the graph and move to −8°C. Density altitude is 9,000 ft.

   b. Density altitude for take-off. Pressure altitude is 4,950 ft. and the temperature is 97°F. Enter the graph at 4,950 ft. and move across to 97°F. Density altitude is 8,200 ft. Note that in the warm air, density altitude is considerably higher than pressure altitude.

F.   Selected Contractions and Acronyms

1.   Contractions are used extensively in surface, radar, and pilot reports and in forecasts. Most of them are known from common usage or can be deciphered phonetically.

2.   The list of contractions below and on page 379 contains only those most likely to give you difficulty.

A

ACLD	above clouds
ACSL	standing lenticular altocumulus
ACYC	anticyclonic
AFDK	after dark
ALQDS	all quadrants
AC	altocumulus
ACCAS	altocumulus castellanus
AS	altostratus
AOA	at or above
AOB	at or below

B

BCKG	backing
BFDK	before dark
BINOVC	breaks in overcast
BL	between layers
BLZD	blizzard
BOVC	base of overcast

C

CBMAM	cumulonimbus mamma
CC	cirrocumulus
CCSL	cirrocumulus standing lenticular
CFP	cold frontal passage
CI	cirrus
CLRS	clear and smooth
CRLCN	circulation
CS	cirrostratus
CU	cumulus
CUFRA	cumulus fractus
CYC	cyclonic

D

DFUS	diffuse
DNSLP	downslope
DP	deep
DTRT	deteriorate
DURC	during climb
DURD	during descent
DWNDFTS	downdraft

E

EMBDD	embedded

F

FNTGNS	frontogenesis (front forming)
FNTLYS	frontolysis (front decaying)
FROPA	frontal passage

G

GFDEP	ground fog estimated depth (feet)

H

HDEP	haze layer estimated depth (feet)
HLSTO	hailstones
HLYR	haze layer aloft

I

ICG	icing
ICGIC	icing in cloud
ICGICIP	icing in cloud in precipitation
ICGIP	icing in precipitation
INTMT	intermittent
INVRN	inversion
IPV	improve
ISOLD	isolated

K

KDEP	smoke layer estimated depth (feet)
KLYR	smoke layer
KOCTY	smoke over city

L

LLWS	low level wind shear
LTG, LTNG	lightning
LTGCC	lightning cloud to cloud
LTGCCCG	lightning cloud to cloud, cloud to ground
LTGCW	lightning cloud to water
LTGIC	lightning in cloud

M

MEGG	merging
MLTLVL	melting level
MNLD	mainland
MOGR	moderate or greater
MRGL	marginal
MSTR	moisture

N

NCWX	no change in weather
NPRS	nonpersistent
NRW	narrow
NS	nimbostratus

O

OAOI	on and off instruments
OAT	outside air temperature
OCFNT	occluded front
OCLD	occlude
OFP	occluded frontal passage
OFSHR	off shore
OI	on instruments
OMTNS	over mountains
ONSHR	on shore
OTAS	on top and smooth
OVRNG	overrunning

P

PDW	priority delayed weather
PRESFR	pressure falling rapidly
PRESRR	pressure rising rapidly
PRIND	present indications are
PRST	persist

Q

QSTNRY	quasistationary
QUAD	quadrant

R

RGD	ragged
RTD	routine delayed weather

S

SC	stratocumulus
SKC, CLR	sky clear
SNOINCR	snow depth increase in past hour
SNRS, SR	sunrise
SNST, SS	sunset
SNWFL	snowfall
SQAL	squall
SQLN	squall line
ST	stratus
STFRA	stratus fractus
STFRM	stratiform
STM	storm

T

TCU	towering cumulus
TOVC	top of overcast
TROP	tropopause
TWRG	towering

U

UDDF	up and down drafts
UPDFTS	updrafts
UPSLP	upslope

V

VLNT	violent
VR	veer

W

WDSPRD	widespread
WFP	warm frontal passage
WK	weak
WRMFNT	warm front
WSHFT	wind shift
WV	wave

3.   Acronyms used in Part III of this book

AC	Convective Outlook Bulletin; identifies a forecast of probable convective storms.
AIRMET	Airman's Meteorological Information; an inflight advisory forecast of conditions possibly hazardous to light aircraft or inexperienced pilots.
ARTCC	Air Route Traffic Control Center, FAA.
CWSU	Center Weather Service Unit, NWS/FAA.
EFAS	En route Flight Advisory Service (Flight Watch), FAA.
FA	Area Forecast; identifies a forecast of general aviation weather over a relatively large area.
FD	Winds and Temperatures Aloft Forecast; a forecast identifier.
FT	Terminal Forecast; identifies a forecast in the U.S. forecast code.
GOES	Geostationary Operational Environmental Satellite.
HIWAS	Hazardous In flight Weather Advisory Service, FAA.
ICAO	International Civil Aviation Organization.
IFSS	International Flight Service Station, FAA.
LAWRS	Limited Aviation Weather Reporting Station, usually a control tower; reports fewer weather elements than a complete SA.
NAWAU	National Aviation Weather Advisory Unit, NWS.
NESDIS	National Environmental Satellite Data and Information Service.
NHC	National Hurricane Center, NWS.
NMC	National Meteorological Center, NWS.
NOAA	National Oceanic and Atmospheric Administration, Department of Commerce.

NOTAM	Notice to Airman.
NSSFC	National Severe Storms Forecast Center, NWS.
NWS	National Weather Service, National Oceanic and Atmospheric Administration, Department of Commerce.
PATWAS	Pilot's Automatic Telephone Weather Answering Service; a self-briefing service.
PIREP	Pilot Weather Report.
RAREP	Radar Weather Report.
SA	Surface Aviation Weather Report; a message identifier.
SAWRS	Supplemental Aviation Weather Report; usually an airline office at a terminal not having NWS or FAA facilities.
SFSS	Satellite Field Service Station.
SIGMET	Significant Meteorological Information; an in flight advisory forecast of weather hazardous to all aircraft.
TAF	Terminal Aerodrome Forecast; identifies a terminal forecast in the ICAO code.
TWEB	Transcribed Weather Broadcast; a self-briefing radio broadcast service.
UA	Pilot Weather Report (PIREP); a message identifier.
WA	AIRMET valid for a specified period, a message identifier.
WS	SIGMET valid for a specified period, a message identifier.
WSFO	Weather Service Forecast Office, NWS.
WSO	Weather Service Office, NWS.
WST	Convective SIGMET; a message identifier.
WW	Service Weather Watch; identifies a forecast of probable severe thunderstorms or tornadoes.

G. Standard Conversions

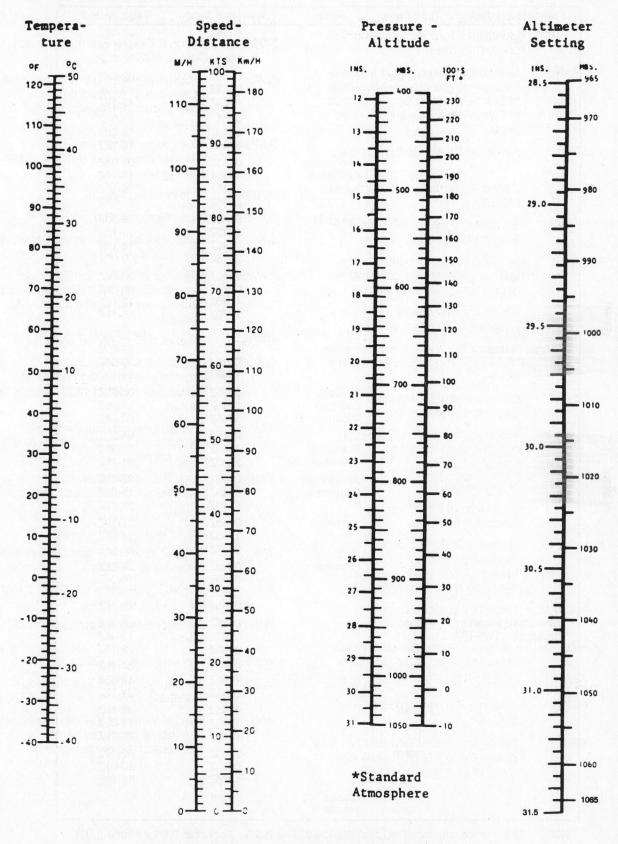

Figure 14-1. Standard Conversion Tables

H. Scheduled Issuance and Valid Time of Forecast Products

Forecast Products	Time Zone	Area	Issuance Time	Valid Period
Terminal Forecast (FT)	Pacific		0245Z	03-03Z
			1045Z	11-11Z
			1945Z	20-20Z
	Mountain		0145Z	02-02Z
			0945Z	10-10Z
			1845Z	19-19Z
	Central Eastern		0045Z	01-01Z
			0845Z	09-09Z
			1745Z	18-18Z
		Anchorage Fairbanks Juneau	0345Z	04-04Z
			1045Z	11-11Z
			1545Z	16-16Z
			2045Z	21-21Z
		Honolulu	0545Z	06-06Z
			1145Z	12-12Z
			1745Z	18-18Z
			2345Z	00-00Z
	ICAO	All	2340Z	00-00Z
			0540Z	06-06Z
			1140Z	12-12Z
			1740Z	18-18Z
	In-flight	All		See Chapter 9
Area Forecast (FA)		Boston Miami	0145Z	02-20Z
			0945Z	10-04Z
			1845Z	19-13Z
		Chicago Dallas-Ft. Worth	0245Z	03-21Z
			1045Z	11-05Z
			1945Z	20-14Z
		San Francisco Salt Lake City	0345Z	04-22Z
			1145Z	12-06Z
			2045Z	21-15Z
		Anchorage Fairbanks	0645Z	07-01Z
			1445Z	15-09Z
			2245Z	23-17Z
		Juneau	0645Z	07-10Z
			1345Z	14-08Z
			2245Z	23-17Z
		Honolulu	0345Z	04-22Z
			0945Z	10-04Z
			1545Z	16-10Z
			2145Z	22-16Z
Transcribed Weather Broadcast (TWEB)	Pacific		0445Z	05-20Z
			1245Z	13-04Z
			2145Z	22-13Z
	Mountain		0345Z	04-19Z
			1145Z	12-03Z
			2045Z	21-12Z
	Central		0245Z	03-18Z
			1045Z	11-02Z
			1945Z	20-11Z
	Eastern		0145Z	02-17Z
			0945Z	10-01Z
			1845Z	19-10Z
		Alaska Hawaii	None	

NOTE: All times are referenced to Local Standard Time (LST). For Local Daylight Time (LDT), subtract 1 hour.

END OF CHAPTER

ADDITIONAL USEFUL INFORMATION

The following five appendices contain information that will facilitate your access to additional weather forecasts and briefings.

      A. FAA Glossary of Weather Terms
      B. AM WEATHER and the Weather Channel
      C. Automated Flight Service Stations
      D. Direct User Access Terminal (DUAT)
      E. New International Aerodrome Meteorological Codes

Appendix A is a glossary of weather terms reproduced from the FAA's *Aviation Weather* (AC 00-6A). Appendices B, C, and D provide details of weather forecasts available on television, preflight and in-flight briefings available through Automated Flight Service Stations, and weather briefings and flight planning that are accessible anywhere by a personal computer with a modem. Appendix E contains discussion about the new international aerodrome meteorological codes.

APPENDIX A
FAA GLOSSARY OF WEATHER TERMS

The following glossary of weather terms is taken from AC 00-6A *Aviation Weather*.

A

absolute instability—A state of a layer within the atmosphere in which the vertical distribution of temperature is such that an air parcel, if given an upward or downward push, will move away from its initial level without further outside force being applied.

absolute temperature scale—*See* Kelvin Temperature Scale.

absolute vorticity—*See* vorticity.

adiabatic process—The process by which fixed relationships are maintained during changes in temperature, volume, and pressure in a body of air without heat being added or removed from the body.

advection—The horizontal transport of air or atmospheric properties. In meteorology, sometimes referred to as the horizontal component of *convection*.

advection fog—Fog resulting from the transport of warm, humid air over a cold surface.

air density—The mass density of the air in terms of weight per unit volume.

air mass—In meteorology, an extensive body of air within which the conditions of temperature and moisture in a horizontal plane are essentially uniform.

air mass classification—A system used to identify and to characterize the different *air masses* according to a basic scheme. The system most commonly used classifies air masses primarily according to the thermal properties of their *source regions*: "tropical" (T); "polar" (P); and "Arctic" or "Antarctic" (A). They are further classified according to moisture characteristics as "continental" (c) or "maritime" (m).

air parcel—*See* parcel.

albedo—The ratio of the amount of electromagnetic *radiation* reflected by a body to the amount incident upon it, commonly expressed in percentage; in meteorology, usually used in reference to *insolation* (solar radiation); i.e., the albedo of wet sand is 9, meaning that about 9% of the incident insolation is reflected; albedoes of other surfaces range upward to 80–85 for fresh snow cover; average albedo for the earth and its atmosphere has been calculated to range from 35 to 43.

altimeter—An instrument which determines the altitude of an object with respect to a fixed level. *See* pressure altimeter.

altimeter setting—The value to which the scale of a *pressure altimeter* is set so as to read true altitude at field elevation.

altimeter setting indicator—A precision *aneroid barometer* calibrated to indicate directly the altimeter setting.

altitude—Height expressed in units of distance above a reference plane, usually above mean sea level or above ground.

(1) **corrected altitude**—Indicated altitude of an aircraft altimeter corrected for the temperature of the column of air below the aircraft, the correction being based on the estimated departure of existing temperature from standard atmospheric temperature; an approximation of true altitude.

(2) **density altitude**—The altitude in the standard atmosphere at which the air has the same density as the air at the point in question. An aircraft will have the same performance characteristics as it would have in a standard atmosphere at this altitude.

(3) **indicated altitude**—The altitude above mean sea level indicated on a *pressure altimeter* set at current local *altimeter setting*.

(4) **pressure altitude**—The altitude in the standard atmosphere at which the pressure is the same as at the point in question. Since an altimeter operates solely on pressure, this is the uncorrected altitude indicated by an altimeter set at standard sea level pressure of 29.92 inches or 1013 millibars.

(5) **radar altitude**—The altitude of an aircraft determined by radar-type radio altimeter; thus the actual distance from the nearest terrain or water feature encompassed by the downward directed radar beam. For all practical purposes, it is the "actual" distance above a ground or inland water surface or the true altitude above an ocean surface.

(6) **true altitude**—The exact distance above mean sea level.

altocumulus—White or gray layers or patches of cloud, often with a waved appearance; cloud elements appear as rounded masses or rolls; composed mostly of liquid water droplets which may be supercooled; may contain ice crystals at subfreezing temperatures.

altocumulus castellanus—A species of middle cloud of which at least a fraction of its upper part presents some vertically developed, cumuliform protuberances (some of which are taller than they are wide, as castles) and which give the cloud a crenelated or turreted appearance; especially evident when seen from the side; elements usually have a common base arranged in lines. This cloud indicates instability and turbulence at the altitudes of occurrence.

anemometer—An instrument for measuring *wind speed*.

aneroid barometer—A *barometer* which operates on the principle of having changing atmospheric pressure bend a metallic surface which, in turn, moves a pointer across a scale graduated in units of pressure.

angel—In radar meteorology, an *echo* caused by physical phenomena not discernible to the eye; they have been observed when abnormally strong temperature and/or moisture *gradients* were known to exist; sometimes attributed to insects or birds flying in the radar beam.

anomalous propagation (sometimes called AP)—In radar meteorology, the greater than normal bending of the radar beam such that *echoes* are received from ground *targets* at distances greater than normal *ground clutter*.

anticyclone—An area of high atmospheric pressure which has a closed circulation that is anticyclonic, i.e., as viewed from above, the circulation is clockwise in the Northern Hemisphere, counterclockwise in the Southern Hemisphere, undefined at the Equator.

anvil cloud—Popular name given to the top portion of a *cumulonimbus* cloud having an anvil-like form.

APOB—A *sounding* made by an aircraft.

Arctic air—An air mass with characteristics developed mostly in winter over Arctic surfaces of ice and snow. Arctic air extends to great heights, and the surface temperatures are basically, but not always, lower than those of *polar air*.

Arctic front—The šurface of discontinuity between very cold (Arctic) air flowing directly from the Arctic region and another less cold and, consequently, less dense air mass.

astronomical twilight—*See* twilight.

atmosphere—The mass of air surrounding the Earth.

atmospheric pressure (also called barometric pressure)—The pressure exerted by the atmosphere as a consequence of gravitational attraction exerted upon the "column" of air lying directly above the point in question.

atmospherics—Disturbing effects produced in radio receiving apparatus by atmospheric electrical phenomena such as an electrical storm. Static.

aurora—A luminous, radiant emission over middle and high latitudes confined to the thin air of high altitudes and centered over the earth's magnetic poles. Called "aurora borealis" (northern lights) or "aurora australis" according to its occurrence in the Northern or Southern Hemisphere, respectively.

attenuation—In radar meteorology, any process which reduces power density in radar signals.

　(1) **precipitation attenuation**—Reduction of power density because of absorption or reflection of energy by precipitation.

　(2) **range attenuation**—Reduction of radar power density because of distance from the antenna. It occurs in the outgoing beam at a rate proportional to 1/range². The return signal is also attenuated at the same rate.

B

backing—Shifting of the wind in a counterclockwise direction with respect to either space or time; opposite of *veering*. Commonly used by meteorologists to refer to a cyclonic shift (counterclockwise in the Northern Hemisphere and clockwise in the Southern Hemisphere).

backscatter—Pertaining to radar, the energy reflected or scattered by a *target;* an *echo*.

banner cloud (also called cloud banner)—A banner-like cloud streaming off from a mountain peak.

barograph—A continuous-recording *barometer*.

barometer—An instrument for measuring the pressure of the atmosphere; the two principle types are *mercurial* and *aneroid*.

barometric altimeter—*See* pressure altimeter.

barometric pressure—Same as *atmospheric pressure*.

barometric tendency—The change of barometric pressure within a specified period of time. In aviation weather observations, routinely determined periodically, usually for a 3-hour period.

beam resolution—*See* resolution.

Beaufort scale—A scale of wind speeds.

black blizzard—Same as *duststorm*.

blizzard—A severe weather condition characterized by low temperatures and strong winds bearing a great amount of snow, either falling or picked up from the ground.

blowing dust—A type of *lithometeor* composed of dust particles picked up locally from the surface and blown about in clouds or sheets.

blowing sand—A type of *lithometeor* composed of sand picked up locally from the surface and blown about in clouds or sheets.

blowing snow—A type of *hydrometeor* composed of snow picked up from the surface by the wind and carried to a height of 6 feet or more.

blowing spray—A type of *hydrometeor* composed of water particles picked up by the wind from the surface of a large body of water.

bright band—In radar meteorology, a narrow, intense *echo* on the *range-height indicator* scope resulting from water-covered ice particles of high reflectivity at the melting level.

Buys Ballot's law—If an observer in the Northern Hemisphere stands with his back to the wind, lower pressure is to his left.

C

calm—The absence of wind or of apparent motion of the air.

cap cloud (also called cloud cap)—A standing or stationary cap-like cloud crowning a mountain summit.

ceiling—In meteorology in the U.S., (1) the height above the surface of the base of the lowest layer of clouds or *obscuring phenomena* aloft that hides more than half of the sky, or (2) the *vertical visibility* into an *obscuration*. *See* summation principle.

ceiling balloon—A small balloon used to determine the height of a cloud base or the extent of vertical visibility.

ceiling light—An instrument which projects a vertical light beam onto the base of a cloud or into surface-based obscuring phenomena; used at night in conjunction with a *clinometer* to determine the height of the cloud base or as an aid in estimating the vertical visibility.

ceilometer—A cloud-height measuring system. It projects light on the cloud, detects the reflection by a photo-electric cell, and determines height by triangulation.

Celsius temperature scale (abbreviated C)—A temperature scale with zero degrees as the melting point of pure ice and 100 degrees as the boiling point of pure water at standard sea level atmospheric pressure.

Centigrade temperature scale—Same as *Celsius temperature scale*.

chaff—Pertaining to radar, (1) short, fine strips of metallic foil dropped from aircraft, usually by military forces, specifically for the purpose of jamming radar; (2) applied loosely to *echoes* resulting from chaff.

change of state—In meteorology, the transformation of water from one form, i.e., solid (ice), liquid, or gaseous (water vapor), to any other form. There are six possible transformations designated by the five terms following:

(1) **condensation**—The change of water vapor to liquid water.

(2) **evaporation**—The change of liquid water to water vapor.

(3) **freezing**—The change of liquid water to ice.

(4) **melting**—The change of ice to liquid water.

(5) **sublimation**—The change of (a) ice to water vapor or (b) water vapor to ice. *See* latent heat.

Chinook—A warm, dry *foehn* wind blowing down the eastern slopes of the Rocky Mountains over the adjacent plains in the U.S. and Canada.

cirriform—All species and varieties of *cirrus*, *cirrocumulus*, and *cirrostratus* clouds; descriptive of clouds composed mostly or entirely of small ice crystals, usually transparent and white; often producing *halo* phenomena not observed with other cloud forms. Average height ranges upward from 20,000 feet in middle latitudes.

cirrocumulus—A *cirriform* cloud appearing as a thin sheet of small white puffs resembling flakes or patches of cotton without shadows; sometimes confused with *altocumulus*.

cirrostratus—A *cirriform* cloud appearing as a whitish veil, usually fibrous, sometimes smooth; often produces *halo* phenomena; may totally cover the sky.

cirrus—A *cirriform* cloud in the form of thin, white feather-like clouds in patches or narrow bands; have a fibrous and/or silky sheen; large ice crystals often trail downward a considerable vertical distance in fibrous, slanted, or irregularly curved wisps called mares' tails.

civil twilight—*See* twilight.

clear air turbulence (abbreviated CAT)—Turbulence encountered in air where no clouds are present; more popularly applied to high level turbulence associated with *wind shear*.

clear icing (or clear ice)—Generally, the formation of a layer or mass of ice which is relatively transparent because of its homogeneous structure and small number and size of air spaces; used commonly as synonymous with *glaze*, particularly with respect to aircraft icing. Compare with *rime icing*. Factors which favor clear icing are large drop size, such as those found in *cumuliform* clouds, rapid accretion of supercooled water, and slow dissipation of *latent heat* of fusion.

climate—The statistical collective of the weather conditions of a point or area during a specified interval of time (usually several decades); may be expressed in a variety of ways.

climatology—The study of *climate*.

clinometer—An instrument used in weather observing for measuring angles of inclination; it is used in conjunction with a *ceiling light* to determine cloud height at night.

cloud bank—Generally, a fairly well-defined mass of cloud observed at a distance; it covers an appreciable portion of the horizon sky, but does not extend overhead.

cloudburst—In popular teminology, any sudden and heavy fall of *rain*, almost always of the *shower* type.

cloud cap—*See* cap cloud.

cloud detection radar—A vertically directed radar to detect cloud bases and tops.

cold front—Any non-occluded *front* which moves in such a way that colder air replaces warmer air.

condensation—*See* change of state.

condensation level—The height at which a rising *parcel* or layer of air would become saturated if lifted adiabatically.

condensation nuclei—Small particles in the air on which water vapor condenses or sublimates.

condensation trail (or contrail) (also called vapor trail)—A cloud-like streamer frequently observed to form behind aircraft flying in clear, cold, humid air.

conditionally unstable air—Unsaturated air that will become unstable on the condition it becomes saturated. *See* instability.

conduction—The transfer of heat by molecular action through a substance or from one substance in contact with another; transfer is always from warmer to colder temperature.

constant pressure chart—A chart of a constant pressure surface; may contain analyses of height, wind, temperature, humidity, and/or other elements.

continental polar air—*See* polar air.

continental tropical air—*See* tropical air.

contour—In meteorology, (1) a line of equal height on a constant pressure chart; analogous to contours on a relief map; (2) in radar meteorology, a line on a radar scope of equal *echo* intensity.

contouring circuit—On weather radar, a circuit which displays multiple contours of *echo* intensity simultaneously on the *plan position indicator* or *range-height indicator* scope. *See* contour (2).

contrail—Contraction for *condensation trail*.

convection—(1) In general, mass motions within a fluid resulting in transport and mixing of the properties of that fluid. (2) In meteorology, atmospheric motions that are predominantly vertical, resulting in vertical transport and mixing of atmospheric properties; distinguished from *advection*.

convective cloud—*See* cumuliform.

convective condensation level (abbreviated CCL)—The lowest level at which condensation will occur as a result of *convection* due to surface heating. When condensation occurs at this level, the layer between the surface and the CCL will be thoroughly mixed, temperature *lapse rate* will be dry adiabatic, and *mixing ratio* will be constant.

convective instability—The state of an unsaturated layer of air whose *lapse rates* of temperature and moisture are such that when lifted adiabatically until the layer becomes saturated, convection is spontaneous.

convergence—The condition that exists when the distribution of winds within a given area is such that there is a net horizontal inflow of air into the area. In convergence at lower levels, the removal of the resulting excess is accomplished by an upward movement of air; consequently, areas of low-level convergent winds are regions favorable to the occurrence of clouds and precipitation. Compare with *divergence*.

Coriolis force—A deflective force resulting from earth's rotation; it acts to the right of wind direction in the Northern Hemisphere and to the left in the Southern Hemisphere.

corona—A prismatically colored circle or arcs of a circle with the sun or moon at its center; coloration is from blue inside to red outside (opposite that of a *halo*); varies in size (much smaller) as opposed to the fixed diameter of the halo; characteristic of clouds composed of water droplets and valuable in differentiating between middle and cirriform clouds.

corposant—*See* St. Elmo's Fire.

corrected altitude (approximation of true altitude)—*See* altitude.

cumuliform—A term descriptive of all convective clouds exhibiting vertical development in contrast to the horizontally extended *stratiform* types.

cumulonimbus—A cumuliform cloud type; it is heavy and dense, with considerable vertical extent in the form of massive towers; often with tops in the shape of an *anvil* or massive plume; under the base of cumulonimbus, which often is very dark, there frequently exists *virga*, precipitation and low ragged clouds (*scud*), either merged with it or not; frequently accompanied by lightning, thunder, and sometimes hail; occasionally produces a tornado or a waterspout; the ultimate manifestation of the growth of a cumulus cloud, occasionally extending well into the stratosphere.

cumulonimbus mamma—A *cumulonimbus* cloud having hanging protuberances, like pouches, festoons, or udders, on the under side of the cloud; usually indicative of severe turbulence.

cumulus—A cloud in the form of individual detached domes or towers which are usually dense and well defined; develops vertically in the form of rising mounds of which the bulging upper part often resembles a cauliflower; the sunlit parts of these clouds are mostly brilliant white; their bases are relatively dark and nearly horizontal.

cumulus fractus—*See* fractus.

cyclogenesis—Any development or strengthening of cyclonic circulation in the atmosphere.

cyclone—(1) An area of low atmospheric pressure which has a closed circulation that is cyclonic, i.e., as viewed from above, the circulation is counterclockwise in the Northern Hemisphere, clockwise in the Southern Hemisphere, undefined at the Equator. Because cyclonic circulation and relatively low atmospheric pressure usually coexist, in common practice the terms cyclone and low are used interchangeably. Also, because cyclones often are accompanied by inclement (sometimes destructive) weather, they are frequently referred to simply as storms. (2) Frequently misused to denote a *tornado*. (3) In the Indian Ocean, a *tropical cyclone* of hurricane or typhoon force.

D

deepening—A decrease in the central pressure of a pressure system; usually applied to a *low* rather than to a *high*, although technically, it is acceptable in either sense.

density—(1) The ratio of the mass of any substance to the volume it occupies—weight per unit volume. (2) The ratio of any quantity to the volume or area it occupies, i.e., population per unit area, *power density*.

density altitude—*See* altitude.

depression—In meteorology, an area of low pressure; a *low* or *trough*. This is usually applied to a certain stage in the development of a *tropical cyclone*, to migratory lows and troughs, and to upper-level lows and troughs that are only weakly developed.

dew—Water condensed onto grass and other objects near the ground, the temperatures of which have fallen below the initial dew point temperature of the surface air, but is still above freezing. Compare with *frost*.

dew point (or dew-point temperature)—The temperature to which a sample of air must be cooled, while the *mixing ratio* and barometric pressure remain constant, in order to attain saturation with respect to water.

discontinuity—A zone with comparatively rapid transition of one or more meteorological elements.

disturbance—In meteorology, applied rather loosely: (1) any low pressure or cyclone, but usually one that is relatively small in size; (2) an area where weather, wind, pressure, etc., show signs of cyclonic development; (3) any deviation in flow or pressure that is associated with a disturbed state of the weather, i.e., cloudiness and precipitation; and (4) any individual circulatory system within the primary circulation of the atmosphere.

diurnal—Daily, especially pertaining to a cycle completed within a 24-hour period, and which recurs every 24 hours.

divergence—The condition that exists when the distribution of winds within a given area is such that there is a net horizontal flow of air outward from the region. In divergence at lower levels, the resulting deficit is compensated for by subsidence of air from aloft; consequently the air is heated and the relative humidity lowered making divergence a warming and drying process. Low-level divergent regions are areas unfavorable to the occurrence of clouds and precipitation. The opposite of *convergence*.

doldrums—The equatorial belt of calm or light and variable winds between the two tradewind belts. Compare *intertropical convergence zone*.

downdraft—A relative small scale downward current of air; often observed on the lee side of large objects restricting the smooth flow of the air or in precipitation areas in or near *cumuliform* clouds.

drifting snow—A type of *hydrometeor* composed of snow particles picked up from the surface, but carried to a height of less than 6 feet.

drizzle—A form of *precipitation*. Very small water drops that appear to float with the air currents while falling in an irregular path (unlike *rain*, which falls in a comparatively straight path, and unlike *fog* droplets which remain suspended in the air).

dropsonde—A *radiosonde* dropped by parachute from an aircraft to obtain *soundings* (measurements) of the atmosphere below.

dry adiabatic lapse rate—The rate of decrease of temperature with height when unsaturated air is lifted adiabatically (due to expansion as it is lifted to lower pressure). *See* adiabatic process.

dry bulb—A name given to an ordinary thermometer used to determine temperature of the air; also used as a contraction for *dry-bulb temperature*. Compare *wet bulb*.

dry-bulb temperature—The temperature of the air.

dust—A type of *lithometeor* composed of small earthen particles suspended in the atmosphere.

dust devil—A small, vigorous *whirlwind*, usually of short duration, rendered visible by dust, sand, and debris picked up from the ground.

duster—Same as *duststorm*.

duststorm (also called duster, black blizzard)—An unusual, frequently severe weather condition characterized by strong winds and dust-filled air over an extensive area.

D-value—Departure of true altitude from pressure altitude (*see* altitude); obtained by algebraically subtracting true altitude from pressure altitude; thus it may be plus or minus. On a constant pressure chart, the difference between actual height and *standard atmospheric* height of a constant pressure surface.

E

echo—In radar, (1) the energy reflected or scattered by a *target;* (2) the radar scope presentation of the return from a target.

eddy—A local irregularity of wind in a larger scale wind flow. Small scale eddies produce turbulent conditions.

estimated ceiling—A ceiling classification applied when the ceiling height has been estimated by the observer or has been determined by some other method; but, because of the specified limits of time, distance, or precipitation conditions, a more descriptive classification cannot be applied.

evaporation—*See* change of state.

extratropical low (sometimes called extratropical cyclone, extratropical storm)—Any *cyclone* that is not a *tropical cyclone*, usually referring to the migratory frontal cyclones of middle and high latitudes.

eye—The roughly circular area of calm or relatively light winds and comparatively fair weather at the center of a well-developed *tropical cyclone*. A *wall cloud* marks the outer boundary of the eye.

F

Fahrenheit temperature scale (abbreviated F)—A temperature scale with 32 degrees as the melting point of pure ice and 212 degrees as the boiling point of pure water at standard sea level atmospheric pressure (29.92 inches or 1013.2 millibars).

Fall wind—A cold wind blowing downslope. Fall wind differs from *foehn* in that the air is initially cold enough to remain relatively cold despite compressional heating during descent.

filling—An increase in the central pressure of a pressure system; opposite of *deepening;* more commonly applied to a low rather than a high.

first gust—The leading edge of the spreading downdraft, *plow wind*, from an approaching thunderstorm.

flow line—A *streamline*.

foehn—A warm, dry downslope wind; the warmness and dryness being due to adiabatic compression upon descent; characteristic of mountainous regions. *See* adiabatic process, Chinook, Santa Ana.

fog—A *hydrometeor* consisting of numerous minute water droplets and based at the surface; droplets are small enough to be suspended in the earth's atmosphere indefinitely. (Unlike *drizzle*, it does not fall to the surface; differs from cloud only in that a cloud is not based at the surface; distinguished from haze by its wetness and gray color.)

fractus—Clouds in the form of irregular shreds, appearing as if torn; have a clearly ragged appearance; applies only to stratus and cumulus, i.e., *cumulus* fractus and *stratus* fractus.

freezing—*See* change of state.

freezing level—A level in the atmosphere at which the temperature is 0° C (32° F).

front—A surface, interface, or transition zone of discontinuity between two adjacent *air masses* of different densities; more simply the boundary between two different air masses. *See* frontal zone.

frontal zone—A *front* or zone with a marked increase of density gradient; used to denote that fronts are not truly a "surface" of discontinuity but rather a "zone" of rapid transition of meteorological elements.

frontogenesis—The initial formation of a *front* or *frontal zone*.

frontolysis—The dissipation of a *front*.

frost (also hoarfrost)—Ice crystal deposits formed by sublimation when temperature and dew point are below freezing.

funnel cloud—A *tornado* cloud or *vortex* cloud extending downward from the parent cloud but not reaching the ground.

G

glaze—A coating of ice, generally clear and smooth, formed by freezing of supercooled water on a surface. *See* clear icing.

gradient—In meteorology, a horizontal decrease in value per unit distance of a parameter in the direction of maximum decrease; most commonly used with pressure, temperature, and moisture.

ground clutter—Pertaining to radar, a cluster of *echoes*, generally at short range, reflected from ground *targets*.

ground fog—In the United States, a *fog* that conceals less than 0.6 of the sky and is not contiguous with the base of clouds.

gust—A sudden brief increase in wind; according to U.S. weather observing practice, gusts are reported when the variation in wind speed between peaks and lulls is at least 10 knots.

H

hail—A form of *precipitation* composed of balls or irregular lumps of ice, always produced by convective clouds which are nearly always *cumulonimbus*.

halo—A prismatically colored or whitish circle or arcs of a circle with the sun or moon at its center; coloration, if not white, is from red inside to blue outside (opposite

that of a *corona*); fixed in size with an angular diameter of 22° (common) or 46° (rare); characteristic of clouds composed of ice crystals; valuable in differentiating between *cirriform* and forms of lower clouds.

haze—A type of *lithometeor* composed of fine dust or salt particles dispersed through a portion of the atmosphere; particles are so small they cannot be felt or individually seen with the naked eye (as compared with the larger particles of *dust*), but diminish the visibility; distinguished from *fog* by its bluish or yellowish tinge.

high—An area of high barometric pressure, with its attendant system of winds; an *anticyclone*. Also high pressure system.

hoar frost—*See* frost.

humidity—Water vapor content of the air; may be expressed as *specific humidity*, *relative humidity*, or *mixing ratio*.

hurricane—A *tropical cyclone* in the Western Hemisphere with winds in excess of 65 knots or 120 km/h.

hydrometeor—A general term for particles of liquid water or ice such as rain, fog, frost, etc., formed by modification of water vapor in the atmosphere; also water or ice particles lifted from the earth by the wind such as sea spray or blowing snow.

hygrograph—The record produced by a continuous-recording *hygrometer*.

hygrometer—An instrument for measuring the water vapor content of the air.

I

ice crystals—A type of *precipitation* composed of unbranched crystals in the form of needles, columns, or plates; usually having a very slight downward motion, may fall from a cloudless sky.

ice fog—A type of fog composed of minute suspended particles of ice; occurs at very low temperatures and may cause *halo* phenomena.

ice needles—A form of *ice crystals*.

ice pellets—Small, transparent or translucent, round or irregularly shaped pellets of ice. They may be (1) hard grains that rebound on striking a hard surface or (2) pellets of snow encased in ice.

icing—In general, any deposit of ice forming on an object. *See* clear icing, rime icing, glaze.

indefinite ceiling—A ceiling classification denoting *vertical visibility* into a surface based obscuration.

indicated altitude—*See* altitude.

insolation—Incoming solar *radiation* falling upon the earth and its atmosphere.

instability—A general term to indicate various states of the atmosphere in which spontaneous *convection* will occur when prescribed criteria are met; indicative of turbulence. *See* absolute instability, conditionally unstable air, convective instability.

intertropical convergence zone—The boundary zone between the trade wind system of the Northern and Southern Hemispheres; it is characterized in maritime climates by showery precipitation with cumulonimbus clouds sometimes extending to great heights.

inversion—An increase in temperature with height—a reversal of the normal decrease with height in the *troposphere;* may also be applied to other meteorological properties.

isobar—A line of equal or constant barometric pressure.

iso echo—In radar circuitry, a circuit that reverses signal strength above a specified intensity level, thus causing a void on the scope in the most intense portion of an echo when maximum intensity is greater than the specified level.

isoheight—On a weather chart, a line of equal height; same as *contour* (1).

isoline—A line of equal value of a variable quantity, i.e., an isoline of temperature is an *isotherm,* etc. *See* isobar, isotach, etc.

isoshear—A line of equal *wind shear.*

isotach—A line of equal or constant wind speed.

isotherm—A line of equal or constant temperature.

isothermal—Of equal or constant temperature, with respect to either space or time; more commonly, temperature with height; a zero *lapse rate.*

J

jet stream—A quasi-horizontal stream of winds 50 knots or more concentrated within a narrow band embedded in the westerlies in the high *troposphere.*

K

katabatic wind—Any wind blowing downslope. *See* fall wind, foehn.

Kelvin temperature scale (abbreviated K)—A temperature scale with zero degrees equal to the temperature at which all molecular motion ceases, i.e., absolute zero (0° K = −273° C); the Kelvin degree is identical to the Celsius degree; hence at standard sea level pressure, the melting point is 273° K and the boiling point 373° K.

knot—A unit of speed equal to one nautical mile per hour.

L

land breeze—A coastal breeze blowing from land to sea, caused by temperature difference when the sea surface is warmer than the adjacent land. Therefore, it usually blows at night and alternates with a *sea breeze,* which blows in the opposite direction by day.

lapse rate—The rate of decrease of an atmospheric variable with height; commonly refers to decrease of temperature with height.

latent heat—The amount of heat absorbed (converted to kinetic energy) during the processes of change of liquid water to water vapor, ice to water vapor, or ice to liquid water; or the amount released during the reverse processes. Four basic classifications are:

(1) **latent heat of condensation**—Heat released during change of water vapor to water.

(2) **latent heat of fusion**—Heat released during change of water to ice or the amount absorbed in change of ice to water.

(3) **latent heat of sublimation**—Heat released during change of water vapor to ice or the amount absorbed in the change of ice to water vapor.

(4) **latent heat of vaporization**—Heat absorbed in the change of water to water vapor; the negative of latent heat of condensation.

layer—In reference to sky cover, clouds or other obscuring phenomena whose bases are approximately at the same level. The layer may be continuous or composed of detached elements. The term "layer" does not imply that a clear space exists between the layers or that the clouds or *obscuring phenomena* composing them are of the same type.

lee wave—Any stationary wave disturbance caused by a barrier in a fluid flow. In the atmosphere when sufficient moisture is present, this wave will be evidenced by *lenticular clouds* to the lee of mountain barriers; also called *mountain wave* or *standing wave.*

lenticular cloud (or lenticularis)—A species of cloud whose elements have the form of more or less isolated, generally smooth lenses or almonds. These clouds appear most often in formations of orographic origin, the result of *lee waves,* in which case they remain nearly stationary with respect to the terrain (standing cloud), but they also occur in regions without marked orography.

level of free convection (abbreviated LFC)—The level at which a *parcel* of air lifted dry-adiabatically until saturated and moist-adiabatically thereafter would become warmer than its surroundings in a conditionally unstable atmosphere. *See* conditional instability and adiabatic process.

lifting condensation level (abbreviated LCL)—The level at which a *parcel* of unsaturated air lifted dry-adiabatically would become saturated. Compare *level of free convection* and *convective condensation level.*

lightning—Generally, any and all forms of visible electrical discharge produced by a *thunderstorm.*

lithometeor—The general term for dry particles suspended in the atmosphere such as dust, haze, smoke, and sand.

low—An area of low barometric pressure, with its attendant system of winds. Also called a barometric depression or *cyclone.*

M

mammato cumulus—Obsolete. *See* cumulonimbus mamma.

mare's tail—*See* cirrus.

maritime polar air (abbreviated mP)—*See* polar air.

maritime tropical air (abbreviated mT)—*See* tropical air.

maximum wind axis—On a constant pressure chart, a line denoting the axis of maximum wind speeds at that constant pressure surface.

mean sea level—The average height of the surface of the sea for all stages of tide; used as reference for elevations throughout the U.S.

measured ceiling—A ceiling classification applied when the ceiling value has been determined by instruments or the known heights of unobscured portions of objects, other than natural landmarks.

melting—*See* change of state.

mercurial barometer—A *barometer* in which pressure is determined by balancing air pressure against the weight of a column of mercury in an evacuated glass tube.

meteorological visibility—In U.S. observing practice, a main category of *visibility* which includes the subcategories of *prevailing visibility* and *runway visibility*. Meteorological visibility is a measure of horizontal visibility near the earth's surface, based on sighting of objects in the daytime or unfocused lights of moderate intensity at night. Compare *slant visibility, runway visual range, vertical visibility*. *See* surface visibility, tower visibility, and sector visibility.

meteorology—The science of the *atmosphere*.

microbarograph—An aneroid *barograph* designed to record atmospheric pressure changes of very small magnitudes.

millibar (abbreviated mb.)—An internationally used unit of pressure equal to 1,000 dynes per square centimeter. It is convenient for reporting *atmospheric pressure*.

mist—A popular expression for drizzle or heavy fog.

mixing ratio—The ratio by weight of the amount of water vapor in a volume of air to the amount of dry air; usually expressed as grams per kilogram (g/kg).

moist-adiabatic lapse rate—*See* saturated-adiabatic lapse rate.

moisture—An all-inclusive term denoting water in any or all of its three states.

monsoon—A wind that in summer blows from sea to a continental interior, bringing copious rain, and in winter blows from the interior to the sea, resulting in sustained dry weather.

mountain wave—A *standing wave* or *lee wave* to the lee of a mountain barrier.

N

nautical twilight—*See* twilight.

negative vorticity—*See* vorticity.

nimbostratus—A principal cloud type, gray colored, often dark, the appearance of which is rendered diffuse by more or less continuously falling rain or snow, which in most cases reaches the ground. It is thick enough throughout to blot out the sun.

noctilucent clouds—Clouds of unknown composition which occur at great heights, probably around 75 to 90 kilometers. They resemble thin *cirrus*, but usually with a bluish or silverish color, although sometimes orange to red, standing out against a dark night sky. Rarely observed.

normal—In meteorology, the value of an element averaged for a given location over a period of years and recognized as a standard.

numerical forecasting—*See* numerical weather prediction.

numerical weather prediction—Forecasting by digital computers solving mathematical equations; used extensively in weather services throughout the world.

O

obscuration—Denotes sky hidden by surface-based *obscuring phenomena* and *vertical visibility* restricted overhead.

obscuring phenomena—Any *hydrometeor* or *lithometeor* other than clouds; may be surface based or aloft.

occlusion—Same as *occluded front*.

occluded front (commonly called occlusion, also called frontal occlusion)—A composite of two fronts as a *cold front* overtakes a *warm front* or *quasi-stationary front*.

orographic—Of, pertaining to, or caused by mountains as in orographic clouds, orographic lift, or orographic precipitation.

ozone—An unstable form of oxygen; heaviest concentrations are in the stratosphere; corrosive to some metals; absorbs most ultraviolet solar radiation.

P

parcel—A small volume of air, small enough to contain uniform distribution of its meteorological properties, and large enough to remain relatively self-contained and respond to all meteorological processes. No specific dimensions have been defined, however, the order of magnitude of 1 cubic foot has been suggested.

partial obscuration—A designation of sky cover when part of the sky is hidden by surface based *obscuring phenomena*.

pilot balloon—A small free-lift balloon used to determine the speed and direction of winds in the upper air.

pilot balloon observation (commonly called PIBAL)—A method of winds-aloft observation by visually tracking a *pilot balloon*.

plan position indicator (PPI) scope—A radar indicator scope displaying range and azimuth of *targets* in polar coordinates.

plow wind—The spreading downdraft of a *thunderstorm*; a strong, straight-line wind in advance of the storm. *See* first gust.

polar air—An air mass with characteristics developed over high latitudes, especially within the subpolar highs. Continental polar air (cP) has cold surface temperatures, low moisture content, and, especially in its source regions, has great stability in the lower layers. It is shallow in comparison with *Arctic air.* Maritime polar (mP) initially possesses similar properties to those of continental polar air, but in passing over warmer water it becomes unstable with a higher moisture content. Compare *tropical air.*

polar front—The semipermanent, semicontinuous *front* separating air masses of tropical and polar origins.

positive vorticity—*See* vorticity.

power density—In radar meteorology the amount of radiated energy per unit cross sectional area in the radar beam.

precipitation—Any or all forms of water particles, whether liquid or solid, that fall from the atmosphere and reach the surface. It is a major class of *hydrometeor,* distinguished from cloud and *virga* in that it must reach the surface.

precipitation attenuation—*See* attenuation.

pressure—*See* atmospheric pressure.

pressure altimeter—An *aneroid barometer* with a scale graduated in altitude instead of pressure using *standard atmospheric* pressure-height relationships; shows indicated altitude (not necessarily true altitude); may be set to measure altitude (indicated) from any arbitrarily chosen level. *See* altimeter setting, altitude.

pressure altitude—*See* altitude.

pressure gradient—The rate of decrease of pressure per unit distance at a fixed time.

pressure jump—A sudden, significant increase in *station pressure.*

pressure tendency—*See* barometric tendency.

prevailing easterlies—The broad current or pattern of persistent easterly winds in the Tropics and in polar regions.

prevailing visibility—In the U.S., the greatest horizontal visibility which is equaled or exceeded throughout half of the horizon circle; it need not be a continuous half.

prevailing westerlies—The dominant west-to-east motion of the atmosphere, centered over middle latitudes of both hemispheres.

prevailing wind—Direction from which the wind blows most frequently.

prognostic chart (contracted PROG)—A chart of expected or forecast conditions.

pseudo-adiabatic lapse rate—*See* saturated-adiabatic lapse rate.

psychrometer—An instrument consisting of a *wet-bulb* and a *dry-bulb* thermometer for measuring wet-bulb and dry-bulb temperature; used to determine water vapor content of the air.

pulse—Pertaining to radar, a brief burst of electromagnetic radiation emitted by the radar; of very short time duration. *See* pulse length.

pulse length—Pertaining to radar, the dimension of a radar pulse; may be expressed as the time duration or the length in linear units. Linear dimension is equal to time duration multiplied by the speed of propagation (approximately the speed of light).

Q

quasi-stationary front (commonly called stationary front)—A *front* which is stationary or nearly so; conventionally, a front which is moving at a speed of less than 5 knots is generally considered to be quasi-stationary.

R

RADAR (contraction for radio detection and ranging)—An electronic instrument used for the detection and ranging of distant objects of such composition that they scatter or reflect radio energy. Since *hydrometeors* can scatter radio energy, *weather radars,* operating on certain frequency bands, can detect the presence of precipitation, clouds, or both.

radar altitude—*See* altitude.

radar beam—The focused energy radiated by radar similar to a flashlight or searchlight beam.

radar echo—*See* echo.

radarsonde observation—A *rawinsonde observation* in which winds are determined by radar tracking a balloon-borne target.

radiation—The emission of energy by a medium and transferred, either through free space or another medium, in the form of electromagnetic waves.

radiation fog—*Fog* characteristically resulting when radiational cooling of the earth's surface lowers the air temperature near the ground to or below its initial dew point on calm, clear nights.

radiosonde—A balloon-borne instrument for measuring pressure, temperature, and humidity aloft. Radiosonde observation—a *sounding* made by the instrument.

rain—A form of *precipitation;* drops are larger than *drizzle* and fall in relatively straight, although not necessarily vertical, paths as compared to drizzle which falls in irregular paths.

rain shower—*See* shower.

range attenuation—*See* attenuation.

range-height indicator (RHI) scope—A radar indicator scope displaying a vertical cross section of *targets* along a selected azimuth.

range resolution—*See* resolution.

RAOB—A *radiosonde* observation.

rawin—A *rawinsonde* observation.

rawinsonde observation—A combined winds aloft and radiosonde observation. Winds are determined by tracking the *radiosonde* by radio direction finder or radar.

refraction—In radar, bending of the *radar beam* by variations in atmospheric density, water vapor content, and temperature.

(1) **normal refraction**—Refraction of the radar beam under normal atmospheric conditions; normal radius of curvature of the beam is about 4 times the radius of curvature of the Earth.

(2) **superrefraction**—More than normal bending of the radar beam resulting from abnormal vertical gradients of temperature and/or water vapor.

(3) **subrefraction**—Less than normal bending of the radar beam resulting from abnormal vertical gradients of temperature and/or water vapor.

relative humidity—The ratio of the existing amount of water vapor in the air at a given temperature to the maximum amount that could exist at that temperature; usually expressed in percent.

relative vorticity—*See* vorticity.

remote scope—In radar meteorology a "slave" scope remoted from weather *radar*.

resolution—Pertaining to radar, the ability of radar to show discrete *targets* separately, i.e., the better the resolution, the closer two targets can be to each other, and still be detected as separate targets.

(1) **beam resolution**—The ability of radar to distinguish between targets at approximately the same range but at different azimuths.

(2) **range resolution**—The ability of radar to distinguish between targets on the same azimuth but at different ranges.

ridge (also called ridge line)—In meteorology, an elongated area of relatively high atmospheric pressure; usually associated with and most clearly identified as an area of maximum anticyclonic curvature of the wind flow (*isobars, contours,* or *streamlines*).

rime icing (or rime ice)—The formation of a white or milky and opaque granular deposit of ice formed by the rapid freezing of supercooled water droplets as they impinge upon an exposed aircraft.

rocketsonde—A type of *radiosonde* launched by a rocket and making its measurements during a parachute descent; capable of obtaining *soundings* to a much greater height than possible by balloon or aircraft.

roll cloud (sometimes improperly called rotor cloud)—A dense and horizontal roll-shaped accessory cloud located on the lower leading edge of a *cumulonimbus* or less often, a rapidly developing *cumulus;* indicative of turbulence.

rotor cloud (sometimes improperly called *roll cloud*)—A turbulent cloud formation found in the lee of some large mountain barriers, the air in the cloud rotates around an axis parallel to the range; indicative of possible violent turbulence.

runway temperature—The temperature of the air just above a runway, ideally at engine and/or wing height, used in the determination of density *altitude;* useful at airports when critical values of density altitude prevail.

runway visibility—The *meteorological visibility* along an identified runway determined from a specified point on the runway; may be determined by a *transmissometer* or by an observer.

runway visual range—An instrumentally derived horizontal distance a pilot should see down the runway from the approach end; based on either the sighting of high intensity runway lights or on the visual contrast of other objects, whichever yields the greatest visual range.

S

St. Elmo's Fire (also called corposant)—A luminous brush discharge of electricity from protruding objects, such as masts and yardarms of ships, aircraft, lightning rods, steeples, etc., occurring in stormy weather.

Santa Ana—A hot, dry, *foehn* wind, generally from the northeast or east, occurring west of the Sierra Nevada Mountains especially in the pass and river valley near Santa Ana, California.

saturated adiabatic lapse rate—The rate of decrease of temperature with height as saturated air is lifted with no gain or loss of heat from outside sources; varies with temperature, being greatest at low temperatures. *See* adiabatic process and dry-adiabatic lapse rate.

saturation—The condition of the atmosphere when actual *water vapor* present is the maximum possible at existing temperature.

scud—Small detached masses of stratus *fractus* clouds below a layer of higher clouds, usually *nimbostratus*.

sea breeze—A coastal breeze blowing from sea to land, caused by the temperature difference when the land surface is warmer than the sea surface. Compare *land breeze*.

sea fog—A type of *advection fog* formed when air that has been lying over a warm surface is transported over a colder water surface.

sea level pressure—The *atmospheric pressure* at *mean sea level*, either directly measured by stations at sea level or empirically determined from the *station pressure* and temperature by stations not at sea level; used as a common reference for analyses of surface pressure patterns.

sea smoke—Same as *steam fog*.

sector visibility—*Meteorological visibility* within a specified sector of the horizon circle.

sensitivity time control—A radar circuit designed to correct for range *attenuation* so that echo intensity on the scope is proportional to reflectivity of the *target* regardless of range.

shear—*See* wind shear.

shower—*Precipitation* from a *cumuliform* cloud; characterized by the suddenness of beginning and ending, by the rapid change of intensity, and usually by rapid change in the appearance of the sky; showery precipitation may be in the form of rain, ice pellets, or snow.

slant visibility—For an airborne observer, the distance at which he can see and distinguish objects on the ground.

sleet—*See* ice pellets.

smog—A mixture of *smoke* and *fog*.

smoke—A restriction to visibility resulting from combustion.

snow—Precipitation composed of white or translucent ice crystals, chiefly in complex branched hexagonal form.

snow flurry—Popular term for snow *shower*, particularly of a very light and brief nature.

snow grains—*Precipitation* of very small, white opaque grains of ice, similar in structure to *snow* crystals. The grains are fairly flat or elongated, with diameters generally less than 0.04 inch (1 mm.).

snow pellets—*Precipitation* consisting of white, opaque approximately round (sometimes conical) ice particles having a snow-like structure, and about 0.08 to 0.2 inch in diameter; crisp and easily crushed, differing in this respect from *snow grains*; rebound from a hard surface and often break up.

snow shower—*See* shower.

solar radiation—The total electromagnetic *radiation* emitted by the sun. *See* insolation.

sounding—In meteorology, an upper-air observation; a *radiosonde* observation.

source region—An extensive area of the earth's surface characterized by relatively uniform surface conditions where large masses of air remain long enough to take on characteristic temperature and moisture properties imparted by that surface.

specific humidity—The ratio by weight of *water vapor* in a sample of air to the combined weight of water vapor and dry air. Compare *mixing ratio*.

squall—A sudden increase in wind speed by at least 15 knots to a peak of 20 knots or more and lasting for at least one minute. Essential difference between a *gust* and a squall is the duration of the peak speed.

squall line—Any nonfrontal line or narrow band of active *thunderstorms* (with or without *squalls*).

stability—A state of the atmosphere in which the vertical distribution of temperature is such that a *parcel* will resist displacement from its initial level. (*See also* instability.)

standard atmosphere—A hypothetical atmosphere based on climatological averages comprised of numerous physical constants of which the most important are:

(1) A surface *temperature* of 59° F (15° C) and a surface pressure of 29.92 inches of mercury (1013.2 millibars) at sea level;

(2) A *lapse rate* in the troposphere of 6.5° C per kilometer (approximately 2° C per 1,000 feet);

(3) A *tropopause* of 11 kilometers (approximately 36,000 feet) with a temperature of −56.5° C; and

(4) An *isothermal* lapse rate in the stratosphere to an altitude of 24 kilometers (approximately 80,000 feet).

standing cloud (standing lenticular altocumulus)—*See* lenticular cloud.

standing wave—A wave that remains stationary in a moving fluid. In aviation operations it is used most commonly to refer to a *lee wave* or *mountain wave*.

stationary front—Same as *quasi-stationary front*.

station pressure—The actual *atmospheric pressure* at the observing station.

steam fog—Fog formed when cold air moves over relatively warm water or wet ground.

storm detection radar—A weather radar designed to detect *hydrometeors* of precipitation size; used primarily to detect storms with large drops or hailstones as opposed to clouds and light precipitation of small drop size.

stratiform—Descriptive of clouds of extensive horizontal development, as contrasted to vertically developed *cumuliform* clouds; characteristic of stable air and, therefore, composed of small water droplets.

stratocumulus—A low cloud, predominantly *stratiform* in gray and/or whitish patches or layers, may or may not merge; elements are tessellated, rounded, or roll-shaped with relatively flat tops.

stratosphere—The atmospheric layer above the tropopause, average altitude of base and top, 7 and 22 miles respectively; characterized by a slight average increase of temperature from base to top and is very stable; also characterized by low moisture content and absence of clouds.

stratus—A low, gray cloud layer or sheet with a fairly uniform base; sometimes appears in ragged patches; seldom produces precipitation but may produce *drizzle* or *snow grains*. A *stratiform* cloud.

stratus fractus—*See* fractus.

streamline—In meteorology, a line whose tangent is the wind direction at any point along the line. A flowline.

sublimation—*See* change of state.

subrefraction—*See* refraction.

subsidence—A descending motion of air in the atmosphere over a rather broad area; usually associated with *divergence*.

summation principle—The principle states that the cover assigned to a layer is equal to the summation of the sky cover of the lowest layer plus the additional coverage at all successively higher layers up to and including the layer in question. Thus, no layer can be assigned a sky cover less than a lower layer, and no sky cover can be greater than 1.0 (10/10).

superadiabatic lapse rate—A *lapse rate* greater than the *dry-adiabatic lapse rate*. *See* absolute instability.

supercooled water—Liquid water at temperatures colder than freezing.

superrefraction—*See* refraction.

surface inversion—An *inversion* with its base at the surface, often caused by cooling of the air near the surface as a result of *terrestrial radiation*, especially at night.

surface visibility—Visibility observed from eye-level above the ground.

synoptic chart—A chart, such as the familiar weather map, which depicts the distribution of meteorological conditions over an area at a given time.

T

target—In radar, any of the many types of objects detected by radar.

temperature—In general, the degree of hotness or coldness as measured on some definite temperature scale by means of any of various types of thermometers.

temperature inversion—*See* inversion.

terrestrial radiation—The total infrared *radiation* emitted by the Earth and its atmosphere.

thermograph—A continuous-recording *thermometer*.

thermometer—An instrument for measuring *temperature*.

theodolite—An optical instrument which, in meteorology, is used principally to observe the motion of a *pilot balloon*.

thunderstorm—In general, a local storm invariably produced by a *cumulonimbus* cloud, and always accompanied by lightning and thunder.

tornado (sometimes called cyclone, twister)—A violently rotating column of air, pendant from a cumulonimbus cloud, and nearly always observable as "funnel-shaped." It is the most destructive of all small-scale atmospheric phenomena.

towering cumulus—A rapidly growing *cumulus* in which height exceeds width.

tower visibility—*Prevailing visibility* determined from the control tower.

trade winds—Prevailing, almost continuous winds blowing with an easterly component from the subtropical high pressure belts toward the *intertropical convergence zone;* northeast in the Northern Hemisphere, southeast in the Southern Hemisphere.

transmissometer—An instrument system which shows the transmissivity of light through the atmosphere. Transmissivity may be translated either automatically or manually into *visibility* and/or *runway visual range*.

tropical air—An air mass with characteristics developed over low latitudes. Maritime tropical air (mT), the principal type, is produced over the tropical and subtropical seas; very warm and humid. Continental tropical (cT) is produced over subtropical arid regions and is hot and very dry. Compare *polar air*.

tropical cyclone—A general term for a *cyclone* that originates over tropical oceans. By international agreement, tropical cyclones have been classified according to their intensity, as follows:

(1) **tropical depression**—winds up to 34 knots (64 km/h);

(2) **tropical storm**—winds of 35 to 64 knots (65 to 119 km/h);

(3) **hurricane or typhoon**—winds of 65 knots or higher (120 km/h).

tropical depression—*See* tropical cyclone.

tropical storm—*See* tropical cyclone.

tropopause—The transition zone between the *troposphere* and *stratosphere*, usually characterized by an abrupt change of *lapse rate*.

troposphere—That portion of the *atmosphere* from the earth's surface to the *tropopause;* that is, the lowest 10 to 20 kilometers of the atmosphere. The troposphere is characterized by decreasing temperature with height, and by appreciable water vapor.

trough (also called trough line)—In meteorology, an elongated area of relatively low atmospheric pressure; usually associated with and most clearly identified as an area of maximum cyclonic curvature of the wind flow (*isobars, contours*, or *streamlines*); compare with *ridge*.

true altitude—*See* altitude.

true wind direction—The direction, with respect to true north, from which the wind is blowing.

turbulence—In meteorology, any irregular or disturbed flow in the atmosphere.

twilight—The intervals of incomplete darkness following sunset and preceding sunrise. The time at which evening twilight ends or morning twilight begins is determined by arbitrary convention, and several kinds of twilight have been defined and used; most commonly civil, nautical, and astronomical twilight.

(1) **Civil Twilight**—The period of time before sunrise and after sunset when the sun is not more than 6° below the horizon.

(2) **Nautical Twilight**—The period of time before sunrise and after sunset when the sun is not more than 12° below the horizon.

(3) **Astronomical Twilight**—The period of time before sunrise and after sunset when the sun is not more than 18° below the horizon.

twister—In the United States, a colloquial term for *tornado*.

typhoon—A *tropical cyclone* in the Eastern Hemisphere with winds in excess of 65 knots (120 km/h).

U

undercast—A cloud *layer* of ten-tenths (1.0) coverage (to the nearest tenth) as viewed from an observation point above the layer.

unlimited ceiling—A clear sky or a sky cover that does not meet the criteria for a *ceiling*.

unstable—*See* instability.

updraft—A localized upward current of air.

upper front—A *front* aloft not extending to the earth's surface.

upslope fog—Fog formed when air flows upward over rising terrain and is, consequently, adiabatically cooled to or below its initial *dew point*.

V

vapor pressure—In meteorology, the pressure of water vapor in the atmosphere. Vapor pressure is that part of the total atmospheric pressure due to water vapor and is independent of the other atmospheric gases or vapors.

vapor trail—Same as *condensation trail.*

veering—Shifting of the wind in a clockwise direction with respect to either space or time; opposite of backing. Commonly used by meteorologists to refer to an anticyclonic shift (clockwise in the Northern Hemisphere and counterclockwise in the Southern Hemisphere).

vertical visibility—The distance one can see upward into a surface based *obscuration;* or the maximum height from which a pilot in flight can recognize the ground through a surface based obscuration.

virga—Water or ice particles falling from a cloud, usually in wisps or streaks, and evaporating before reaching the ground.

visibility—The greatest distance one can see and identify prominent objects.

visual range—*See* runway visual range.

vortex—In meteorology, any rotary flow in the atmosphere.

vorticity—Turning of the atmosphere. Vorticity may be imbedded in the total flow and not readily identified by a flow pattern.

(a) **absolute vorticity**—the rotation of the Earth imparts vorticity to the atmosphere; absolute vorticity is the combined vorticity due to this rotation and vorticity due to circulation relative to the Earth (relative vorticity).

(b) **negative vorticity**—vorticity caused by anticyclonic turning; it is associated with downward motion of the air.

(c) **positive vorticity**—vorticity caused by cyclonic turning; it is associated with upward motion of the air.

(d) **relative vorticity**—vorticity of the air relative to the Earth, disregarding the component of vorticity resulting from Earth's rotation.

W

wake turbulence—*Turbulence* found to the rear of a solid body in motion relative to a fluid. In aviation terminology, the turbulence caused by a moving aircraft.

wall cloud—The well-defined bank of vertically developed clouds having a wall-like appearance which form the outer boundary of the *eye* of a well-developed *tropical cyclone.*

warm front—Any non-occluded *front* which moves in such a way that warmer air replaces colder air.

warm sector—The area covered by warm air at the surface and bounded by the *warm front* and *cold front* of a *wave cyclone.*

water equivalent—The depth of water that would result from the melting of snow or ice.

waterspout—*See* tornado.

water vapor—Water in the invisible gaseous form.

wave cyclone—A *cyclone* which forms and moves along a front. The circulation about the cyclone center tends to produce a wavelike deformation of the front.

weather—The state of the *atmosphere,* mainly with respect to its effects on life and human activities; refers to instantaneous conditions or short term changes as opposed to *climate.*

weather radar—Radar specifically designed for observing weather. *See* cloud detection radar and storm detection radar.

weather vane—A *wind vane.*

wedge—Same as *ridge.*

wet bulb—Contraction of either *wet-bulb temperature* or *wet-bulb thermometer.*

wet-bulb temperature—The lowest *temperature* that can be obtained on a *wet-bulb thermometer* in any given sample of air, by evaporation of water (or ice) from the muslin wick; used in computing *dew point* and *relative humidity.*

wet-bulb thermometer—A thermometer with a muslin-covered bulb used to measure wet-bulb temperature.

whirlwind—A small, rotating column of air; may be visible as a dust devil.

willy-willy—A *tropical cyclone* of hurricane strength near Australia.

wind—Air in motion relative to the surface of the earth; generally used to denote horizontal movement.

wind direction—The direction from which wind is blowing.

wind speed—Rate of wind movement in distance per unit time.

wind vane—An instrument to indicate wind direction.

wind velocity—A vector term to include both *wind direction* and *wind speed.*

wind shear—The rate of change of *wind velocity* (direction and/or speed) per unit distance; conventionally expressed as vertical or horizontal wind shear.

X—Y—Z

zonal wind—A west wind; the westerly component of a wind. Conventionally used to describe large-scale flow that is neither cyclonic nor anticyclonic.

END OF APPENDIX

APPENDIX B
AM WEATHER AND THE WEATHER CHANNEL

> Please take a few minutes to study each of the concepts listed above and anticipate/imagine what they are and how they relate to the other listed concepts.

A. AM WEATHER

1. AM WEATHER is a comprehensive weather service available through public television across the U.S. and southern portions of Canada.

    a. This 15-min. program gives viewers vital weather information in the following sequence:

        1) Satellite imagery sequences
        2) Jet stream
        3) National and local radar
        4) National map of current weather
        5) 12, 24, 48, and 72-hr. forecast maps
        6) IFR/MVFR/VFR flying weather
        7) Icing and turbulence forecasts
        8) Winds aloft
        9) Severe weather watches, warnings, and advisories

    b. The program is transmitted seven times each weekday (Monday through Friday) via satellite to participating Public Broadcast System (PBS) stations and updated throughout the morning.

        1) The first line report is transmitted at 6:45 a.m. ET.

        2) Each of the more than 300 stations carrying the program determines the local broadcast schedule for its area.

            a) Note that while the program is updated throughout the morning, most stations broadcast the show only once a day.

            b) Check your local television listings to confirm the exact broadcast time in your area.

2. AM WEATHER meteorologists are permanently assigned to the program by the National Weather Service (NWS) and the National Environmental Satellite Data and Information Service (NESDIS).

    a. Thus, all forecasts are products of the NWS.

    b. The assigned meteorologists enhance the forecasts by making them more visually appealing to the viewer.

    c. On Friday, an extended forecast for the weekend is made by the meteorologists, which is normally their own analysis.

3. AM WEATHER is produced by Maryland Public Television, 11767 Owings Mills Boulevard, Owings Mills, Maryland 21117; telephone (410) 356-5000.

B. Using AM WEATHER

1.   AM WEATHER is a concise but informative 15 min. of weather information.

    a.   It is an excellent way to begin your day with the current weather and forecast conditions.

    b.   The format gives you the big picture of the national weather and areas of concern.

        1)   The charts and symbols used are the same as described throughout Part III, beginning on page 211.

        2)   AM WEATHER adds colors to depict areas of MVFR and IFR to help you to quickly identify those areas.

2.   While AM WEATHER uses charts, graphics, and terminology familiar to pilots, it is also used by other groups of people (e.g., boaters, travelers, farmers, etc.).

3.   Watch AM WEATHER to start your day, and then watch The Weather Channel (see page 402) for updated radar imagery and surface maps.

    a.   Use AM WEATHER to complement your weather briefing with an FSS specialist.

    b.   If you saw an area of concern while watching AM WEATHER, ask the FSS specialist for updates and/or clarifications to satisfy your needs.

    c.   This approach will give you the information you need to make a competent go/no go decision.

4.   The disadvantage of AM WEATHER is that most stations broadcast only once a day, Monday through Friday.

AM WEATHER Station Time Listing

ALABAMA
Birmingham
7:15 WBIQ/10
Demopolis
7:15 WIIQ/41
Dozier
7:15 WDIQ/2
Florence
7:15 WFIQ/36
Huntsville
7:15 WHIQ/25
Louisville
7:15 WGIQ/43
Mobile
7:15 WEIQ/42
Montgomery
7:15 WAIQ/26
Mt. Cheaha
7:15 WCIQ/7

ARIZONA
Chinle
6:45 **and** 7:45
K51AV/51
Phoenix/Tempe
5:45 KAET/8
Tsaile
6:45 **and** 7:45
K40AP/40
Tucson
7:00 KUAT/6
Window Rock
6:45 **and** 7:45
K44BB/44

ARKANSAS
Arkadelphia
5:45 KETG/9
Fayetteville
5:45 KAFT/13
Jonesboro
5:45 KTEJ/19
Little Rock
5:45 KETS/2
Mountain View
5:45 KEMV/6

CALIFORNIA
Cedarville
5:45 **and** 6:45
K13IU/13
Eureka
7:45 KEET/13
Fresno
6:45 KVIT/18
Los Angeles
5:15 KCET/28
Sacramento/Stockton
6:00 **and** 6:45
KVIE/6
San Diego
6:15 KPBS/15
San Francisco
6:15 KQED/9
San Jose
6:15 KTEH/54

San Mateo/San Francisco
5:45 KCSM/60
Santa Rosa/Rohnert Park
6:45 KRCB/22

CANADA
Ottawa
7:15 Cable 4
Ottawa
7:15 Cable 16

COLORADO
Denver
6:15 KRMA/6
Pueblo/Colorado Springs
6:15 KTSC/8/15

FLORIDA
Ft. Myers
6:45 WSFP/30
Gainesville
6:45 WUFT/5
Jacksonville
6:45 WJCT/7
Miami/Ft. Lauderdale
6:45 WPBT/2
Orlando
7:15 WMFE/24
Panama City
5:45 WFSG/56
Pensacola
6:45 WSRE/23
Tallahassee
6:45 WFSU/11
Tampa
6:45 WEDU/3
West Palm Beach
6:45 WXEL/42

GEORGIA
Albany/Pelham
6:45 WABW/14
Americus/Dawson
6:45 WACS/25
Atlanta
7:45 WPBA/30
Atlanta/Athens
6:45 WGTV/8
Augusta/Wrens
6:45 WCES/20
Columbus/Warm Springs
6:45 WJSP/28
Dalton/Chatsworth
6:45 WCLP/18
Macon/Cochran
6:45 WDCO/15
Savannah/Pembroke
6:45 WVAN/9
Valdosta/Waycross
6:45 WXGA/8

IDAHO
Boise
7:45 KAID/4
Coeur D'Alene
6:45 KCDT/26
Moscow
6:45 KUID/12

Pocatello
7:45 KISU/10
Twin Falls
7:45 KIPT/13

ILLINOIS
Carbondale
6:15 WSIU/8
Chicago
6:15 WTTW/11
Moline
7:45 WQPT/24
Olney
6:15 WUSI/16
Peoria
7:15 WTVP/47
Urbana
6:15 WILL/12

INDIANA
Bloomington
7:15 WTIU/30
Fort Wayne
6:45 WFWA/39
Indianapolis
6:45 WFYI/20
Muncie
6:45 WIPB/49
South Bend/Elkhart
6:45 WNIT/34
Vincennes
7:15 WVUT/22

IOWA
Council Bluffs
6:45 KBIN/32
Des Moines
6:45 KDIN/11
Fort Dodge
6:45 KTIN/21
Iowa City
6:45 KIIN/12
Mason City
6:45 KYIN/24
Red Oak
6:45 KHIN/36
Sioux City
6:45 KSIN/27
Waterloo
6:45 KRIN/32

KANSAS
Concordia
7:15 K64BS/64
Hoxie
7:15 K69DB/69
Iola-Moran
6:45 K30AL/30
Lakin/Garden City
7:15 KSWK/3
Phillipsburg
7:15 K66CD/66
Quinter
7:15 K12KH/12
Salina/Hays
7:15 KOOD/9

Topeka
6:45 KTWU/11
Wichita/Hutchinson
6:45 KPTS/8

KENTUCKY
Louisville
6:45 WKPC/15

LOUISIANA
Alexandria
6:15 KLPA/25
Baton Rouge
6:15 WLPB/27
Lafayette
6:15 KLPB/24
Lake Charles
6:15 KLTL/18
Monroe
6:15 KLTM/13
Shreveport
6:15 KLTS/24

MAINE
Augusta/Lewiston/
Portland
6:45 WCBB/10
Calais
6:45 WMED/13
Orono/Bangor
6:45 WMEB/12
Portland/Biddeford
6:45 WMEA/26
Presque Isle
6:45 WMEM/10

MARYLAND
Annapolis
6:45 WMPT/22
Baltimore
6:45 WMPB/67
Frederick
6:45 WFPT/62
Hagerstown
6:45 WWPB/31
Oakland
6:45 WGPT/36
Salisbury
6:45 WCPB/28

MASSACHUSETTS
Boston
6:45 WGBX/44
Springfield
6:45 WGBY/57

MICHIGAN
Alpena
6:45 WCML/6
Cadillac
6:45 WCMV/27
Detroit
7:15 WTVS/56
East Lansing
6:45 WKAR/23
Flint
6:45 WFUM/28

Grand Rapids/Allendale
7:15 WGVU/35
     Kalamazoo
7:15 WGVK/52
     Leland
6:45 W69AV/69
     Manistee
6:45 WCMW/21
     Marquette
7:15 WNMU/13
     Mount Pleasant
6:45 WCMU/14
     Traverse City
6:45 W46AD/46

MINNESOTA
     Appleton
6:45 KWCM/10
     Austin
7:45 KSMQ/15
     Bemidji
6:30 and 6:45
     KAWE/9
     Brainerd
6:30 and 6:45
     KAWB/22
     Duluth
6:45 WDSE/8
     St. Paul/Minneapolis
6:15 KTCA/2
     St. Paul/Minneapolis
5:45 and 6:45
7:15 and 7:45
8:15 and 8:45
     KTCI/17

MISSISSIPPI
     Biloxi
6:45 WMAH/19
     Booneville
6:45 WMAE/12
     Bude
6:45 WMAU/17
     Greenwood
6:45 WMAO/23
     Jackson
6:45 WMPN/29
     Meridian
6:45 WMAW/14
     Mississippi State
6:45 WMAB/2
     Oxford
6:45 WMAV/18

MISSOURI
     Joplin
5:45 KOZJ/26
     Kansas City
6:15 KCPT/19
     Sedalia/Warrensburg
7:45 KMOS/6
     Springfield
5:45 KOZK/21
     St. Louis
6:15 and 6:45
     KETC/9

MONTANA
     Boulder
6:45 and 7:45
     K27CD/27
     Bozeman
6:45 KUSM/9
     Bridger
6:45 and 7:45
     K63EA/63
     Browning
6:45 and 7:45
     K57FL/57
     Chinook
6:45 and 7:45
     K59AS/59
     Colstrip
6:45 and 7:45
     K28CB/28
     Hamilton
6:45 and 7:45
     K21AN/21
     Pablo
6:45 and 7:45
     K25CL/25
     Plains
6:45 and 7:45
     K21CA/21
     St. Ignatius
6:45 and 7:45
     K28CF/28
     Stevensville
6:45 and 7:45
     K67EX/67
     Thompson Falls
6:45 and 7:45
     K36BW/36
     White Hall
6:45 and 7:45
     K52CE/52
     White Sulphur Spring
6:45 and 7:45
     K57AL/57

NEBRASKA
     Alliance
6:45 KTNE/13
     Hastings
6:45 KHNE/29
     Kearney/Lexington
6:45 KLNE/3
     Lincoln
6:45 KUON/12
     Norfolk
6:45 KXNE/19
     North Platte
6:45 KPNE/9
     Omaha
6:45 KYNE/26
     O'Neill/Ainsworth
6:45 KMNE/7
     Valentine/Cherry County
6:45 KRNE/12

NEVADA
     Battle Mountain
5:45 and 6:45
     K32CA/32
     Elko
5:45 and 6:45
     K14AO/14
     Ely
5:45 and 6:45
     K14AL/14
     Eureka
5:45 and 6:45
     K47DG/47
     Fallon
5:45 and 6:45
     K25AK/25
     Incline Village
5:45 and 6:45
     K14AJ/14
     Las Vegas
7:15 KLVX/10
     Lovelock
5:45 and 6:45
     K14AK/14
     Reno
6:45 KNPB/5
     Tonopah
5:45 and 6:45
     K17AH/17
     Winnemucca
5:45 and 6:45
     K15AL/15

NEW HAMPSHIRE
     Durham/Manchester
7:45 WENH/11
     Keene
7:45 WEKW/52
     Littleton
7:45 WLED/49

NEW JERSEY
     Camden
8:45 WNJS/23
     Montclair
8:45 WNJM/50
     New Brunswick
8:45 WNJB/58
     Trenton
8:45 WNJT/52

NEW MEXICO
     Albuquerque
5:45 KNME/5
     Las Cruces
6:45 KRWG/22
     Mescalero
6:45 and 7:45
     K02KQ/2
     Mescalero
6:45 and 7:45
     K02KR/2
     Portales
6:45 KENW/3
     Shiprock
6:45 and 7:45
     K48AW/48

NEW YORK
     Binghamton
6:45 WSKG/46
     Buffalo
6:45 WNED/17
     Garden City
7:15 WLIW/21
     Norwood/St. Lawrence Co.
7:15 WNPI/18
     Rochester
6:45 WXXI/21
     Schenectady
6:45 WMHT/17
     Syracuse
6:45 WCNY/24
     Watertown
7:15 WNPE/16

NORTH CAROLINA
     Asheville
6:45 WUNF/33
     Chapel Hill
6:45 WUNC/4
     Charlotte
6:45 WTVI/42
     Columbia
6:45 WUND/2
     Concord/Charlotte
6:45 WUNG/58
     Greenville
6:45 WUNK/25
     Jacksonville
6:45 WUNM/19
     Linville
6:45 WUNE/17
     Roanoke Rapids
6:45 WUNP/36
     Wilmington
6:45 WUNJ/39
     Winston-Salem
6:45 WUNL/26

NORTH DAKOTA
     Bismarck
6:45 KBME/3
     Dickinson
6:45 KDSE/9
     Fargo
6:45 KFME/13
     Grand Forks
6:45 KGFE/2
     Minot
6:45 KSRE/6
     Williston
6:45 KWSE/4

OHIO
     Akron
6:45 WEAO/49
     Alliance
6:45 WNEO/45
     Athens
6:45 Cable 7
     Bowling Green
6:45 WBGU/27

Cincinnati
6:45 WCET/48
Cleveland
6:45 WVIZ/25
Dayton
6:45 WPTD/16
Oxford
6:45 WPTO/14
Toledo
6:45 WGTE/30

OKLAHOMA
Elk City/Cheyenne
7:15 KWET/12
McAlester
7:15 KOET/3
Oklahoma City
7:15 KETA/13
Oklahoma City
8:45 KTLC/43
Tulsa
7:15 KOED/11

OREGON
Bend
6:00 **and** 6:45
KOAB/3
Corvallis
6:00 **and** 6:45
KOAC/7
Eugene
6:00 **and** 6:45
KEPB/28
Klamath Falls
7:00 **and** 7:45
KFTS/22
Medford
7:00 **and** 7:45
KSYS/8
Pendleton/Baker/
La Grande
6:00 **and** 6:45
KTVR/13
Portland
6:00 **and** 6:45
KOPB/10

PENNSYLVANIA
Bethlehem/Allentown
7:15 WLVT/39
Erie
6:45 WQLN/54
Harrisburg
6:45 WITF/33
Philadelphia
6:45 WHYY/12
Scranton/Wilkes-Barre
6:45 WVIA/44
State College/
University Park
6:45 WPSX/3

SOUTH CAROLINA
Allendale/Barnwell
6:45 WEBA/14
Beaufort
6:45 WJWJ/16
Charleston
6:45 WITV/7
Columbia
6:45 WRLK/35
Conway
6:45 WHMC/23
Florence
6:45 WJPM/33
Greenville
6:45 WNTV/29
Greenwood
6:45 WNEH/38
Rock Hill
6:45 WNSC/30
Spartanburg
6:45 WRET/49
Sumter
6:45 WRJA/27

SOUTH DAKOTA
Aberdeen
6:15 KDSD/16
Brookings
6:15 KESD/8
Eagle Butte/Faith
6:15 KPSD/13
Martin/Pine Ridge
6:15 KZSD/8
Mobridge
6:15 KQSD/11
Pierre
6:15 KTSD/10
Rapid City
6:15 KBHE/9
Yankton/Vermillion
6:15 KUSD/2

TENNESSEE
Chattanooga
6:45 WTCI/45
Cookeville
6:45 WCTE/22
Lexington/Jackson/Martin
7:45 WLJT/11
Nashville
6:45 WDCN/8

TEXAS
Amarillo
6:15 KACV/2
Austin
6:45 KLRU/18
College Station
6:45 KAMU/15

Corpus Christi
6:45 KEDT/16
Dallas/Ft. Worth
5:24 **and** 6:15
KERA/13
Harlingen
7:45 KMBH/60
Houston
6:45 KUHT/8
Kileen
6:45 KNCT/46
Lubbock
8:15 KTXT/5
Odessa
7:15 KOCV/36
San Antonio
6:15 KLRN/9
Waco
6:15 KCTF/34

UTAH
Provo
5:15 KBYU/11
Salt Lake City
6:15 KUED/7

VERMONT
Burlington
7:15 WETK/33
Rutland
7:15 WVER/28
St. Johnsbury
7:15 WVTB/20
Windsor
7:15 WVTA/41

VIRGIN ISLANDS
St. Thomas
8:45 WTJX/12

VIRGINIA
Falls Church
8:40 WNVT/53
Charlottesville
8:15 WHTJ/41
Harrisonburg
6:45 WVPT/51
Marion
6:45 WMSY/52
Norfolk
6:45 WHRO/15
Norton
6:45 WSBN/47
Richmond
8:15 WCVE/23
Richmond
8:15 WCVW/57
Roanoke
6:45 WBRA/15

WASHINGTON
Richland
6:45 KTNW/31
Seattle
5:45 KCTS/9
Spokane
6:45 KSPS/7
Yakima
6:45 KYVE/47

WEST VIRGINIA
Beckley
6:45 WSWP/9
Huntington
6:45 WPBY/33
Morgantown
6:45 WNPB/24

WISCONSIN
Green Bay
6:15 WPNE/38
LaCrosse
6:15 WHLA/31
Madison
6:15 WHA/21
Menomonie/Eau Clair
6:15 WHWC/28
Milwaukee
6:15 WMVS/10
Park Falls
6:15 WLEF/36
Wausau
6:15 WHRM/20

WYOMING
Riverton
6:45 KCWC/4

C. The Weather Channel

1. The Weather Channel (TWC) is the only all-weather network on cable television that broadcasts local, regional, national, and international (mostly limited to Europe) weather reports and forecasts 24 hr. a day, 7 days a week.

   a. TWC began broadcasting in May 1982, and today it is available to over 53 million cable subscribers.

2. TWC's diversified programming is designed to meet almost everyone's weather needs.

3. TWC employs more than 65 staff meteorologists who analyze information from the NWS and other sources to produce local reports and forecasts for the entire nation.

   a. These forecasts are those made by the staff at TWC, not by the NWS.

4. TWC produces over 1,000 computer-generated graphics each day to display current and forecast weather conditions. These graphics help make the presentation more viewer-friendly.

   a. Local radar reports and surface observations are updated hourly, and sometimes more often.

   b. With this information current surface analysis charts are also reviewed and updated hourly.

5.   Regularly scheduled programs are presented each hour as shown in the TWC clock below.

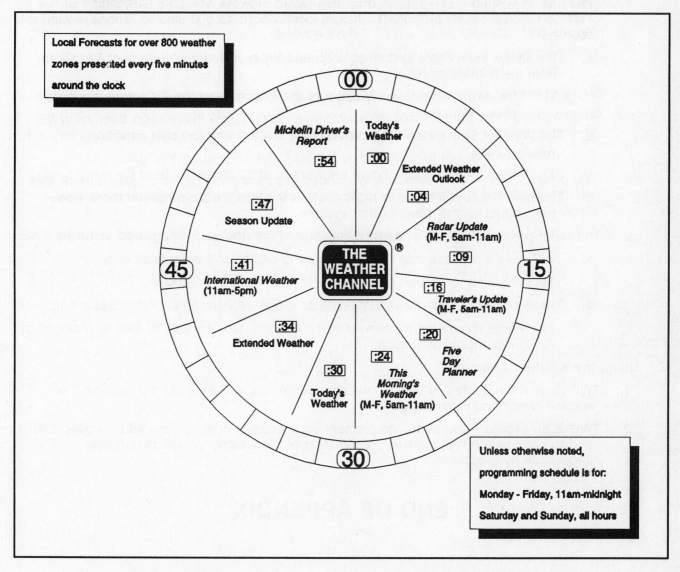

Local Forecasts for over 800 weather zones presented every five minutes around the clock

Today's Weather
:00

Michelin Driver's Report
:54

Extended Weather Outlook
:04

Radar Update
(M-F, 5am-11am)
:09

Season Update
:47

THE WEATHER CHANNEL ®

Traveler's Update
(M-F, 5am-11am)
:16

International Weather
(11am-5pm)
:41

Five Day Planner
:20

Extended Weather
:34

This Morning's Weather
(M-F, 5am-11am)
:24

Today's Weather
:30

Unless otherwise noted, programming schedule is for:
Monday - Friday, 11am-midnight
Saturday and Sunday, all hours

a.   TWC also provides a variety of specialty and seasonal programming.

D.   **Local Weather**

    1.   TWC has developed a computerized system called THE WEATHER STAR, which allows TWC to simultaneously broadcast different local reports to 800 weather zones around the country.

        a.   This allows each cable system to automatically receive local surface observations from the nearest observation site.

            1)   This information normally appears at the bottom of the TV screen during other presentations.

        b.   The Weather Star system also provides the current and forecast conditions for each weather zone.

            1)   These forecasts are shown an average of 12 times per hour (or every 5 min.).

        c.   The part that is of interest to most pilots is the local radar image, which is also transmitted on the Weather Star system.

    2.   The radar presentations are enhanced versions of the raw radar images provided by NWS.

        a.   TWC uses a process that eliminates ground clutter and other false echoes.

            1)   This cleanup allows accurate radar imaging and contouring.

        b.   These enhancements make TWC's radar image very useful in preflight planning.

            1)   Radar shows the movement of the echoes over the last 90 min. to provide you with more information.

E.   **Using the Weather Channel**

    1.   TWC does not provide any aviation weather segments, but it constantly updates the surface charts and radar imagery.

    2.   TWC is an excellent source to complement your weather briefing from an FSS specialist and to provide you with a visual presentation of the weather, which is currently impossible by telephone.

END OF APPENDIX

APPENDIX C
AUTOMATED FLIGHT SERVICE STATIONS

Please take a few minutes to study each of the concepts listed above and anticipate/imagine what they are and how they relate to the other listed concepts.

A. **Flight Service Stations (FSSs)** -- Flight Service Stations (FSSs) are air traffic facilities which provide a variety of services to pilots.

1. Automated FSSs (AFSSs) have replaced most of the older FSSs, usually with one AFSS per state.

   a. Nonautomated FSSs were at a large number of airports around the country.

   b. The same services are available at AFSSs that were available at the FSSs, but with a much faster and more complete dissemination of information nationwide.

2. AFSS personnel are responsible for providing emergency, in-flight, and preflight services.

   a. Emergency situations are those where life or property is in danger and include:

      1) VFR search and rescue services
      2) Assistance to lost aircraft and other aircraft in emergency situations

   b. In-flight services are those provided to or affecting aircraft in flight or operating on the airport surface.

      1) NAVAID monitoring and restoration
      2) Local Airport Advisories (LAA)
      3) Delivery of ATC clearances, advisories, or requests
      4) Issuance of military flight advisory messages
      5) En Route Flight Advisory Service (EFAS) or Flight Watch
      6) Issuance of NOTAMs
      7) Transcribed or live weather broadcasts
      8) Weather observations
      9) PIREPs
      10) Radio pilot briefings

   c. Preflight services are those which are provided prior to actual departure, usually by telephone.

      1) Pilot briefings
      2) Flight plan filing and processing
      3) Aircraft operational reservations

B. **Transcribed or Live Broadcasts**

1. A variety of reports and forecasts are broadcast over selected VOR and NDB frequencies.

2. Unscheduled broadcasts are made upon receipt of special weather reports, PIREPS, NOTAMS, weather advisories, radar reports, military training route data, alert notices, and other information considered necessary to enhance safety and efficiency of flight.

3. Transcribed Weather Broadcasts (TWEBs) are broadcast continuously over selected NDB and VOR frequencies.

    a. A TWEB is weather information presented in a route-of-flight format.

4. Pilot's Automatic Telephone Weather Answering Service (PATWAS) is provided by nonautomated FSSs, and is a recorded telephone briefing service with the forecast for the local area, usually within a 50-NM radius of the station.

    a. A few selected stations also include route forecasts similar to the TWEB.

5. Telephone Information Briefing Service (TIBS) is provided by AFSSs, and provides weather and/or aeronautical information for a variety of routes and/or areas.

    a. TIBS may provide any of the following:

        1) Area and/or route briefings

            a) Surface observations
            b) Terminal forecasts
            c) Winds aloft

        2) Airspace procedures
        3) Special announcements

    b. A pilot who calls an AFSS by touch-tone telephone can select from different TIBS briefing areas and/or routes.

        1) The information contained in a TIBS is the same as in a standard briefing.

6. Scheduled weather broadcasts are made at H+15 in Alaska only, and include surface reports from a number of airports in the vicinity of the station.

7. Hazardous In-flight Weather Advisory Service (HIWAS) is a continuous broadcast service of in-flight weather advisories, i.e., SIGMETs, Convective SIGMETs, AIRMETs, CWAs, and AWWs over selected VORs.

C. **Pilot Briefing**

1. AFSSs are equipped with computer terminals and graphic displays that can generate or display any desired report, forecast, chart, and/or satellite imagery.

2. A preflight briefing may be any one of the following types:

    a. A standard briefing is a summary and interpretation of all available data concerning an intended flight.

    b. An abbreviated briefing is intended to supplement mass disseminated data (e.g., TWEB, TIBS, etc.), update a previous briefing, or provide the pilot with only specifically requested information.

    c. An outlook briefing is provided when the proposed departure is 6 hr. or more from the time of the briefing.

D. **In-Flight Services**

1. AFSSs provide services to aircraft on a "first come, first served" basis, except for the following, which receive priority:

    a. Aircraft in distress
    b. Lifeguard (air ambulance) aircraft
    c. Search and rescue aircraft

2. The following types of radio communications are generally made by AFSSs:

   a. Authorized transmission of messages necessary for ATC or safety
   b. Routine radio contacts, e.g., in-flight briefings
   c. ATC clearances, advisories, and requests
   d. Flight progress reports upon receipt of pilot position reports

3. AFSSs regularly monitor NAVAIDs within their area for proper operation.

   a. When a malfunction is indicated or reported by several aircraft, the AFSS attempts to restore the NAVAID or discontinue its use and issue an appropriate NOTAM.

4. Local Airport Advisory (LAA) is a terminal service provided by an FSS physically located on an airport without an operating control tower.

   a. LAA provides information to arriving and departing aircraft concerning wind direction and speed, favored runway, altimeter setting, pertinent known traffic, and pertinent known field conditions.

      1) This information is advisory in nature and is not an ATC clearance.

   b. LAA may also provide control of airport lighting, e.g., rotating beacon, runway lights, etc., after dark and at times of reduced visibility.

   c. If Class D or E airspace (formerly known as a control zone) exists at the surface of an airport served by an LAA, the FSS there will provide special VFR clearances to aircraft, when appropriate.

5. En Route Flight Advisory Service (EFAS), or Flight Watch, provides en route aircraft with timely and pertinent weather data tailored to a specific altitude and route using the most current available information.

   a. Briefings are intended to apply to the en route phase of flight (i.e., between climbout and descent to land).

   b. When conditions dictate, information is provided on weather for alternate routes and/or altitudes.

   c. EFAS may not be used for in-flight services, i.e., flight plan filing, position reporting, or full route (preflight) briefing.

   d. PIREPs are solicited from pilots who contact Flight Watch.

E. **Emergency Services**

   1. An emergency can be either a distress or urgent situation.

      a. Distress is a condition of being threatened by serious and/or imminent danger and of requiring immediate assistance.

      b. Urgency is a condition of being concerned about safety and of requiring timely but not immediate assistance, i.e., a potential distress condition.

   2. FSSs may provide assistance to aircraft in emergency situations by

      a. Assisting in orienting a lost pilot,

      b. Coordinating communications between the aircraft, ATC, and search and rescue crews, as appropriate,

      c. Monitoring frequencies 121.5 and 243.0 for emergency radio calls and ELT signals.

   3. Direction Finder Service (DF Steer) allows an FSS to determine an aircraft's bearing from the station using the two-way VHF radio communication transmitter on board the aircraft.

      a. Under emergency conditions where a standard instrument approach cannot be executed, an FSS can provide DF guidance and instrument approach service.

   4. Using VOR or ADF facilities as appropriate to the aircraft's equipment, an FSS can assist a pilot in determining his/her position, and provide guidance to the airport if desired.

F. **Flight Data**

1. FSSs accept domestic and international IFR, VFR, and DVFR flight plans and forward the flight plan data to the appropriate agencies.

2. Data from IFR flight plans are transmitted to the ARTCC as part of the IFR fight plan proposal.

   a. Search and rescue information from the flight plan are retained in the FSS and are available upon request.

3. VFR flight plans are forwarded to the FSSs serving the departure and the destination points.

   a. When an aircraft on a VFR flight plan changes destination, the FSS will forward a notification to the original and new destination stations.

   b. When an aircraft changes its ETA, the information is passed by FSS to the destination station.

4. Military flight plans are handled by FSS in much the same way as civilian flight plans.

5. For security control of air traffic, flight data and position reports from DVFR and IFR aircraft operating within an ADIZ are forwarded by FSS to ARTCC.

6. FSSs disseminate law enforcement alert messages, stolen aircraft summaries, and aircraft lookout alerts within FSS facilities and offices, but not to the general public.

7. FSSs coordinate all pertinent information received from pilots prior to and during parachute jumping activity with other affected ATC facilities.

G. **International Operations**

1. AFSSs record and relay flight plans to and from Canada and Mexico to the appropriate stations.

   a. When the pilot requests Customs flight notification service (by including ADCUS in the remarks), the FSS will notify the appropriate Customs office.

      1) This service is available at airports so indicated in the *A/FD*.

   b. Round-robin flight plans to Mexico cannot be accepted.

2. AFSSs also relay International Civil Aeronautics Organization (ICAO) messages in four categories.

   a. Distress or urgency messages, or other messages concerning known or suspected emergencies, such as radio communications failures.

   b. Movement and control

      1) Flight plans, amendments, and cancellations
      2) Clearances and flow control
      3) Requests and position reports

   c. Flight information

      1) Traffic information
      2) Weather information
      3) NOTAMs

   d. Technical messages

H. **Search and Rescue (SAR) Operations**

1. An aircraft on a VFR or DVFR flight plan is considered to be overdue when it fails to arrive 30 min. after its ETA, and communications or location cannot be established.

   a. An aircraft not on a flight plan is considered to be overdue when it is reported to be at least 1 hr. late at its destination.

      b.   In either case, the actions taken by an FSS are the same.

  2.   As soon as a VFR aircraft becomes overdue, the destination FSS will attempt to locate the aircraft by checking all adjacent flight plan area airports.

      a.   Appropriate approach control and ARTCC facilities are also checked.

      b.   If these measures do not locate the aircraft, a request is made for SAR information from the departure FSS.

      c.   The departure FSS then checks locally for any information about the aircraft.

          1)   If the aircraft is located, a notification is sent to the destination FSS.

          2)   If not, all SAR information from the flight plan is sent, including any remarks which may be pertinent to the search.

  3.   If the aircraft has not been located within 30 min. after it becomes overdue, an Information Request (INREQ) is sent to all FSSs, Flight Watch stations, and ARTCCs along the route, as well as the Rescue Coordination Center (RCC).

      a.   Upon receipt of an INREQ, a facility will check its records and all area airports along the proposed route of flight.

  4.   If the aircraft is not located within 1 hr. after transmission of the INREQ, an Alert Notice (ALNOT) is transmitted to all facilities within the search area.

      a.   The search area is normally 50 mi. on either side of the proposed route of flight, from the last reported position to the destination.

      b.   Upon receipt of an ALNOT, each station within the search area conducts a search of airports within the area which were not checked previously.

      c.   ALNOTs are also broadcast over VOR and NDB voice capable frequencies.

  5.   When lake, island, mountain, or swamp reporting service programs have been established and a pilot requests the service, the FSS will establish contact every 10 min. with the aircraft while it is crossing the hazardous area.

      a.   If contact with the aircraft is lost for more than 15 min., Search and Rescue is notified.

      b.   Hazardous Area Reporting Service and chart depictions are published in the *AIM*.

I. Aviation Weather Services

  1.   Surface aviation observations (SA, SP, RS) are filed at scheduled and unscheduled intervals with stations having sending capability for dissemination.

  2.   FSSs actively solicit PIREPs when one or more of the following conditions are reported or forecast.

      a.   Ceilings at or below 5,000 ft.
      b.   Visibilities 5 SM or less
      c.   Thunderstorms and related phenomena
      d.   Turbulence of moderate degree or greater
      e.   Icing of light degree or greater
      f.   Wind shear

  3.   Radar Weather Reports (SD) are collected by the National Meteorological Center (NMC) in Maryland, and transmitted to FSSs.

  4.   Winds and Temperature Aloft Forecasts (FD) are computer-prepared and issued by the NMC to FSSs.

  5.   Terminal Forecasts (FT) for selected U.S. airports are prepared by National Weather Service Forecast Offices and forwarded for distribution.

6. Area Forecasts (FA) are issued by the National Aviation Weather Advisory Unit in Kansas City, MO and disseminated to all FSSs.

7. Severe weather forecasts are filed by the NWS and distributed to all FSSs.

8. SIGMETs and AIRMETs are issued by the National Aviation Weather Advisory Unit in Kansas City, MO to provide notice of potentially hazardous weather conditions.

9. The TWEB and synopsis for selected routes are prepared by National Weather Service Forecast Offices and distributed to the appropriate FSSs.

10. A Meteorological Impact Statement (MIS) is an unscheduled planning forecast, intended for ARTCC specialists responsible for making flow control related decisions.

    a. It enables these specialists to include the impact of expected weather conditions in making such decisions.

11. A Center Weather Advisory (CWA) is issued by the Central Weather Service unit to reflect adverse weather conditions in existence at the time of issuance or conditions beginning within the next 2 hr.

J. Using the AFSS

1. A universal toll-free number has been established for AFSSs throughout the country.

    a. When you dial (800) WX-BRIEF [(800) 992-7433], you are switched automatically to the AFSS that serves your area.

        1) You are answered by a recording giving instructions for both touch-tone and rotary dial telephones.

        2) Touch-tone users can elect to talk to a briefer or any of the direct-access services, e.g., TIBS, or "fast-file" flight plan filing.

        3) If you are using a rotary or pulse dial telephone, you will be switched automatically to a briefer.

    b. So that your preflight briefing can be tailored to your needs, give the briefer the following information:

        1) Your qualifications, e.g., student, private, commercial, and whether instrument rated

        2) The type of flight, either VFR or IFR

        3) The aircraft's tail number

        4) The aircraft type

        5) Your departure point

        6) Your proposed route of flight

        7) Your destination

        8) Your proposed flight altitude(s)

        9) Your estimated time of departure

        10) Your estimated time en route

    c. At a minimum, a standard briefing should include the following information in sequence:

        1) Adverse conditions -- significant meteorological and aeronautical information that might influence you to alter your proposed route of flight, or even cancel your flight entirely. Expect the briefer to emphasize conditions that are particularly significant, e.g., low-level wind shear, embedded thunderstorms, reported icing, or frontal zones.

2)   VFR Flight Not Recommended. When VFR flight is proposed and conditions are present or forecast, surface or aloft, that in the briefer's judgment would make flight under VFR doubtful, the briefer will describe the conditions, affected locations, and announce, "VFR flight is not recommended."

  a)   This is advisory in nature.

  b)   You are responsible to make a final decision as to whether the flight can be conducted safely.

3)   Synopsis. A brief statement describing the type, location, and movement of weather systems and/or air masses which may affect the proposed flight.

4)   Current Conditions. Reported weather conditions applicable to the flight will be summarized from all available sources.

  a)   This summary is omitted if the proposed time of departure is over 2 hr. later, unless you request it.

5)   En Route Forecast. Conditions for the proposed route are summarized in logical order; i.e., departure/climbout, en route, and descent.

6)   Destination Forecast. At the planned ETA, any significant changes within 1 hr. before and after the planned arrival are included.

7)   Winds Aloft. Forecast winds aloft will be summarized for the proposed route and altitude.

8)   NOTAMs. Information from any NOTAM (D) or NOTAM (L) pertinent to the proposed flight, and pertinent FDC NOTAMs within approximately 400 mi. of the FSS providing the briefing.

  a)   NOTAM (D) and FDC NOTAMs which have been published in the Notices to Airmen publication are not included, unless requested by you.

9)   ATC Delays. Any known ATC delays and flow control advisories which might affect the proposed flight.

10)   The following may be obtained on your request:

  a)   Information on military training routes (MTR) and military operations area (MOA) activity within the flight plan area and a 100 NM extension around the flight plan area

  b)   Approximate density altitude data

  c)   Information regarding such items as air traffic services and rules, customs/immigration procedures, ADIZ rules, etc.

  d)   LORAN-C NOTAMs

  e)   Other assistance as required

d. Request an abbreviated briefing when you need information to supplement mass disseminated data (e.g., weather channel or TIBS), update a previous briefing, or when you only need one or two specific items.

1)   Provide the briefer with appropriate background information, the time you received the previous information, and/or the specific items needed.

2)   The briefer can then limit the briefing to the information that you have not received, and/or changes in conditions since your previous briefing.

  a)   To the extent possible, the briefing will be given in the same sequence as a Standard Briefing.

3)   If you request only one or two specific items, the briefer will advise you if adverse conditions are present or forecast.

   a)   Details on these conditions will be provided at your request.

   e.   You should request an outlook briefing whenever your proposed time of departure is 6 hr. or more from the time of the briefing.

   1)   The briefer will provide available forecast data applicable to the proposed flight for planning purposes only.

   2)   You should obtain a standard briefing prior to departure in order to obtain such items as current conditions, updated forecasts, winds aloft, and NOTAMs.

2.   To provide in-flight services such as weather updates, filing a flight plan, etc., a number of sources exist.

   a.   Transcribed weather broadcasts such as TWEB or HIWAS are available on certain VOR and NDB frequencies, as indicated in the identifier boxes on your navigational charts.

   1)   To hear such a broadcast, listen on the appropriate frequency through your VOR or ADF receiver.

   b.   To obtain current weather along your route of flight or to file a PIREP, contact Flight Watch (i.e., EFAS) on 122.0 MHz below FL 180 and as published at and above FL 180.

   c.   For weather outlooks and to file, activate, or extend a flight plan, contact the nearest FSS.

   1)   Use the frequency shown on your navigational chart atop the identifier box of the VOR nearest you.

3.   Following any briefing, feel free to ask for any information that you or the briefer may have missed.

   a.   It helps to save your questions until the briefing has been completed.

   b.   This enables the briefer to present the information in a logical sequence and reduces the chance of important items being overlooked.

K.   **Visiting the AFSS**

1.   You should visit an AFSS and tour the facility. This is an excellent way for you to learn what services are available and how you can best use them.

   a.   AFSSs are open 24 hr., but usually the best time to visit is during normal business hours.

   b.   You do not need to call ahead, but you may want to so you do not visit during a normal peak time or when staffing is reduced.

2.   Your tour will be given by a specialist and usually begins at a telephone briefing station.

   a.   The specialist is able to call up the various weather reports and forecasts for the departure, en route, and destination phases of a flight.

   b.   Various charts (e.g., surface analysis chart) and a sequence of satellite pictures (either NOAA or GOES) can be displayed on another monitor.

   c.   Here, also, the specialist will input the data for a flight plan. The specialist can also view information on aircraft that have already filed, either departing from or arriving in the AFSS's area.

3. You will also see the radio communication station used to contact aircraft.

a. Here you will learn why it is very important that you identify the frequency on which you are transmitting and your aircraft's location.

1) Remember, most states will only have one AFSS, so it is important for the specialist to know where you are in order to select the proper remote transmitter to talk to you.

b. At these positions, the specialist can provide the same services as those positions that answer incoming telephone calls, in addition to accepting PIREPs.

c. At a separate station will be the specialist who provides the En Route Flight Advisory Service (EFAS) for aircraft en route.

4. As you tour the facility, you may also observe the other services that may be provided at the AFSS. These may include the VHF Direction Finding (VHF/DF) equipment to help lost pilots and information on how weather observations are taken.

5. By touring the AFSS and talking with specialists, you will gain a better understanding of how the AFSS supports pilot operations.

a. By understanding the strengths and weaknesses of the AFSS, you can maximize the services provided by the AFSS, and you can provide assistance (i.e., by making a PIREP).

L. **Kavouras Weather Graphics**

1. A company in Minneapolis called Kavouras is the FAA's contractor to provide weather graphics to AFSSs. It provides both hardware and software.

2. The capability of each computer console includes four graphics menus:

a. Current menu -- analysis charts
b. Forecast menu -- forecast charts
c. Satellite menu -- satellite imagery
d. Radar menu -- radar data

3. A final menu provides maps, charts, etc., that are produced at the specific request of a particular AFSS, e.g., Bahamas maps for Florida AFSSs.

4. A paragraph on each chart follows.

a. Current Menu (Analysis Charts)

1) Weather Depiction

a) The Weather Depiction Analysis is a contoured and shaded depiction of MVFR and IFR areas. This chart gives the user a general overview of the country in terms of ceiling and visibility. Major synoptic features such as highs, lows, and fronts are included to aid in interpreting the chart. Individual station data and models are not shown on this chart. The Weather Depiction Analysis is quite valuable in assessing the large-scale picture in the vicinity of a route of flight; however, the latest surface observations at the destination airport must also be checked.

2) North American Surface

a) The North American Surface Analysis depicts isobars, high and low pressure centers, and fronts. Individual station data and station models are not shown. The analysis depicts synoptic features -- those of fairly large scale. The placement of frontal features is determined by the Kavouras meteorologist using computer-generated surface plots which are hand analyzed every 2 to 3 hr. Before any features are placed, they are compared to the previous 3-hourly position, and also to recent trends noticed over the last 12 hr.

3)   National Radar Summary

    a)   The National Radar Summary Analysis is a composite of 211 National Weather Service, military, and ARTCC radars. The chart depicts precipitation areas using the standard VIP scale of six intensity levels. Echo top data, movement, "Out for Maintenance," and "Not Available" sites are also included.

VIP Level	Contour Color	Intensity Level
1	Light green	Light
2	Dark green	Moderate
3	Light yellow	Heavy
4	Dark yellow	Very heavy
5	Light red	Intense
6	Dark red	Extreme

4)   Upper Air (850, 700, 500 mb)

    a)   Upper air charts at 850, 700, and 500 mb display height contours in decameters, temperature in degrees Celsius and relative moisture content. The following table gives an approximate relationship between millibar level and altitude:

MB	Altitude (ft.)
500	18,000
700	10,000
850	5,000

The actual altitude of these levels varies significantly with season and latitude. All levels are lower in winter and in northern latitudes since the atmosphere is colder and more compact (denser).

5)   Upper Air (300, 200 mb)

    a)   Upper air charts at 300 mb and 200 mb display height contours in decameters, and wind speeds in knots. The 300-mb level is at approximately 30,000 ft., and 200 mb is at approximately 39,000 ft. The exact altitude of these levels varies significantly with season and latitude.

6)   Freezing Level

    a)   The freezing level chart displays the height of the lowest freezing level in thousands of feet above the surface. Data are derived from NWS balloon soundings, taken twice daily at 0000Z and 1200Z. The map is computer-generated, with a contour color of red and an interval of 4,000 ft.

7)   Lifted Index/K Index

    a)   The Lifted/K Index chart is a measure of atmospheric stability. Data are derived from National Weather Service radiosondes launched at 0000Z and 1200Z. Two values are displayed for every radiosonde site. The top value is the Lifted Index, with negative values indicating an unstable atmosphere and positive values indicating a more stable atmosphere. The lower value is the "K" Index. The larger this number, the greater the likelihood of precipitation.

8)   Precipitable Water

a)   Precipitable water is a measure of the amount of liquid water in a vertical column of air. This chart displays a contoured analysis of precipitable water, which can be correlated with precipitation total. Higher values indicate the atmosphere is holding a greater amount of moisture and imply that more significant precipitation is possible from the air mass, given that other conditions are favorable.

9)   Average Relative Humidity (SFC -- 500 mb)

a)   The Surface to 500 mb Relative Humidity Chart gives the average humidity in the lower 18,000 ft. of the atmosphere. Relative humidity values are contoured every 10% in red.

10)   Winds Aloft (FL 040, FL 140, FL 240, FL 340)

a)   Winds aloft charts are an analysis of wind flow at various levels in the atmosphere. Data are displayed using the conventional wind barb format. The data are derived from NWS radiosondes, with the analysis computer-generated. Barbs are yellow and temperatures are black for FL 240 and FL 340 (temperatures are not reported for FL 040 and FL 140).

b.   Forecast Menu (Forecast Charts)

1)   North American Surface

a)   The North American Surface Forecast depicts high and low pressure centers, fronts, and precipitation. The chart depicts synoptic features, those of fairly large scale. Therefore, weak pressure centers and features produced by terrain are generally ignored.

2)   Low Level Significant Weather

a)   Low Level Significant Weather Forecasts display important weather features from the surface to 24,000 ft. Freezing levels, shaded regions of MVFR and IFR, and turbulence are provided.

3)   Winds/Temperatures Aloft (FD Winds)

a)   Forecast winds aloft are generated twice daily by the National Weather Service in Suitland, Maryland. Forecasts are available for 12-, 24-, 36-, and 48-hr. time periods. At this time, only the 12-hr. chart (actually the 9- to 18-hr. forecast) is displayed since it is the most pertinent and often the most accurate. These data are reproduced without any changes by Kavouras, Inc. Wind data are displayed using the conventional wind barb format in yellow. Forecast temperatures are shown in degrees Celsius and colored black.

FL 390 . . .	approximately 200 mb
FL 340 . . .	approximately 250 mb
FL 300 . . .	approximately 300 mb
FL 240 . . .	approximately 400 mb
FL 180 . . .	approximately 500 mb
FL 120 . . .	approximately 650 mb
FL 090 . . .	approximately 700 mb
FL 060 . . .	approximately 800 mb

4)   U.S. High Level Significant Wx

a)   High level significant weather forecasts cover events occurring at levels above 24,000 ft. Forecasts are available four times daily, covering a 6-hr. time period (3 hr. either side of 00Z and 12Z). Each chart displays jet stream axes with altitude and wind maximums, tropopause heights, areas of broken thunderstorm coverage, areas of moderate or greater turbulence, and surface fronts.

5)   36-Hour Thickness/Sea Level Pressure

a)   This gives a depiction of how the atmosphere will look 36 hr. in the future. Sea level pressure is depicted, along with frontal features. The thickness of the lower half (approximately) of the atmosphere is also displayed.

c.   GOES Menu (GOES Satellite Imagery)

1)   GOES (Geostationary Operational Environmental Satellite) imagery is available to Automated Flight Service Stations every half hour. Kavouras receives the data directly from the satellite via its communications facility in Minneapolis. The transmission schedule is dictated partly by NOAA and partly by Kavouras.

d.   Radar Menu (NWS Radar Data)

1)   National Radar Composite

a)   The national radar composite consisting of the National Weather Service "S" and "C" band radars is available at 5-min. intervals. This composite is assembled at Kavouras' central processing facility in Minneapolis. Option 1 displays the animation of the most recent images. Option 6 displays the latest image received by the system. The ANIMATE UP and DOWN keys can be used to change the speed of the animation loop when option 1 is chosen, or to single step through the images when option 6 is chosen.

2)   Regional Radar Composite

a)   Twenty regional radar composites are assembled and transmitted every 5 min. Each AFSS facility will receive one region, likely one which correlates with the most air traffic handled by the facility. Each region is about the size of five or six states and gives the advantage of seeing high resolution radar data on a larger basemap.

3)   Single Site Radar Imagery

a)   Data from two single site radars can be combined and either animated or viewed singly. The options are either to have one radar site with two ranges, or to have two individual radar sites with one range from each. It is up to the AFSS facility to choose whether two radar sites or two ranges from a single site are needed.

END OF APPENDIX

APPENDIX D
DIRECT USER ACCESS TERMINAL (DUAT)

Please take a few minutes to study each of the concepts listed above and anticipate/imagine what they are and how they relate to the other listed concepts.

A. **DUAT Introduction** -- The FAA's Direct User Access Terminal (DUAT) computerized weather briefing and flight planning system provides pilots with the most up-to-date and reliable briefing information possible.

   1. DUAT is accessed over toll-free 800-number telephone lines, and there are no usage fees for the basic service.

   2. The service is provided under contract from the FAA by two companies, Data Transformation Corporation (DTC) and GTE (formerly Contel Federal Systems).

   3. You should try both DTC and GTE DUAT systems before deciding which one you will use regularly.

      a. This is a personal decision, i.e., choosing the service that you feel is the most comfortable.

      b. You, of course, may also use both systems.

   4. Both systems are being upgraded continuously.

B. **Accessing DUAT** -- To access DUAT, you need an IBM-compatible or Apple (e.g., MacIntosh) personal computer, a modem, a telephone line, and one of the many telecommunications programs on the market.

   1. You may need assistance setting up your computer modem. Consult a friend or computer store personnel.

   2. Set your telecommunications program to use 8 bits, one stop bit, no parity, full duplex, and echo off to access either service.

      a. DTC's modem line number is (800) 245-3828.
      b. GTE's modem line number is (800) 767-9989.

         1) GTE's DUAT can also be accessed through SprintNet, a national telecommunications service that provides local telephone numbers for remote services.

   3. Both services also offer customer assistance lines that are open 24 hr. a day.

      a. DTC's help line is (800) 243-3828.
      b. GTE's help line is (800) 345-3828.

   4. In the following discussion you will see 14 DUAT screens, 7 from DTC and 7 from GTE. We recommend both vendors to you equally.

C. **Signing Up** -- Once you establish a connection, both systems walk you through a sign-up procedure that confirms your authorization to use the system by asking you to enter your access code and password, as shown below.

<table>
<tr><td align="center">DTC's Sign-Up Procedure</td><td align="center">GTE's Sign-Up Procedure</td></tr>
</table>

```
Welcome to Data Transformation's

DIRECT USER ACCESS TERMINAL SYSTEM

*************************************
* Message Space for News and Info  *
*   on new DUAT System Features     *
*************************************

Transaction 2433828 7/27/92 1245 (UTC)

If you do not have an Access Code
Press <ENTER>:

Enter Access Code. . . . . . . . . ?  _
Enter Password  . . . . . . . . . . ?  _

    For HELP enter a ? at any prompt.
```

```
Session number:  26155

Enter DUAT access code  -or-  last name:

If you are NOT a pilot, press RETURN at
the next prompt.

Enter pilot certification number:
```

1. Anyone can use the service to obtain weather information. However, only pilots with current medicals and balloon and glider pilots can access the flight planning function for the purpose of filing a flight plan.

2. If you do not have an access code and a password and you are a licensed pilot, follow the instructions on the screen and DUAT will assign you an access code (which you can change if you wish) and ask you to choose a password.

    a. These can then be used in all subsequent DUAT briefings.

D. **Storing Your Aircraft Profile** -- After signing onto DUAT, or anytime during a briefing, you can store a profile of your aircraft that is retained permanently in the DUAT system.

1. This aircraft profile consists of search and rescue information such as tail number, type of aircraft, aircraft color, etc.

2. When you use DUAT to file a flight plan, this information is entered automatically in the appropriate spaces unless you wish to change it.

E. **DUAT Main Menu** -- The DUAT main menu allows you to access a weather briefing, plan a flight, and use various other options, as shown below.

<table>
<tr><td align="center">DTC's Main Menu</td><td align="center">GTE's Main Menu</td></tr>
</table>

```
       FAA FUNCTIONS MAIN MENU

Weather Briefing (and Expanded Text) . . 1
Flight Planning . . . . . . . . . . . . . . . . . . . . 2
Encode Function . . . . . . . . . . . . . . . . . 3
Decode Function . . . . . . . . . . . . . . . . . 4
Obtain/Change Access Code/Password . 5

Information:  Congressional Phones . . . . I
System Parameters Menu . . . . . . . . . . . S
Value Added Services . . . . . . . . . . . . . V
DTC's Catalog . . . . . . . . . . . . . . . . . A

Exit DUAT system  . . . . . . . . . . . . . . . X

Selection . . . . ?
```

```
          DUAT MAIN MENU

Weather Briefing                      1
Flight Plan                           2
Encode                                3
Decode                                4
Modify Screen Width/Length            5
Golden Eagle Services (tm)            6
Service Information                   7
Extended Decode                       8
FAA/NWS Contractions                  9

Select function (or 'Q' to quit):
```

1. Most of the options are functions sponsored by the FAA and are free of charge to you.

    a. DTC's Value Added Services and GTE's Golden Eagle Services are charged to you per minute of usage, and allow you access to weather and radar charts.

2. You are allowed 20 min. on the FAA functions, after which the system terminates your briefing automatically.

    a. If you have not completed your briefing by that time, you must sign back on and begin again.

F. **Weather Briefing** -- If you select a weather briefing from the main menu, the screen displays the Weather Briefing Menu, as shown below.

GTE's Weather Briefing Menu

Route Briefing	1
Local Briefing	2
Selected Location Weather	3
Plain Language Weather	4
State/Collective Weather	5
Defined Radius Weather	6

1. Both DUAT services offer similar options here.

    a. The route options are used to retrieve weather products along a proposed route of flight.

        1) You are required to enter your departure point, time of departure, altitude, destination, route of flight, and estimated time en route.

        2) If a Preferred IFR Route is available, you have a choice of selecting it or entering your own route.

    b. The local options are used to retrieve weather products available at or near a specific location.

        1) You are prompted for the location identifier and your time of departure.

    c. Selected WX products (DTC) and Selected Location Weather (GTE) allow you to request specific weather products from specified weather reporting locations.

    d. Regional and State Collectives allow you to request weather products by state or geographical regions.

    e. DTC's General Weather Products provides public forecasts, as shown below.

DTC's General Weather Menu

Local Forecast	1
Extended Forecast	2
Zone Forecast	3
Recreational Forecast	4
Radar Narrative Summary	5
Marine/Coastal Products	6
Severe/Special Advisories	7
Return to Main Menu	M

2.   As with an FSS briefer, DUAT allows you the option of a standard, abbreviated, or outlook briefing.

    a.   Selecting a Route or Local briefing provides you with a standard briefing, including all the reports and forecasts commonly provided by a briefer.

    b.   Any of the Selected or Collective Weather options provides you with an abbreviated briefing by allowing you to select from any of the following weather products.

GTE's Weather Data Types

```
o Enter a sequence of weather data types separated by
    spaces (e.g., SA FT FD)
  Valid types are:
    SA      -- Hourly Surface Observations
    WT      -- SA Weather Trends (3 hours)
    FT      -- Terminal Forecasts
    NO      -- NOTAMs (Notice to Airmen)
    NS      -- NOTAM Summaries (Notice to Airmen)
    FDC     -- FDC NOTAMs
    UA      -- Pilot Reports
    SD      -- Radar Summaries
    FD      -- FD Winds Aloft Forecast
    FA      -- Area Forecast
    AC      -- Severe Weather Outlook -- only through
                  selected location weather
    CWA     -- Center Weather Advisory
    WS      -- SIGMETs
    WST     -- Convective SIGMETs
    WA      -- AIRMETs
    WW      -- Severe Weather Warnings
    AWW     -- Severe Weather Forecast Alerts
    WW-A    -- Amended Severe Weather Forecasts
o Enter '!' to escape to ENCODE/DECODE function.
o Enter 'Q' to quit.
```

3.   The weather data presented are normally displayed in the standard abbreviated format unless you request the expanded (plain English) version.

    a.   EXAMPLES:

        1)   Standard format:

            LAL SA 1454 40 SCT E120 OVC 8 70/62/1904/014

        2)   Expanded format:

            LAKELAND FL (LAKELAND LINDER REGIONAL) [LAL] HOURLY OBSERVATIONS AT 9:54 A.M. EST: 4,000 SCATTERED, ESTIMATED CEILING 12,000 OVERCAST, VISIBILITY 8 MILES, TEMPERATURE 70, DEWPOINT 62, WIND 190 AT 4, ALTIMETER 30.14

    b.   Requesting weather information more than 6 hr. in advance of your proposed time of departure provides you with an outlook briefing.

G. **Flight Plan Filing** -- The Flight Planning Menu is used to file, amend, or cancel flight plans.

GTE's Flight Plan Menu

File Flight Plan	1
Amend Flight Plan	2
Cancel Flight Plan	3
Flight Planner	4
Modify Flight Planner Profile	5
Flight Planner Users Guide	6

1. Note that if you have not previously entered your access code and password, the system will not allow you to file or amend any flight plan information.

2. When filing a flight plan, you may use any previously stored data on your aircraft, or enter all the values at this time.

   a. After you answer all the flight data prompts, your entire flight plan is displayed and you are prompted for any changes you wish to make.

   b. When all the fields are correct, enter an "F" to file your flight plan.

      1) DUAT displays the ARTCC or FSS to which your flight plan will be sent and the time it will be transmitted.

      2) Up to the time it is sent, you can amend or cancel with DUAT service.

      3) After it is sent, you must contact an FSS.

3. To amend or cancel a flight plan you filed with DUAT, you must enter the aircraft ID number used in the flight plan.

   a. If you attempt to amend or cancel your flight plan after it is forwarded, you are advised to contact the appropriate FSS.

4. DTC's Flow Control Messages provide current advisories issued by the Air Traffic Control Systems Command Center, as shown below.

DTC's Flow Control Message Menu

Advisories	1
Weather Outlook	2
North Atlantic Tracks	3
Miscellaneous Messages . .	4
Return to Main Menu	M

   a. Selection 1 retrieves departure, en route, and arrival delay messages.

   b. Selection 2 retrieves the ATC System Outlook which includes a weather outlook and potential or current problem areas for delays.

   c. Selection 3 retrieves the daily routes for the North Atlantic tracks.

   d. Selection 4 retrieves miscellaneous messages that may concern NOTAMs, missile firing, etc.

5. DTC's Data File and GTE's Modify Flight Planner Profile allow you to store or modify any data on your aircraft profile.

    a. This data may be used when requesting a weather briefing, filing a flight plan, or creating a flight log.

6. DTC's Flight Log and GTE's Flight Planner both create a navigation log of a proposed flight based on your estimated departure time, airspeed, and the forecast winds aloft at your proposed altitude.

    a. To plan a flight, you must supply DUAT with certain information.

        1) Departure point
        2) Destination
        3) Departure time
        4) Route Selection
        5) Aircraft performance information
        6) Cruise altitude

    b. You may input your own route, or allow DUAT to generate one of its own.

        1) DTC will automatically generate the preferred IFR route for you, if one exists.
        2) GTE will generate one of the following routes at your request.

            a) Low-Altitude Airway Auto-Routing selects the shortest path from your origin to the destination using Victor airways.

                i) This may not be possible for certain airports which are very remote from any navigational facility.

            b) Jet Route Auto-Routing selects the shortest path from your origin to the destination using Jet routes.

                i) You will be prompted for departure and arrival routings (e.g., SIDs or STARs).

            c) VOR-Direct Auto-Routing is similar to Low-Altitude Airway Routing.

            d) Direct Routing for LORAN.

            e) Direct Routing for RNAV.

            f) User Selected Routing.

    c. Once you select one of the routing options, DUAT computes and displays the optimal route, along with the distance along the route.

        1) NOTE: DUAT does NOT take into account obstacles, terrain, controlled airspace, and special use airspace.

        2) You must verify the suggested route against your charts to see that it can be flown safely.

d.   If you approve the suggested route, DTC's DUAT generates a flight log as shown
     below, complete with forecast winds aloft and heading and groundspeed
     computations.

DTC's Flight Log

The IFR Low Altitude Preferred Route(s) between GNV and MIA:

1. V157 LBV V529 V035 CURVE
    100 AND BLO : 1300-0300 :
2. Return to Previous Menu

Selection....? 1

===> Data Transformation's Flight Log <===
Altitude--7000 Ft   Air Speed--150 Knots   Departure Time--2200Z

LEG	MAG CRS	MAG HDG	GND SPD	DIST (NM)	ETE (MIN)	WIND	ATE
GNV ARPT	217	225	138	9	0+04	276/023	_____
GNV VORTAC	162	170	157	25	0+09	276/023	_____
OCF VORTAC	171	179	160	72	0+27	290/024	_____
LAL+	153	160	166	77	0+28	290/024	_____
LBV VORTAC	157	163	160	42	0+16	284/018	_____
SWAGS	156	162	160	19	0+07	284/018	_____
XING V035	109	110	167	19	0+07	284/018	_____
CURVE	087	085	167	19	0+07	284/018	_____
MIA ARPT	TOTAL			282	1+45		_____

YOUR ETA = 2345Z

e.   GTE will prompt you for your aircraft's climb, cruise, and descent airspeeds, climb and descent rates, and fuel consumption rates throughout, and then generate the following flight log. Other flight log formats are available in Modify Flight Planner Profile menu.

GTE's Flight Log

From: KGNV -- Gainesville FL
To: KMIA -- Miami FL
Time: Fri Feb 26 22:00 (UTC)

Routing options selected: Automatic low altitude airway.
Flight plan route:
      KGNV OCF V295 BAIRN V267 GREMM KMIA
Flight totals: fuel: 30 gallons, time: 1:37, distance 265.7 nm.

Ident Type/Morse Code Name or Fix/radial/dist Latitude Longitude Alt.	Route Mag KTS Fuel Winds Crs TAS Time Temp Hdg GS Dist	Fuel Time Dist
1. KGNV Apt. Gainesville FL 29:41:24 82:16:18 2	Direct 4.8 291/15 179 130 0:14 +16C 184 136 31	0.0 0:00 266
2. OCF --- -.-. ..-. d113.7 Ocala 29:10:39 82:13:35 70	V295 6.1 284/24 132 155 0:20 +10C 136 176 60	4.8 0:14 235
3. ORL --- .-. .-.. d112.2 Orlando 28:32:34 81:20:06 70	V295 4.1 284/24 165 155 0:14 +10C 172 166 38	10.9 0:34 175
4. BAIRN Int. ORLr162/37 VRBr300 27:56:53 81:06:54 70	V267 8.0 284/24 165 155 0:26 +10C 173 166 73	15.0 0:48 137
5. PHK .-- .... -.- d115.4 Pahokee 26:46:58 80:41:29 70	V267 4.3 277/17 158 161 0:15 +13C 164 169 41	23.0 1:14 64
6. GREMM Int. PHKr155/41 MIAr020 26:09:57 80:22:36 40	Direct 2.4 263/17 171 170 0:08 +17C 177 171 23	27.3 1:29 23
7. KMIA Apt. Miami FL 25:47:35 80:17:25 0		29.7 1:37 0

NOTE: Fuel calculations do not include required reserves.
Flight totals: fuel: 30 gallons, time: 1:37, distance 265.7 nm.
Average groundspeed 165 knots.
Great circle distance is 256.4 nm -- this route is 4% longer.

7. DTC's Special Use Airspace selection provides you with information on Alert Areas, MOAs, Prohibited Areas, Restricted Areas, and Warning Areas within 5 NM of your route of flight.

8. GTE's Flight Planner Users Guide gives complete user-friendly directions on how to use the Flight Planner option.

H. **Encode and Decode Functions** -- The Encode and Decode Functions are used to find the identifier for a known location and to receive information on NAVAIDs and airports for any valid location identifier, respectively.

1. In the Encode Function, all locations that use the name you enter are shown.

DTC's Encode Function

```
ENTER LOCATION, STATE...?  PHILI, PA

IDENT CITY, STATE              AIRPORT/NAVAID/WX RPT
1N3    PHILIPSBURG, PA         ALBERT//
PSB    PHILIPSBURG, PA         MID-STATE/VORTAC/WX


ENTER LOCATION, STATE...?  FERGUSON

IDENT CITY, STATE              AIRPORT/NAVAID/WX RPT
82J    PENSACOLA, FL           FERGUSON//
12M0   WINDSOR, MO             FERGUSON FARMS//
TN09   PHILADELPHIA, TN        FERGUSONS FLYING CIRCUS//
```

2. In the Decode Function, information concerning a given identifier or contraction is provided.

GTE's Decode Function

```
ENTER IDENT(S):  GNV

GNV AIRPORT GAINESVILLE REGIONAL   GAINESVILLE, FL
     FSS: GNV(GAINESVILLE) ARTCC: ZJX(JACKSONVILLE)
GNV NAVAID    GAINESVILLE, FL                              EFAS
     FSS: GNV(GAINESVILLE) ARTCC:
GNV NAVAID GAINESVILLE   GAINESVILLE, FL                   VORTAC
     FSS: GNV(GAINESVILLE) ARTCC: JACKSONVILLE
GNV WEATHER  GAINESVILLE, FL
```

3.   At any prompt that requires entry of a location identifier, both DTC and GTE allow you to retrieve Encode/Decode information.

    a.   Whenever you are prompted for an aircraft type, the escape to Encode/Decode Function is changed to search for aircraft type designators and model numbers, rather than for location identifiers.

    b.   EXAMPLE:

DTC's Aircraft Decode/Encode Function

Aircraft Type/ Special Equipment . . ? ? **PA24**	
DESI MODEL PA24 COMANCHE	**MANUFACTURER** PIPER AIRCRAFT CORP.
Aircraft Type/ Special Equipment . . ? ? **DAUP**	
DESI MODEL HH65 DAUPHINE 2 HR3C DAUPHINE 2	**MANUFACTURER** AEROSPATIALE AEROSPATIALE

I.   **Extra Services for a Fee** -- Beyond the basics, DUAT offers extra services for which you are charged a fee.

1.   GTE offers its Golden Eagle Services, which provide the following graphic displays of weather data.

    a.   Satellite images
    b.   Weather depictions
    c.   Weather radar

2.   DTC offers graphic displays of weather data it receives under contract from Accu-Weather. These include surface analysis, radar plus, and lightning strike.

    a.   In addition, DTC provides access to EAASY Sabre, the American Airlines on-line reservations system.

3.   Both GTE and DTC bill you for extra services by charging the fees to your credit card.

J.   **In Closing** -- DUAT is a great resource for pilots which should be used. DTC and GTE each provide excellent services which vary slightly, but on balance are very competitive. For example, GTE has more flight planning options, whereas DTC provides the following free character maps.

1.   Individual Radar
2.   Composite Radar
3.   Flight Rules
4.   Temperature/Dewpoint
5.   Visibility/Weather

END OF APPENDIX

APPENDIX E
NEW INTERNATIONAL
AERODROME METEOROLOGICAL CODES

OVERVIEW

The Terminal Forecast (FT) is being changed to a Terminal Aerodrome Forecast (TAF) and the Surface Aviation Observation (SA) is being changed to a Meteorological Aviation Routine weather report (METAR). These changes are the result of international standardization of weather reports/forecasts in conjunction with the International Civil Aviation Organization (ICAO) and the World Meteorological Organization.

Until January 1, 1996, supplementary TAFs will be issued for 80 U.S. international airports and METARs will be issued for 250 U.S. international "landing rights" airports. They will be "supplemental" as the official forecasts and reports will continue to be FTs and SAs. On January 1, 1996, the plan is to convert completely to TAFs and METARS and not issue any FTs or SAs.

We will convert the information in this appendix into Part III, Chapter 2, Surface Aviation Weather Reports (pages 225 through 242) and Part III, Chapter 6, Terminal Forecasts (FT and TAF) (pages 255 through 270) as we approach 1996. Recall that all the Gleim books are reprinted on a regular basis and updated as required.

The official NOAA "key" or explanation of the TAF and METAR is reproduced on page 430.

COMPARISON OF METAR AND SA

Order of presentation:

METAR

1. Location Identifier
2. Issuance Time
3. Wind
4. Visibility
5. Runway Visual Range
6. Significant Present, Forecast, and Recent Weather
7. Cloud Amount
8. Cloud Height
9. Cloud Type
10. Temperatures
11. Altimeter Setting
12. Supplementary Information/Remarks

SA

1. Location Identifier
2. Type and Time of Report
3. Sky Condition and Ceiling
4. Visibility
5. Weather and Obstructions to Vision
6. Sea Level Pressure
7. Temperatures
8. Wind
9. Altimeter Setting
10. Coded Data/Remarks

EXAMPLE:

METAR: KPIT 1955Z 22015G25KT 3/4SM R22L/2700FT TRSA OVC010CB 18/16 A2992

SA: PIT SA 1955 M10 OVC 3/4TRW 103/65/61/2215G25/922/R22LVR27 (Remarks)

METAR explanation (see pages 225 through 242 for SA discussion):

1. K precedes airport identifier for those located in the contiguous 48 states.

2. Issuance time: 4-digit time in UTC -- "Z".

3. Wind: First 3 digits mean true-north direction, nearest 10 degrees, (or VaRiaBle); next 2 digits mean (i.e., average) speed and unit KT, (KMH or MPS); as needed, Gust and 2-digit maximum speed; 00000KT for calm; (for reports only, if directions varies 60 degrees or more, Variability appended, e.g., 180V260).

4. Visibility: Prevailing visibility in Statute Miles and fractions; above 6 SM in TAF Plus6SM. (Or, 4-digit sector visibility in meters and as required, lowest value with direction).

5. Runway Visual Range: R; 2-digit runway designator Left, Center, or Right as needed; "/"; Minus or Plus in U.S., 4-digit value; FeeT in U.S., (usually meters elsewhere); Variability (and tendency Down, Up or No change).

6. Significant Present, Forecast and Recent Weather: Table grouped in categories and used in the order listed below.

QUALIFIER

Intensity or Proximity
- Light   "no sign" Moderate   + Heavy
VC Vicinity: but not at aerodrome; in U.S., 5-10SM from center of runway complex
    (elsewhere within 8000m)

Descriptor

MI Shallow	BC Patches	DR Drifting	TS Thunderstorm
BL Blowing	SH Showers	FZ Supercooled/freezing	

WEATHER PHENOMENA

Precipitation

DZ Drizzle	RA Rain	SN Snow	SG Snow grains
IC Diamond dust	PE Ice pellets	GR Hail	GS Small hail/snow pellets
UP Unknown precipitation in automated observations			

Obscuration

BR Mist	FG Fog	FU Smoke	VA Volcanic ash
SA Sand	HZ Haze	PY Spray	DU Widespread dust

Other

SQ Squall	SS Sandstorm	DS Duststorm	PO Well developed
FC Funnel cloud/tornado/waterspout			dust/sand whirls

---

7. Cloud amount: SKy Clear, SCaTtered, BroKeN, and OVerCast; or Vertical Visibility for obscured sky. More than one layer may be reported. CLeaR for "clear below 12 thousand feet" at automated observation sites.

8. Cloud height: 3-digit height in hundreds of feet; Or Vertical Visibility for obscured sky and height "VV004", or unknown height "VV///".

9. Cloud type: Either Towering CUmulus or CumulonimBus, or Vertical Visibility for obscured sky.

10. Temperatures: In degrees Celsius; first 2 digits, temperature "_/_" last 2 digits, dew-point temperature; below zero reported with Minus, e.g., M06.

11. Altimeter setting: Indicator and 4-digits; in U.S., A-inches and hundredths; (Q-hectoPascals).

12. Supplementary information: (Wind Shear in lower layers, LanDinG or TaKeOFf, RunWaY, 2-digit designator; REcent weather of operational significance.) ReMarK: Automated Observation or AOAugmented; wind shear; recent weather and time; and TORNADO, FUNNEL CLOUD, WATERSPOUT.

COMPARISON OF TAF AND FT

EXAMPLE:

> TAF   KPIT 091720Z 1818 22020KT 3SM –SHRA BKN020 FM20 30015G25KT 3SM SHRA OVC015 PROB40 2022 1/2SM TSRA OVC008CB FM23 27008KT 5SM –SHRA BKN020 OVC040 TEMPO 0407 00000KT 1SM –RA FG FM10 22010KT 5SM –SHRA OVC020 BECMG 1315 20010KT P6SM NWS SKC

> FT   PIT FT 091717 C20 BKN 3RW– 2212 CHC C8 OVC 1TRW 3015G25 AFT 20Z. 23Z C40 BKN 2708 OCNL 3/4F 10Z-13Z. 04Z VFR..

Order of presentation:

	TAF	**FT**
	1. Message Type	1. Location Identifier
	2. Location Identifier	2. Message Type
	3. Issuance Time	3. Issuance Time
	4. Valid Period	4. Sky Condition and Ceiling
	5. Wind	5. Visibility
	6. Prevailing Visibility	6. Weather and Obstructions to Vision
	7. Significant Present, Forecast, and Recent Weather	7. Wind
	8. Cloud Amount	8. Remarks
	9. Cloud Height	9. Expected Changes
	10. Cloud Type	10. Categorical Outlook
	11. Remarks	
	12. Expected Changes	

Additional Commentary on TAFs

1. The TAF is similar to the current FT except for the order of presentation discussed above and the following items.

2. Each forecast time after initial forecast is indicated by FM followed by Zulu hour, e.g., FM 20, FM 23, FM 10 in the above TAF example.

3. Also, probability statements regarding certain conditions are preceded by PROB and the percentage probability, such as PROB40 in above TAF which is followed by the time period in four digits which is Zulu hour to Zulu hour, such as PROB40 2022, which means a 40% probability from 20Z to 22Z of 1/2SM TSRA OVC 008CB (½ SM visibility in thunderstorm/rain, ceiling 800 ft. with cumulonimbus).

NOAA KEY TO NEW TAF AND METARS REPRODUCED

Front

KEY to NEW INTERNATIONAL AERODROME FORECAST (TAF) and NEW AVIATION ROUTINE WEATHER REPORT (METAR)

TAF
KPIT 091720Z 1818 22020KT 3SM -SHRA BKN020 FM20 30015G25KT 3SM
SHRA OVC015 PROB40 2022 1/2SM TSRA OVC008CB FM23 27008KT
5SM -SHRA BKN020 OVC040 TEMPO 0407 00000KT 1SM -RA FG FM10
22010KT 5SM -SHRA OVC020 BECMG 1315 20010KT P6SM NSW SKC

METAR
KPIT 1955Z 22015G25KT 3/4SM R22L/2700FT TSRA OVC010CB 18/16 A2992

Forecast	Report	Explanation
TAF	METAR	Message type: TAF-routine and TAF AMD-amended forecast, METAR-hourly and SPECI-special report
KPIT	KPIT	ICAO location indicator
091720Z	1955Z	Issuance time: ALL times in UTC "Z", 2-digit date and 4-digit time for TAF, 4-digit time for METAR
1818		Valid period: first 2 digits begins and last 2 ends forecast
22020KT	22015G25KT	Wind: first 3 digits mean true-north direction, nearest 10 degrees, (or VaRiaBle); next 2 digits mean speed and unit, KT, (KMH or MPS); as needed, Gust and 2-digit maximum speed; 00000KT for calm; (for reports only, if direction varies 60 degrees or more, Variability appended, e.g., 180V260)
3SM	3/4SM	Prevailing visibility: in U.S., Statute Miles & fractions; above 6 miles in TAF Plus6SM. (Or, 4-digit minimum visibility in meters and as required, lowest value with direction)
	R22L/2700FT	Runway Visual Range: R; 2-digit runway designator Left, Center, or Right as needed; "/", Minus or Plus in U.S., 4-digit value, FeeT in U.S., (usually meters elsewhere); Variability (and tendency Down, Up or No change)
-SHRA	TSRA	Significant present, forecast and recent weather: see table
BKN020	OVC010CB	Cloud amount, height and type: SKy Clear, SCaTtered, BroKeN, OVerCast; 3-digit height in hundreds of feet; and either Towering CUmulus or CumulonimBus. Or Vertical Visibility for obscured sky and height "VV004", or unknown height "VV///". More than one layer may be forecast or reported. CLeaR for "clear below 12 thousand feet" at automated observation sites.
	18/16	Temperature: in degrees Celsius; first 2 digits, temperature "/" last 2 digits, dew-point temperature; below zero reported with Minus, e.g., M06
	A2992	Altimeter setting: indicator and 4 digits; in U.S., A-inches and hundredths; (Q-hectoPascals, e.g., Q1013)

Back

KEY to NEW INTERNATIONAL AERODROME FORECAST (TAF) and NEW AVIATION ROUTINE WEATHER REPORT (METAR)

Supplementary information for report: (Wind Shear in lower layers, LanDinG or TakeOFf, RunWaY, 2-digit designator; REcent weather of operational significance.) ReMarK; Automated Observation or AOAugmented; wind shear; recent weather and time; and TORNADO, FUNNEL CLOUD, WATERSPOUT.

FM20	FroM and 2-digit hour: indicates significant change
PROB40 2022	PRObability and 2-digit percent: probable condition during 2-digit beginning and 2-digit ending time period
TEMPO 0407	TEMPOrary: changes expected for less than 1 hour and in total, less than half of 2-digit beginning and 2-digit ending time period
BECMG 1315	BECoMinG: change expected during 2-digit beginning and 2-digit ending time period

Table of Significant Present, Forecast and Recent Weather - Grouped in categories and used in the order listed below; or as needed in TAF, No Significant Weather.

QUALIFIER
Intensity or Proximity
- Light   "no sign" Moderate   + Heavy
VC Vicinity: but not at aerodrome; in U.S., 5-10SM from center of runway complex (elsewhere within 8000m)

Descriptor

MI Shallow	BC Patches	DR Drifting	TS Thunderstorm
BL Blowing	SH Showers	FZ Supercooled/freezing	

WEATHER PHENOMENA
Precipitation

DZ Drizzle	RA Rain	SN Snow	SG Snow grains
IC Diamond dust	PE Ice pellets	GR Hail	GS Small hail/snow pellets
UP Unknown precipitation in automated observations			

Obscuration

BR Mist	FG Fog	FU Smoke	VA Volcanic ash
SA Sand	HZ Haze	PY Spray	DU Widespread dust

Other

SQ Squall	SS Sandstorm	DS Duststorm	PO Well developed dust/sand whirls
FC Funnel cloud/tornado/waterspout			

Explanations in parentheses "( )" indicate different worldwide practices. Ceiling defined as the lowest broken or overcast layer, or the vertical visibility. Automated Observations may be Augmented manually for certain weather phenomena. TAFs exclude temperature, turbulence and icing forecasts and METARs exclude trend forecasts. Although not used in U.S., Ceiling And Visibility OK replaces visibility, weather and clouds if: visibility is 10 kilometers or more; "no cloud below 1500 meters (5000 feet) or below the highest minimum sector altitude, whichever is greater and no cumulonimbus; and no precipitation, thunderstorm, duststorm, sandstorm, shallow fog, or low drifting dust, sand or snow.

March 1993
NOAA/PA 93054   UNITED STATES DEPARTMENT OF COMMERCE
National Oceanic and Atmospheric Administration—National Weather Service

BOOKS AVAILABLE FROM GLEIM PUBLICATIONS, Inc.
WRITTEN EXAM BOOKS

Before pilots take their FAA written tests, they want to understand the answer to every FAA written test question. Gleim's written test books are widely used because they help pilots learn and understand exactly what they need to know to do well on their FAA written test.

Gleim's books contain all of the FAA's airplane questions (nonairplane questions are excluded). We have unscrambled the questions appearing in the FAA written test books and organized them into logical topics. Answer explanations are provided next to each question. Each of our chapters opens with a brief, user-friendly outline of exactly what you need to know to pass the written test. Information not directly tested is omitted to expedite your passing the written test. This additional information can be found in our flight maneuver and reference books and practical test prep books described below.

PRIVATE PILOT AND RECREATIONAL PILOT FAA WRITTEN EXAM ($12.95)

The FAA's written test for either certificate consists of 60 questions out of the 711 questions in our book.

INSTRUMENT PILOT FAA WRITTEN EXAM ($16.95)

The FAA's written test consists of 60 questions out of the 898 questions in our book. Also, those people who wish to become an instrument-rated flight instructor (CFII) or an instrument ground instructor (IGI) must take the FAA's written test of 50 questions from this book.

COMMERCIAL PILOT FAA WRITTEN EXAM ($14.95)

The FAA's written test will consist of 100 questions out of the 564 questions in our book.

FUNDAMENTALS OF INSTRUCTING FAA WRITTEN EXAM ($9.95)

The FAA's written test consists of 50 questions out of the 160 questions in our book. This is required of any person to become a flight instructor or ground instructor. The test only needs to be taken once. For example, if someone is already a flight instructor and wants to become a ground instructor, taking the FOI test a second time is not required.

FLIGHT/GROUND INSTRUCTOR FAA WRITTEN EXAM ($14.95)

The FAA's written test consists of 100 questions out of the 827 questions in our book. To be used for the Certificated Flight Instructor (CFI) written test and those who aspire to the Advanced Ground Instructor (AGI) rating for airplanes. Note that this book also covers what is known as the Basic Ground Instructor (BGI) rating. However, the BGI is **not** useful because it does not give the holder full authority to sign off private pilots to take their written test. In other words, this book should be used for the AGI rating.

AIRLINE TRANSPORT PILOT FAA WRITTEN EXAM ($23.95)

The FAA's written test consists of 80 questions each for the ATP Part 121, ATP Part 135, and the flight dispatcher certificate. This first edition contains a complete answer explanation to each of the 1,346 airplane ATP questions (111 helicopter questions are excluded). This difficult FAA written test is now made simple by Gleim. As with Gleim's other written test books, studying for the ATP will now be a learning and understanding experience rather than a memorization marathon -- at a lower cost and with higher test scores and less frustration!!

FAA PRACTICAL TEST PREP AND REFERENCE BOOKS

Our new Practical Test Prep books are designed to replace the FAA Practical Test Standards reprint booklets which are universally used by pilots preparing for the practical test. These new Practical Test Prep books will help prepare pilots for FAA practical tests as much as the Gleim written exam books prepare pilots for FAA written tests. Each task, objective, concept, requirement, etc., in the FAA's practical test standards is explained, analyzed, illustrated, and interpreted so pilots will be totally conversant with all aspects of their practical tests.

NOW	*Private Pilot FAA Practical Test Prep*	544 pages	($16.95)
AVAILABLE!	*Instrument Pilot FAA Practical Test Prep*	520 pages	($17.95)
	Commercial Pilot FAA Practical Test Prep	432 pages	($14.95)
	Flight Instructor FAA Practical Test Prep	632 pages	($17.95)

PRIVATE PILOT HANDBOOK ($12.95)

A complete private pilot ground school text in outline format with many diagrams for ease in understanding. A complete, detailed index makes it more useful and saves time. It contains a special section on biennial flight reviews.

AVIATION WEATHER AND WEATHER SERVICES ($18.95)

This is a complete rewrite of the FAA's *Aviation Weather 00-6A* and *Aviation Weather Services 00-45D* into a single easy-to-understand book complete with all of the maps, diagrams, charts, and pictures that appear in the current FAA books. Accordingly, pilots who wish to learn and understand the subject matter in these FAA books can do it much more easily and effectively with this book.

MAIL TO: **GLEIM PUBLICATIONS, Inc.**
P.O. Box 12848
University Station
Gainesville, FL 32604

OR CALL: **(800) 87-GLEIM, (904) 375-0772, FAX (904) 375-6940**

Our customer service staff is available to take your calls from 8:00 a.m. to 7:00 p.m.,
Monday through Friday, and 9:00 a.m. to 2:00 p.m., Saturday, Eastern Time.
Please have your VISA/MasterCard ready.

**THE BOOKS WITH
THE RED COVERS**

WRITTEN TEST BOOKS

Private/Recreational Pilot	Seventh	(1993-1995) Edition	$12.95 _____
Instrument Pilot	Fifth	(1993-1995) Edition	16.95 _____
Commercial Pilot	Fifth	(1993-1995) Edition	14.95 _____
Fundamentals of Instructing	Fifth	(1993-1995) Edition	9.95 _____
Flight/Ground Instructor	Fifth	(1993-1995) Edition	14.95 _____
Airline Transport Pilot	Second	(1993-1995) Edition	23.95 _____

HANDBOOKS AND PRACTICAL TEST PREP BOOKS

Aviation Weather and Weather Services	(First Edition)	18.95 _____
Private Pilot Handbook	(Fourth Edition)	12.95 _____
Private Pilot FAA Practical Test Prep	(First Edition)	16.95 _____
Instrument Pilot FAA Practical Test Prep	(First Edition)	17.95 _____
Commercial Pilot FAA Practical Test Prep	(First Edition)	14.95 _____
Flight Instructor FAA Practical Test Prep	(First Edition)	17.95 _____

Shipping ___3.00___

Add applicable sales tax for shipments within the State of Florida

Sales Tax _____

Please call or write for additional charges for outside the 48 contiguous United States
Printed 08/93. Prices subject to change without notice. We ship latest editions.

TOTAL $ _____

1. *We process and ship orders within 1 day of receipt of your order. We generally ship via UPS for the Eastern U.S. and U.S. mail for the Western U.S.*

2. *Please PHOTOCOPY this order form for friends and others.*

3. *No CODs. All orders from individuals must be prepaid and are protected by our unequivocal refund policy.*

   Library and company orders may be on account. Shipping and handling charges will be added to the invoice, and to prepaid telephone orders.

Name _____
    (please print)
Shipping Address _____
    (street address required for UPS)

_____

City _____ State _____ Zip _____

☐ MC/VISA   ☐ Check/M.O.   Daytime Telephone (___)_____

MC/VISA No. __ __ __ __ - __ __ __ __ - __ __ __ __ - __ __ __ __

Expiration Date *(month/year)* _____/_____

Signature _____

064A

**GLEIM PUBLICATIONS, INC. GUARANTEES
THE IMMEDIATE REFUND OF ALL RESALABLE TEXTS RETURNED IN 30 DAYS
SHIPPING AND HANDLING CHARGES ARE NONREFUNDABLE**

P.S. We presume your local FBO or bookstore does not stock the books you are ordering from us directly. If you provide us with a name and address, we will invite them to do so.

AUTHOR'S RECOMMENDATIONS

The Experimental Aircraft Association, Inc. is a very successful and effective nonprofit organization that represents and serves those of us interested in flying, in general, and in sport aviation, in particular. I personally invite you to enjoy becoming a member:

> $35 for a 1-year membership
> $20 per year for individuals under 19 years old
> Family membership available for $45 per year

Membership includes the monthly magazine *Sport Aviation*.

> *Write to:* Experimental Aircraft Association, Inc.
> P.O. Box 3086
> Oshkosh, Wisconsin 54903

> *Or call:* (414) 426-4800
> (800) 843-3612 (in Wisconsin: 1-800-236-4800)

The annual EAA Oshkosh Fly-in is an unbelievable aviation spectacular with over 10,000 airplanes at one airport! Virtually everything aviation-oriented you can imagine! Plan to spend at least 1 day (not everything can be seen in a day) in Oshkosh (100 miles northwest of Milwaukee).

> *Convention dates:* 1994 -- July 29 through August 4
> 1995 -- July 28 through August 3

* *

The National Association of Flight Instructors (NAFI) is a non-profit organization dedicated to raising and maintaining the professional standing of the flight instructor in the aviation community. Members accept the responsibility to practice its profession according to the highest ethical standards. I personally invite you to become a member:

> *Active - Certified flight instructors (civilian and military)*
> Initial membership = $35
> Annual renewal = $30

> *Associate - Individuals and organizations in support of professional flight education*
> Initial membership and Annual renewal $100.

Membership includes the bimonthly *NAFI Foundation Newsletter,* an accidental death and dismemberment policy, and representation with the FAA in Washington.

> *Write to:* National Association of Flight Instructors
> Ohio State University Airport
> P.O. Box 793
> Dublin, OH 43017

> *Or call:* (614) 889-6148

INDEX

See also the Glossary beginning on page 384.

See also the Glossary beginning on page 384.

See also the Glossary beginning on page 384.

See also the Glossary beginning on page 384.

See also the Glossary beginning on page 384.

Please forward your suggestions, corrections, and comments to **Irvin N. Gleim • c/o Gleim Publications, Inc. • P.O. Box 12848 • University Station • Gainesville, Florida • 32604** for inclusion in the next edition of *Aviation Weather and Weather Services*. Please include your name and address so we can properly thank you for your interest.

1. _____

2. _____

3. _____

4. _____

5. _____

6. _____

7. _____

8. _____

9. _____

10. _____

11. _____

12. _____

13. _____

14. _____

15. _____

16. _____

17. _____

Name: _____

Address: _____

City/State/Zip: _____

Telephone: _____